Advances in Cryogenic Engineering – Materials: Proceedings of the International Cryogenic Materials Conference 2015 (ICMC 2015)

IOP Conference Series: Materials Science and Engineering
Volume 102

Tucson, Arizona, USA
28 June - 2 July 2015

ISBN: 978-1-5108-1811-8
ISSN: 1757-8981

Printed from e-media with permission by:

Curran Associates, Inc.
57 Morehouse Lane
Red Hook, NY 12571

Some format issues inherent in the e-media version may also appear in this print version.

Printed by Curran Associates, Inc. (2016)

For permission requests, please contact the Institute of Physics
at the address below.

Institute of Physics
Dirac House, Temple Back
Bristol BS1 6BE UK

Phone: 44 1 17 929 7481
Fax: 44 1 17 920 0979

techtracking@iop.org

Additional copies of this publication are available from:

Curran Associates, Inc.
57 Morehouse Lane
Red Hook, NY 12571 USA
Phone: 845-758-0400
Fax: 845-758-2634
Email: curran@proceedings.com
Web: www.proceedings.com

Advances in Cryogenic Engineering – Materials: Proceedings of the International Cryogenic Materials Conference 2015 (ICMC 2015)

IOP Conference Series: Materials Science and Engineering
Volume 102

Tucson, Arizona, USA
28 June - 2 July 2015

IOP
Institute of Physics

Table of contents

Volume 102

Advances in Cryogenic Engineering Materials: Proceedings of the International Cryogenic Materials Conference (ICMC) 2015
28 June to 2 July 2015, Tucson, AZ, USA

Accepted papers received: 12 November 2015
Published online: 18 December 2015

Preface

Papers

Mechanical Properties of Structural Materials

Cryogenic Materials Testing, Methods, and Properties

Cryogenic Insulation Materials

HTS Cables

BSCCO, MgB_2, and Coated Conductors

IOP Conf. Series: Materials Science and Engineering **102** (2015) 011001 doi:10.1088/1757-899X/102/1/011001

Preface

The 2015 joint Cryogenic Engineering and International Cryogenic Materials Conferences were held from June 28 through July 2 at the JW Marriott Starr Pass Resort & Spa in Tucson, Arizona. As at past conferences, the international scope of these meetings was strongly maintained with 26 countries being represented by 561 attendees who gathered to enjoy the joint technical programs, industrial exhibits, special events, and natural beauty of the surrounding Sonoran Desert.

The program for the joint conferences included a total of 363 presentations in the plenary, oral, and poster sessions. Four plenary talks gave in-depth discussions of the readiness of bulk superconductors for applications, the role of cryogenics in the development of the hydrogen bomb and vice versa, superconducting turboelectric aircraft propulsion and UPS's uses and plans for LNG fuel. Contributed papers covered a wide range of topics including large-scale and small-scale cryogenics, advances in superconductors and their applications. In total, 234 papers were submitted for publication of which 224 are published in these proceedings.

The CEC/ICMC Cryo Industrial Expo displayed the products and services of 38 industrial exhibitors and provided a congenial venue for a reception and refreshments throughout the week as well as the conference poster sessions.

Spectacular panoramic views of Saguaro National Park, the Sonoran Desert and the night time lights of Tucson set the stage for a memorable week in the American Southwest. Conference participants enjoyed scenic hikes and bike rides, exploring Old Town Tucson, hot and spicy southwestern cuisine, a nighttime lightning display and a hailstorm.

Conference Chairs for 2015 were Peter Kittel, Consultant, for CEC and Michael Sumption from The Ohio State University, Materials Science Department for ICMC. Program Chairs were Jonathan Demko from the LeTourneau University for CEC and Timothy Haugan from AFRL/RQQM for ICMC, assisted by the CEC Program Vice Chair, Jennifer Marquardt from Ball Aerospace & Technologies Corp. The Exhibit chair was Richard Dausman of Cryomech, Inc. who also served as Publicity Chair. Finally, Awards Chairs were Ray Radebaugh from the National Institute of Standards and Technology for CEC and David Cardwell from the University of Cambridge, United Kingdom, for ICMC.

On June 28, the Cryogenic Society of America presented three short courses: *Cryocooler Fundamentals and Space Applications*, a full day course given by Dr. Ray Radebaugh and Dr. Ron Ross, *Superconducting Radio Frequency Systems*, a half-day course by Dr. Rong-Li Geng; and *Practical Thermometry and Instrumentation*, a half-day course by Dr. Scott Courts. The courses were well attended and catered to both seasoned professionals and newcomers to cryogenics.

The able leadership of Paula Pair with the hardworking staff from Centennial Conferences provided outstanding conference management and operations. Thank you Paula and staff for a truly outstanding and memorable conference experience! Finally, to the attendees, authors, editors, and reviewers whose combined efforts have contributed to these proceedings, thanks very much to you all.

We look forward to the next CEC/ICMC in Madison, Wisconsin, July 9 – 13, 2017.

Peter Kittel
2015 CEC Conference Chairman

Michael Sumption
2015 ICMC Conference Chairman

IOP Conf. Series: Materials Science and Engineering **102** (2015) 011002 doi:10.1088/1757-899X/102/1/011002

ICMC Awards

ICMC LIFETIME ACHIEVEMENT AWARD 2015

Professor Herbert C. Freyhardt

ICMC has presented its Award for Lifetime Achievement biannually at the CEC/ICMC conference since its 30th anniversary conference in Keystone in 2005 to an individual to "recognize a lifetime's achievement in advancing the knowledge of cryogenic materials".

The sixth recipient of the ICMC Lifetime Achievement Award, following Ted Collings in 2005, David Larbalestier in 2007, Kyoji Tachikawa in 2009, René Flükiger in 2011 and Harold Weber in 2013, is Professor Herbert Freyhardt.

Herbert Freyhardt completed his PhD in 1970 at the University of Göttingen on the "Hardening of Superconductors by Dislocations and Precipitates". He was subsequently appointed to an Associate Professorship at the University of Göttingen, before spending a year as a Visiting Scientist at Argonne National Laboratory in 1973. Herbert returned to Göttingen to lead the Crystal Growth Facility and the Superconductivity Group at the Institut für Metallphysik, where he completed his Habilitation in 1976 with a thesis on "Radiation Induced Flux-Pinning in Type-II Superconductors". He was appointed "Außerplanmäßiger Professor" in 1980 and to a full Professorship in 1984. In 1990 Herbert established and co-founded the "Zentrum für Funktionswerkstoffe" ("Center for Applied Materials Research"), a non-profit organisation in collaboration with the University of Göttingen, and has been managing this enterprise ever since. Herbert retired from the Institut für Materialphysik, University of Göttingen in 2006 to take up a part-time Research Professorship in Mechanical Engineering and the Texas Center for Superconductivity at the University of Houston, a position that he continues to hold.

Herbert has had a remarkable, successful and innovative research career, resulting in more than 250 publications in scientific journals and several patents. He has investigated a diverse number of topics, primarily within solid state physics, ranging from flux pinning and critical currents to composite superconductors, metastable and amorphous superconductors, A15 and Chevrel phases, bulk high temperature superconductors, coated conductors and thin films, as well as many materials aspects of superconductors and applied superconductivity. Of Herbert's many research interests and achievements, two, in particular, have had an enormous impact on their respective fields. The first concerns studies of the interaction between individual flux lines in type-II superconductors and individual metallurgical defects, which was an extremely hot topic in the 1970's and rather controversial after the publication of Labusch's statistical summation theory of elementary pinning forces. As an early indicator of Herbert's modus operandi, he convened a small group of scientists from all around the world to discuss this issue during the "International

IOP Conf. Series: Materials Science and Engineering **102** (2015) 011002 doi:10.1088/1757-899X/102/1/011002

Discussion Meeting on Flux Pinning in Superconductors", near Göttingen, in 1974, effectively stimulating further dedicated research on this topic worldwide. The second contribution of particular significance came with the advent of high temperature superconductivity in the late 1980's. Herbert was amongst the first to realize that the most promising material, YBCO, required highly specialized fabrication techniques to achieve its potential as a conductor for practical applications. As a consequence, Herbert was one of the pioneers to introducing and develop IBAD techniques for the preparation of YBCO tapes and coated conductors and was able to set up an impressive laboratory at the then newly founded "Zentrum für Funktionswerkstoffe". This activity led to another trademark characteristic of Herbert's attitude and approach towards scientific research, which was to target, and subsequently foster, very close cooperation with industry and to support focused, well managed knowledge transfer from academia to industry. In fact, and not coincidentally, much of the technology developed at ZFW has been transferred to a Bruker company.

Herbert has an extremely high and well-respected profile internationally and his research has received worldwide recognition. He realized very early on that bringing scientists together in informal "discussion meetings" is a very efficient way of promoting progress in topics of actual interest and a way of getting to the heart of a particular issue. Herbert instigated a series of international workshops on "Flux Pinning and Critical Currents in Superconductors" following one of these discussion meetings in 1974 that were held subsequently every two to three years rotating between Europe, Japan and the U.S. At the time, these workshops were renowned for attracting the most eminent researchers in the field. During his research visit at Argonne in 1973, Herbert made close personal contacts with leading scientists in the U.S. As a consequence, he was invited to join the ICMC Board in 1981, subsequently to become elected President from 2000 to 2005, Chairman of the Anchorage Conference in 2005 and Co-Chairman of the 2001 off-year conference. He also became a Board Member of the Applied Superconductivity Conference and suggested rotating this major conference between the U.S., Europe and Japan. Although this turned out not to be achievable, it led to Herbert founding the European Conference on Applied Superconductivity (EUCAS), which was held for the first time in Göttingen in 1993 and which has continued to be an increasingly successful biennial meeting ever since, attracting more than 900 attendants for the first time in Vienna in 2005 and more than 1000 from 40 countries in Genova in 2013. There could be no better legacy of Herbert's contribution to the field of applied superconductivity than the continuing success and importance of the EUCAS conference within the scientific community.

Having established what was to become a subject-defining conference in EUCAS, and since the format of these meetings exhibited few of the trademark characteristics of Herbert's favorites, that is discussion meetings or workshops, he founded yet another series on a hot topic in 2000 - the "International Workshops on Coated Conductors for Applications" (CCA). Following the inaugural meeting in Göttingen, CCA now also rotates to countries around the world and typically attracts around 150 specialists in this technologically important area for applications of high temperature superconductors, fostering particularly close contacts between academia and industry. More recently, Herbert has played a major role in the organization of the two-yearly PASREG conference on bulk superconductivity, which he has done with characteristic enthusiasm and commitment. Herbert also serves currently as Co-editor of *Materials, Conductors and Materials Physics* and Deputy Editor-in-Chief of the global Superconductivity News Forum (SNF), which is published jointly by IEEE-CSC and ESAS.

From Herbert's broad and impressive range of activities aimed at serving the community and promoting the field of superconductivity, it is not surprising to note that numerous institutions worldwide have been eager to welcome him as a Visiting Scientist, which he has always been ready to accept, underlining further his worldwide reputation. Significantly, Herbert received the IEEE Award for Continuing Significant Contributions to the Field of Applied Superconductivity in 2006, which is further evidence of the esteem within which he is held by his peers.

IOP Conf. Series: Materials Science and Engineering **102** (2015) 011002 doi:10.1088/1757-899X/102/1/011002

Herbert's high quality research combined with tireless efforts to serve the community and to advance the field of superconductivity, and particularly regarding its industrial application, make Professor Herbert Freyhardt an excellent candidate for receiving the 2015 ICMC Lifetime Achievement Award. The citation reads:

"The Lifetime Achievement Award is presented for his contribution to the development of type II superconducting materials for practical applications, to the understanding of the effects of flux pinning on the applied properties of these materials and for his general outstanding contribution to the superconductivity community during a long and distinguished career"

It is both a privilege and pleasure for the ICMC Board, on behalf of the worldwide cryogenic materials community, to present the 2015 Lifetime Achievement Award to Professor Herbert C. Freyhardt.

David Cardwell
Chair, ICMC Awards Committee
Tucson, June 2015

ICMC BEST PAPER AWARDS

Awards for the best papers of the 2013 Conference, *Advances in Cryogenic Engineering, Vol. 60 (1574)*, were presented at the Conference in Tucson, Arizona, June 28 – July 2, 2015. From the large number of papers nominated by the technical editors, the committee selected two papers for the award. The awards were presented before the Monday morning plenary session at the 2015 Tucson Conference and the following authors were acknowledged for their outstanding contributions:

Best Superconducting Materials Paper

J. Hänisch, K. Iida, F. Kurth, T. Thersleff, S. Trommler, E. Reich, R. Hühne, L. Schultz, and B. Holzapfel

for their paper

"The Effect of 45° Grain Boundaries and associated Fe particles on J_c and resistivity in $Ba(Fe_{0.9}Co_{0.1})_2As_2$ Thin Films"
Adv. Cryog. Eng., Vol. 60 (2014), pp. 260-267

and

Best Structural Materials Paper

D.W. Cheng, D.R. Dietderich, A. Godeke, A.R. Hafalia, M. Marchevsky, S.O. Prestemon, P.K. Roy, G. Sabbi, and C.A. Swenson

for their paper

"Characterization of Insulating Coatings for Wind-and-React Coil Fabrication"
Adv. Cryog. Eng., Vol. 60 (2014), pp. 178-185

IOP Conf. Series: Materials Science and Engineering **102** (2015) 011003 doi:10.1088/1757-899X/102/1/011003

2015 International Cryogenic Materials Conference Committee

CONFERENCE AND PROGRAM

Wilfried Goldacker, *President* – Karlsruhe Institute of Technology, Germany
Michael Sumption, *Conference Chair* – The Ohio State University
Timothy Haugan, *Program Chair* – AFRL / RQQM

BOARD OF DIRECTORS

U. (Balu) Balachandran – Argonne National Laboratory
David A. Cardwell – University of Cambridge, United Kingdom
Lance D. Cooley – Fermi National Accelerator Laboratory
David Evans – Advanced Cryogenic Materials Ltd, United Kingdom
Herbert C. Freyhardt – University of Houston / University of Göttingen (retired) and ZFW gGmbH Göttingen, Germany
Wilfried Goldacker – Karlsruhe Institute of Technology, Germany
K. Theodore Hartwig, Jr. – Texas A&M University
Timothy Haugan – AFRL /RQQM
Eric E. Hellstrom – Applied Superconductivity Center / NHMFL / FSU
Juan Knaster – Fusion For Energy, Spain
Yanwei Ma – Institute of Electrical Engineering, Chinese Academy of Science, China
Thomas H. Nicol – Fermi National Accelerator Laboratory
Richard P. Reed – Cryogenic Materials, Inc.
Kenichi Sato – Sumitomo Electric Industries Ltd., Japan
Jun-ichi Shimoyama – University of Tokyo, Japan
Michael Sumption – The Ohio State University
Herman H. J. ten Kate – CERN, Switzerland
Marcel ter Brake – University of Twente, The Netherlands
Robert P. Walsh – National High Magnetic Field Laboratory / FSU

AWARDS COMMITTEE

David A. Cardwell, *Chair* – University of Cambridge, United Kingdom
Eric E. Hellstrom – Applied Superconductivity Center / NHMFL / FSU

ICMC Technical Editors

U. (Balu) Balachandran
Chief Technical Editor
Argonne National Laboratory
9700 S. Cass Avenue, Building #212
Argonne, IL 60439
USA
Email: balu@anl.gov

Lance Cooley
Fermi National Accelerator Laboratory
PO Box 500, MS 316
Batavia, IL 60510
USA
Email: ldcooley@fnal.gov

K. Ted Hartwig
Texas A&M University
319 Mechanical Engineering Office Bldg.
College Station, TX 77843-3123
USA
Email: thartwig@tamu.edu

Eric Hellstrom
ASC / NHMFL / FSU
2031 E. Paul Dirac Drive
Tallahassee, FL 32310
USA
Email: hellstrom@asc.magnet.fsu.edu

Takanobu Kiss
Kyushu University
744 Motooka, Nish-ku
Fukuoka, 819-0395
Japan
Email: kiss@sc.kyushu-u.ac.jp

Managing Editor
Annett D'Antonio
Centennial Conferences
908 Main Street, Suite 230
Louisville, CO 80027
USA
Email: cecicmc@centennialconferences.com

Peter Lee
ASC / NHMFL / FSU
2031 E. Paul Dirac Drive
Tallahassee, FL 32310
Email: lee@asc.magnet.fsu.edu

Tatiana Prikhna
Institute for Superhard Materials of the National Academy of Science of Ukraine
2, Avtozavodskaya Str.
Kiev, 04074
Ukraine
Email: Prikhna@mail.ru

Wolfgang Stautner
GE Global Research
One Research Circle
Niskayuna, NY 12309
USA
Email: Stautner@research.ge.com

Michael Sumption
The Ohio State University
Materials Science Department
CSMM, 477 Watts Hall
2041 College Road
Columbus, OH 43210
USA
Email: Sumption.3@osu.edu

Robert Walsh
NHMFL/FSU
1800 E. Paul Dirac Dr.
Tallahassee, FL 32310-3706
USA
Email: walsh@magnet.fsu.ed

IOP Conf. Series: Materials Science and Engineering **102** (2015) 011005 doi:10.1088/1757-899X/102/1/011005

Acknowledgments

The CEC and ICMC Boards wish to thank our supporter and sponsors,

SUPPORTER

- **Fermilab**

SPONSORS

- **Air Liquide**
- **Cryomech, Inc.**
- **Demaco**
- **IEEE Council on Superconductivity**
- **Linde**
- **TechSource**
- **Visit Tucson**

who have contributed to the 2015 CEC/ICMC Conference.
Their contributions helped ensure the success of the 2015 Conference.

Producing the **Open Access** ***IOP Conference Series: Materials Science and Engineering (MSE)*** requires the dedicated effort of many individuals. The quality of the publication relies on the dedication of those who took time from their busy schedules to review and edit papers. The Boards wish to specifically acknowledge the efforts of the managing editor, Annett D'Antonio of Centennial Conferences and the chief technical editors, John Weisend (CEC) and Balu Balachandran (ICMC). Their contributions to ensure a quality publication in a timely manner are greatly appreciated. The chief technical editors are indebted to the individual technical sub-editors for their time spent with reviewers and authors to get the final manuscripts ready for publication. The Boards also wish to thank all students and post-doctoral researchers from a number of universities for their service with the initial format checking of the papers submitted for peer review. Balu Balachandran, the Chief Technical Editor of ICMC, would like to thank Dick Reed for the monumental task of preparing the materials and subject indexes for volumes 22 through 60 of Advances in Cryogenic Engineering (Materials). Those indexes are included in volume 102 of *IOP Conference Series: Materials Science and Engineering*.

CEC/ICMC 2015 was managed by Paula Pair and Centennial Conferences. The Boards recognize this is a tremendous team effort and in addition to Paula would like to especially acknowledge all her staff – Jim Berhalter, Troy Christensen, Annett D'Antonio, Johanna Hild, Alicia Orcutt, Christina Tomhave and Joey Wilkerson – for the tremendous job they did.

Last, but not least, the Conference would not have been possible without the effort and dedication of the attendees in preparing their presentations and papers, traveling to Tucson, Arizona, and showing the professionalism and enthusiasm that makes the CEC/ICMC such a great Conference time after time.

Peer review statement

All papers published in this volume of *IOP Conference Series: Materials Science and Engineering* have been peer reviewed through processes administered by the proceedings Editors. Reviews were conducted by expert referees to the professional and scientific standards expected of a proceedings journal published by IOP Publishing.

Effect of microstructure on high-cycle fatigue properties of Alloy718 plates

Y Ono[1], T Yuri[1], N Nagashima[1], T Ogata[1], and N Nagao[2]

[1] National Institute for Materials Science, 1-2-1 Sengen, Tsukuba, Ibaraki 305-0047 Japan
[2] Japan Aerospace Exploration Agency, 2-1-1 Sengen, Tsukuba, Ibaraki 305-8505 Japan

Email: ONO.Yoshinori@nims.go.jp

Abstract. Effect of microstructure on high-cycle fatigue properties of Alloy718 were investigated at 77 K by using samples with three different microstructures; fine-grained (FG), coarse-grained (CG) and bimodal-grained (BG) ones. The BG sample consisted of FG and CG microstructural regions and grain sizes of those regions were close to those of the FG and the CG samples, respectively. High-cycle fatigue strength of the FG sample was higher than that of the CG sample. High-cycle fatigue strength of the BG sample was clearly lower than that of the FG sample and almost the same as that of the CG one. Flat area (facet) was found at fatigue crack initiation site in all specimens. Facet size was similar to the grain size and found to be almost same in the CG and the BG samples. Observations of the microstructure beneath the fatigue crack initiation site of the BG sample revealed that the facet corresponds to transgranular cracking in the course grain, meaning that fatigue crack initiated at the coarse grain in the BG sample. It is deduced that the high-cycle fatigue strength of Alloy 718 with the BG microstructure is strongly affected by that of the CG region in that material.

1. Introduction

Ni-base superalloy 718 (Alloy718) is a precipitation-strengthened material mainly deriving from γ'-Ni_3Al ($L1_2$) and γ'' -Ni_3Nb (DO_{22}) precipitates [1]. This alloy has been widely used in gas turbines and jet engines with temperature range up to 923 K, because of its high temperature strength and corrosion resistance. This alloy is also used for some cryogenic applications in aerospace industry, since this alloy exhibits high strength at cryogenic temperatures and good weldability [2-4].

It was reported that Alloy718 with finer grains achieves better high-cycle fatigue properties [5]. Delta-phase (δ: Ni_3Nb) plays an important role to obtain fine grains [6]. If distribution of Nb is not homogenized properly, sufficient amount of δ doesn't precipitate in the Nb-lean region at high temperature. It causes the growth of γ grains there and the formation of bimodal-grained (BG) microstructure that consists of fine-grained (FG) and coarse-grained (CG) regions [6].

In this study, Alloy718 with BG microstructure had been obtained actually. It's very interesting to comprehend the effect of certain amount of coarse grains on mechanical properties of Alloy 718 with BG microstructure. Therefore, we also prepared samples that have homogeneously FG and CG microstructures, respectively. Then we investigated high-cycle fatigue properties of three samples and observed fracture surfaces and microstructures of fatigue-fractured specimens, and discussed the effect of microstructure on high-cycle fatigue properties.

Table 1. Chemical compositions of Alloy 718 plates used in this study (mass%).

	C	Si	Mn	P	S	Cu	Ni	Cr	Mo	Co	Al	Ti	Nb	B	Ta	Fe
Plate 1 (Bimodal-grained)	0.03	0.05	0.16	0.002	<0.001	0.02	52.57	18.12	2.85	0.21	0.43	1.02	5.09	0.005	<0.01	bal.
Plate 2 (Normal-grained)	0.03	0.02	0.13	0.007	<0.001	0.01	53.49	18.27	2.92	0.13	0.43	1.06	5.16	0.004	<0.01	bal.

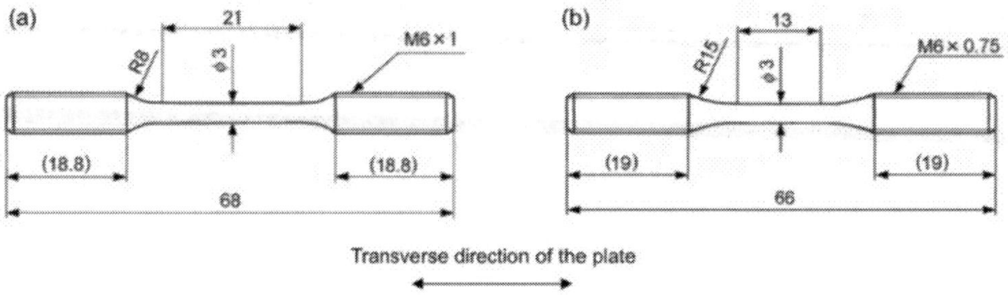

Figure 1. Dimensions and sampling direction of the specimens for tensile test (a) and high-cycle fatigue test (b). All dimensions are shown in mm.

2. Experimental procedure

Two plates of Alloy718 used in this study had been produced in accordance with Aerospace Materials Specification 5596. They had bimodal-grained (BG) and normal-grained microstructures, respectively. The chemical compositions of those plates are shown in table 1. The BG plate was hot-rolled at a reduction of 80%. After the rolling processes, it was solution-treated at 1228 K for 3.6 ks, followed by air-cooling. Regarding the plate for obtaining the FG and the CG microstructures, it was hot-rolled at a reduction of 90% and then subjected to the solution-treatment at the temperatures of 1228 K and 1318 K for 1.8 ks to control the γ grain size. Finally, three solution-treated samples were double-aged at 993 K for 28.8 ks and 893 K for 28.8 ks, followed by air-cooling. The processing details are summarised briefly in table 2.

Figure 1 shows the dimensions and sampling direction of tensile (a) and high-cycle fatigue (b) test specimens. The longitudinal direction of those specimens is parallel to transverse direction (TD) of the plate. Tensile tests and fatigue tests were carried out at 77 K. Tensile tests were performed at an initial strain rate of 4.2×10^{-4} s^{-1}. Fatigue tests were conducted up to 10^7 cycles using sinusoidal waveform loading and uniaxial tension-compression loading under stress ratio (R) of -1. Test frequency was 10

Table 2. Processing details for three samples evaluated in this study.

Plate	Reduction ratio in hot-rolling process (%)	Sample	Solution treatment	Aging
Plate 1 (Bimodal-grained)	80	Bimodal-grained (BG)	1228 K, 3.6 ks → Air-cooling	993 K for 28.8 ks → 893 K for 28.8 ks → Air-cooling
Plate 2 (Normal-grained)	90	Fine-grained (FG)	1228 K, 1.8 ks → Air-cooling	
		Coarse-grained (CG)	1318 K, 1.8 ks → Air-cooling	

Hz at 77 K. The microstructure was observed with an optical microscope and a scanning electron microscope (SEM). Fracture surfaces were observed with the SEM.

3. Results and discussion

Figure 2 shows microstructures of the FG (a), the CG (b) and the BG (c), (d) samples. The mean grain size of the FG sample was 40 μm. The solution-treatment temperature of 1228 K for the FG sample was lower than δ solvus, which was reported around 1283 K [7]. Therefore, plate-like δ was observed mainly along grain boundaries and pinned the grain boundaries effectively during the solution treatment. The mean grain size of the CG sample was 100 μm. The solution-treatment temperature of 1318 K for this sample was higher than δ solvus, resulting in coarsening the grains easily. On the other hand, the BG sample was obtained after the almost same heat treatment as that of the FG sample. It consisted of FG and CG regions (c). Mean grain size of each region was 30 μm and 100 μm, respectively. Figure 2(d) is a backscatter image of the CG region and a dark-etched area observed in the BG sample (c). Dark-etched areas shown in figure 2(c) correspond to grains indicated by arrows in figure 2(d). Variation of contrast was easily found within those grains. This contrast is derived from residual strain, meaning that they seem to be recovered (unrecrystallized) grains. Nb-enriched MC-type carbides were observed in all samples [8].

Tensile properties of the samples are summarized in table 2. The 0.2% proof stress and tensile strength of the FG sample are higher than those of the CG sample due to grain refinement. Regarding the BG sample, the 0.2% proof stress is close to that of the CG sample. Tensile strength of it lies in between those of the FG and the CG samples.

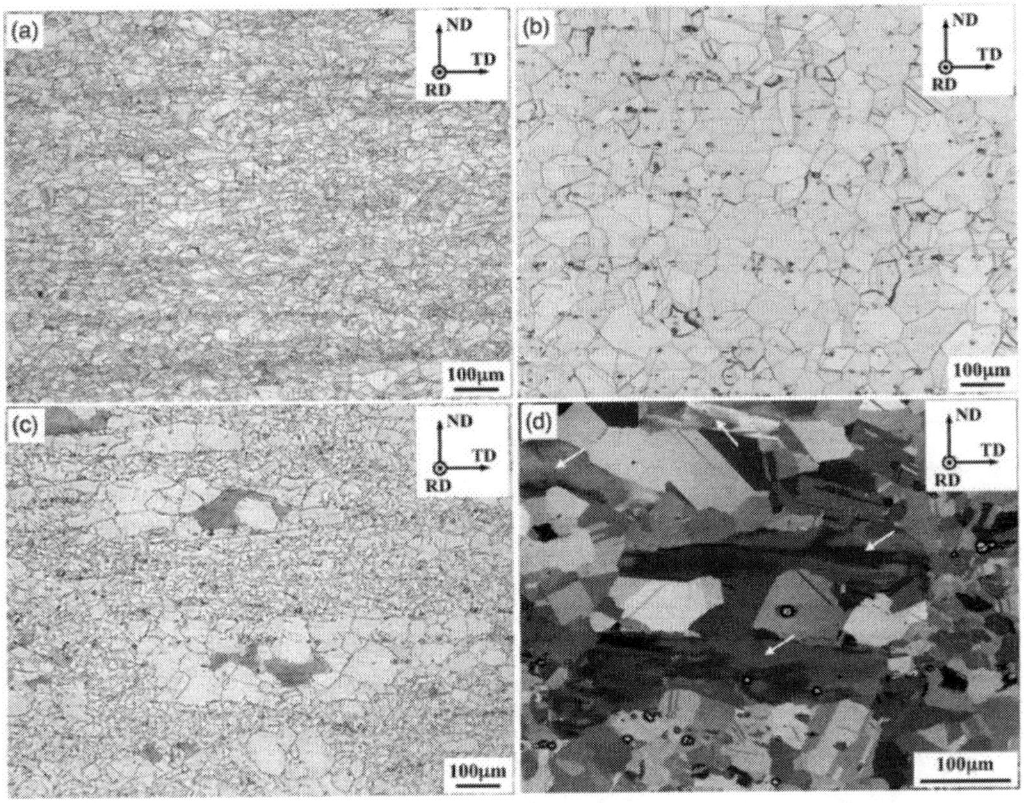

Figure 2. Optical images of the FG (a), the CG (b) and the BG (c) samples and a backscatter electron image of the BG sample (d).

Table 3. Tensile properties of the BG, FG and CG samples of Alloy 718 plates at 77 K.

Microstructure	0.2% proof stress $\sigma_{0.2}$ / MPa	Tensile strength σ_B / MPa	Elongation δ (%)	Reduction of area ϕ (%)	Number of specimens tested
Bimodal-grained (30 μm, 100 μm)	1,310	1,758	21	30	2
Fine-grained (40 μm)	1,393	1,817	26	30	3
Coarse-grained (100 μm)	1,331	1,661	27	50	2

Figure 3 shows the results of the high-cycle fatigue tests at 77 K. High-cycle fatigue strength of the FG sample was higher than that of the CG sample, as the 0.2% proof stress and tensile strength are. On the other hand, high-cycle fatigue strength of the BG sample was clearly lower than that of the FG sample and almost the same as that of the CG sample. It is known empirically that there is a good correlation between high-cycle fatigue strength and tensile strength [9]. However, the results obtained in this study don't meet this empirical law.

Figure 4 shows a whole image of fracture surface (a), a magnified image of the fracture surface around fatigue crack initiation site (b), a magnified image of the area surrounded by broken line in (b) (c) and an optical image of microstructure beneath the fracture surface in cross section along A-A in (c) of the BG sample (d). In this specimen, flat area (facet) was found at fatigue crack initiation site, as indicated by an arrow in figure 6 (b). Each facet size was similar to the grain size of each specimen. In the BG sample, the facet size was close to 100 μm the same as that of the CG sample. Figure 6 (d) shows that the facet corresponds to transgranular cracking in a coarse grain.

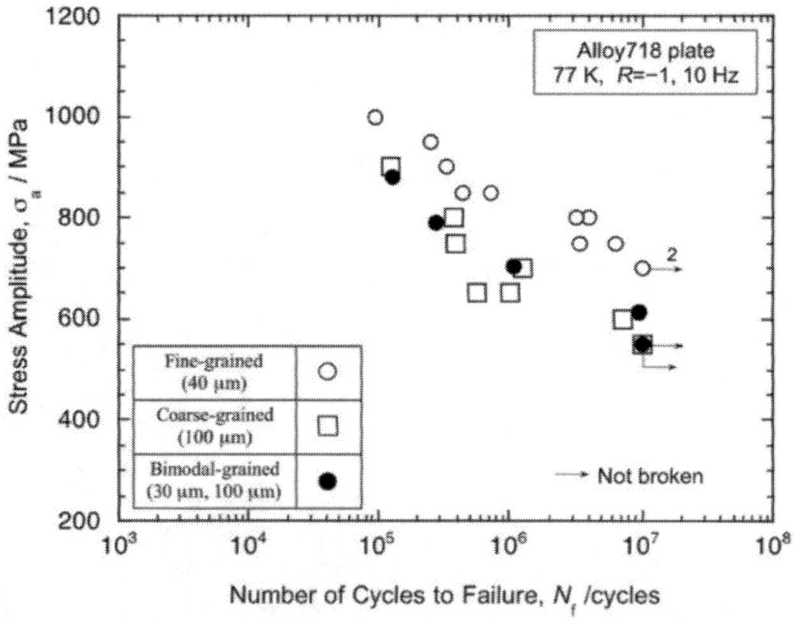

Figure 3. *S-N* diagrams of the FG, the CG and the BG samples obtained at 77 K.

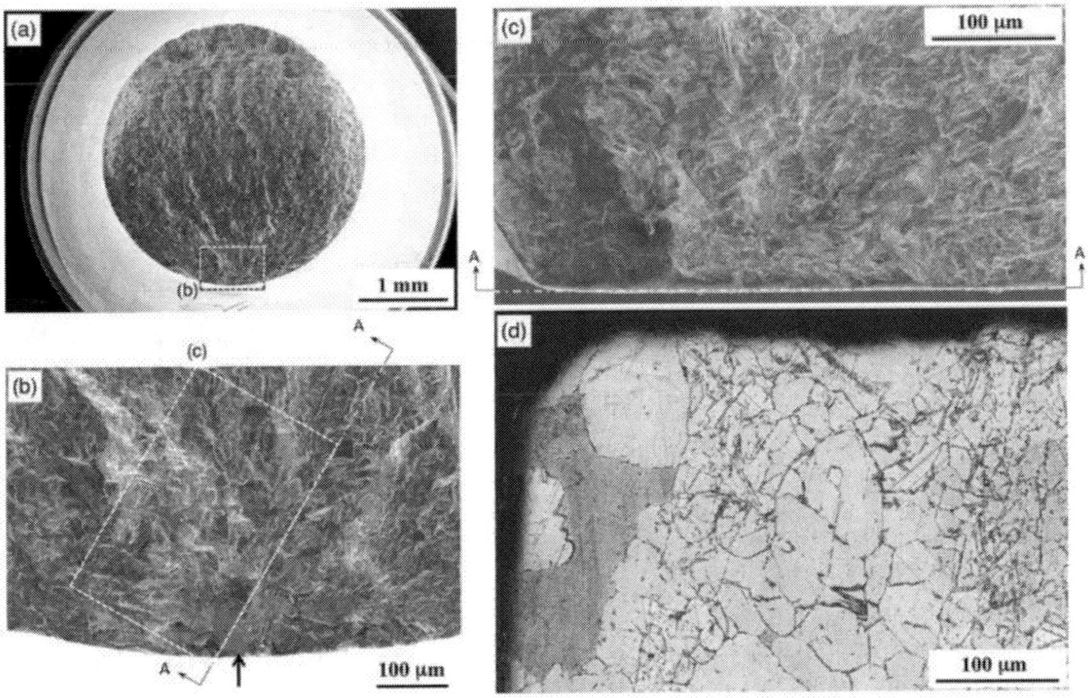

Figure 4. Images of the BG sample fatigue-tested at 77 K (σ_a = 615 MPa, N_f = 9.39×10^6 cycles). SEM images showing whole fracture surface (a), the fracture surface around fatigue crack initiation site that is indicated by an arrow (b), the area surrounded by broken line in (b)(c). An optical image of microstructure beneath the fracture surface in cross section along A-A in (c)(d).

Facet was observed in all fatigue-fractured specimens. NbC wasn't confirmed at fatigue crack initiation site in the specimens used in this study, though it was identified in the Alloy718 fatigue-tested at R=0.01 [8]. Regarding the BG samples, fatigue crack initiated only at the coarse grain. It was reported that fatigue crack initiates much easier in CG sample than in FG sample [5]. Thus, the high-cycle fatigue strength of Alloy 718 with the BG microstructure seems to be strongly affected by that of the CG region.

4. Summary

Effect of microstructure on high-cycle fatigue properties of Alloy718 were investigated at 77 K by using three samples with different microstructures; fine-grained (FG), coarse-grained (CG) and bimodal-grained (BG) ones. High-cycle fatigue strength of the FG sample was higher than that of the CG sample. High-cycle fatigue strength of the BG sample was clearly lower than that of the FG sample and almost the same as that of the CG sample. Fatigue crack initiated at a coarse grain in the BG sample. It is deduced that the high-cycle fatigue strength of the BG sample is strongly affected by that of the CG region, resulting in lowering the high-cycle fatigue properties.

References

[1] Cozar R, Pineau A 1973 Metall. Trans. A **4**, 47
[2] Tobler R L, 1976 *Cryogenics*, **16**, 669.
[3] Hasegawa K, Ando K, Kitade S, Sakamoto M, Fukushima Y, and Okita K, 1995 *Trans. Jpn. Soc. Mech. Eng.*, A **61**, 1707, in Japanese.

[4] Bae S H, Kwon S I, Yoon J G, Lee J H, Do J H, Kim I S, Choi B G, Jo C Y and Hong H U, 2014 *Metallurgical and materials transactions*. **45** A, 537.

[5] Kawagoishi N, Maemura E, Chen Q, Goto M and Morino K, 2008 *Trans. Jpn. Soc. Mech. Eng.*, A **74**, 1000, in Japanese.

[6] Donachie M J, et al. 2003 *SUPERALLOYS - A Technical Guide (2nd Edition)*, ASM International, Ohio, pp. 94-106.

[7] Mosser P E, Leconte G, Leray J, Lasalmonie A, Honnorat Y, 1989 *Superalloy 718-Metallurgy and Applications*, ed. Loria E A (The Minerals, Metals & Materials Society), 179.

[8] Ono Y, Yuri T, Sumiyoshi H, Takeuchi E, Matsuoka S, Ogata T, 2004 *Materials Transactions*, **45**, 342.

[9] Murakami Y, 2002 *Metal fatigue: Effects of small defects and nonmetallic inclusions* Oxford, Elsevier Science Ltd.

IOP Conf. Series: Materials Science and Engineering **102** (2015) 012002 doi:10.1088/1757-899X/102/1/012002

Tensile and fatigue qualification testing of ITER-CS conduit alloy JK2LB

R. P. Walsh[1], D.M. McRae[1], K. Han[1], and N. N. Martovetsky[2]

[1]National High Magnetic Field Laboratory
Florida State University
Tallahassee, FL 32310

[2]US ITER Project, Magnet Group
Oak Ridge, TN 37830 - 6483

Abstract. The ITER Central Solenoid (CS) coils utilize cable-in-conduit conductor (CICC) and the conduit alloy is JK2LB. The production grade conduit alloy (and it's welds) must meet strict requirements for strength, toughness, fatigue crack resistance, and fabricability. The conduit alloy must retain good mechanical properties after additional fabrication steps such as welding, coil winding strain and exposure to the Nb3Sn superconductor's reaction heat treatment. Here we present data from cryogenic tensile, fracture toughness, fatigue crack growth rate, and axial fatigue tests of JK2LB alloy and conduit butt welds, before and after the exposure to the reaction heat treatment. The tests of specimens removed directly from the conduit provide confirmation of the materials properties and the effect of the cold work and aging. The 4 K fatigue performance is extremely important to the reliability of the CS and is covered both by axial cyclic fatigue tests and the fatigue crack growth rate measurements.

1. Introduction

JK2LB is a high-manganese austenitic steel, developed by the Japan Atomic Energy Agency (JAEA) and Kobe Steel, which has been selected for use as the conduit alloy in the ITER Central Solenoid. The steel is a low thermal expansion alloy developed for Nb_3Sn CICC applications [1]. The alloy has a carefully controlled chemistry and microstucture in order to obtain the desired mechanical properties.

The goal of this characterization program is to bolster the existing database for JK2LB and increase the understanding of the material's structure and performance in addition to qualification testing to satisfy the basic minimum requirements.

2. Summary

Production grade conduit material was supplied by ITER to NHMFL in the compacted state for qualification testing reported here. Additional processing (cold work and heat treatment) of about (10) 1 m lengths of conduit are performed at NHMFL for characterization tests the production grade of JK2LB in the as-near-to in-service condition as possible.

Table 1: JK2LB chemistry in wt.%

Alloy	C	Si	Mn	P	S	Cr	Ni	Mo	N	B	Fe
Specification	<0.03	<0.5	20.5-22.5	<0.015	<0.015	12.0-14.0	8.0-10.0	0.5-1.5	0.09-0.15	0.001-0.004	Bal
Heat TB9249	0.012	0.26	21.85	0.007	0.002	12.81	9.19	0.97	0.11	0.0037	Bal
Heat TB9249	0.012	0.27	21.35	0.007	0.002	12.62	9.14	0.98	0.12	0.003	Bal

At 295 K, the base and weld metals must have minimum yield strength (YS), tensile strength (TS) and elongation of 300 MPa, 550 MPa and 30% respectively. The conduit undergoes fabrication steps during magnet construction (welding and deformation processes) that affect the mechanical properties and are include for the qualification testing. The properties of the annealed conduit prior to the magnet processing steps are supplied by the manufacturer - KOBE Steel, (YS = 332 MPa, TS = 569 MPa and Elong. = 45%). Here we investigate how the addition of cold work, welding and an aging heat treatment influence the strength, ductility, and fatigue performance of the conduit. The 295 K tests of the conduit- base and weld metal, in the as-near-to in-service condition, confirm a change in 295 K properties with the NHMFL measurements of (base,weld); (YS = 435 MPa, 466 MPa), (TS = 618 MPa, 619 MPa) and (Elongation = 44%, 34%).

At 4 K mechanical properties specification includes a fracture toughness requirement in addition to the tensile specification. The 4 K requirements are for the conduit to have minimum YS > 850 MPa, TS > 1150 MPa, Elongation > 25%, and K_{IC} > 130 MPa*m$^{0.5}$. Tests of the conduit base metal and welds, in the as-near-to in-service condition, results in average 4 K properties of (base, weld); (YS =1076 MPa, 1031 MPa), (TS = 1456 MPa, 1432 MPa), (Elongation = 42%, 34%) and (K_{IC}=175 MPa*m^$^{0.5}$,K_{IC}=244 MPa*m^$^{0.5}$), which easily satisfy the ITER requirements.

Additionally, the 4 K fatigue properties of the conduit base and weld metal, have been characterized in axial fatigue test to generate cyclic stress vs cycles to failure (S-n) curves. The results from these tests plus 4 K fatigue crack growth rate tests provide the data that allows designers to estimate the conduit's service lifetime and reliability.

In summary, the production grade JK2LB conduit base metal and welds, before and after the final processing steps, are found to have excellent mechanical properties that easily satisfy the ITER 295 K and 4 K requirements.

3. Material Information

The JK2LB's target chemistry and the production grade heat analyses are shown in Table 1. Piece lengths of the seamless square tube with a round hole are butt welded together at mid-length to provide the approximately 1 m lengths for qualification testing. The conduit was initially fabricated to slightly oversized dimensions (51.3 mm across flats, 35 mm inside dia.) for ease of cable insertion (not applicable for this exercise). For this program the manufacturer provides the conduit after performing next fabrication step; compaction, on the hollow tubes that would normally be done after cable insertion. Prior to the compaction step which adds 4 to 5 % cold work strain, the conduit was solution annealed (1100C, 5 min., WQ) and has an average ASTM grain size = 5.

The material preparation procedures for the addition of CW and the heat treatment are briefly described. The 1 m conduit test sections are subjected to bending strains with a 3 pt. bending jig. The bending is done in four steps shown in Figure 1. At Step 3, two different bending sequences can be identified as DB for Double Bend and RB for Reverse Bend. The material preparation and conditions are described below;

3.1 Compaction - the process step used to achieve final conduit cross-section dimensions after cable insertion. Compaction reduces the conduit's nominal outside dimensions approximately 2 to 2.5 mm, and is performed prior to receipt at NHMFL, the material condition is; As-Compacted (AC).

3.2 Welding - the procedure is a combination of automated and manual TIG process and is documented elsewhere [1]. The filler metal is high Mn steel JK2LB and the material referred to as the Weld. Welding is done prior to compaction and the welds are ground on the outside surface to the nominal conduit dimensions. The inside weld surface is not treated (ground) and is left in the as-welded condition.

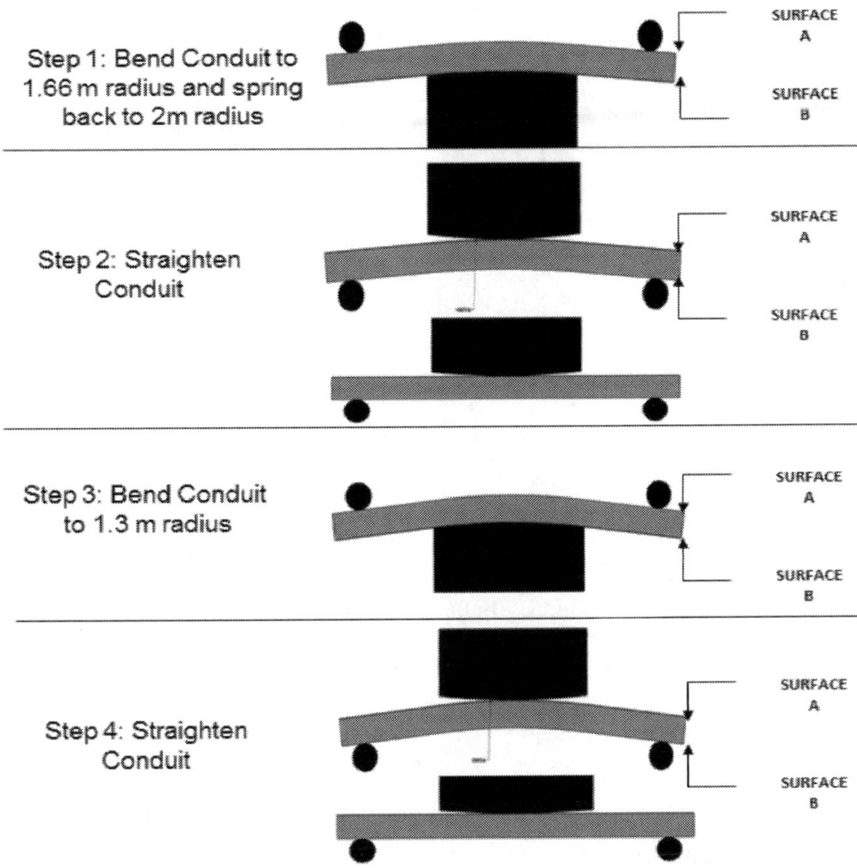

FIGURE 1. Schematic showing the material preparation steps done to simulate subsequent magnet winding operations after compaction and prior to the aging heat treatment.

3.3 Cold Work (CW) - Pre-coiling and coil winding introduce cold work. To simulate the cold work (prior to heat treatment), the (AC) conduit sections were formed in a 3 pt bend method over 2 different radii mandrels as shown in Figure 1. Step 1 bending simulates pre-coiling at 1.66 m radius. Step 3 bending simulates the minimum winding radius of 1.3 m of the CS coils. One of the two CW conditions is achieved with a Double Bend and the other CW condition is achieved with a Reverse Bend. The difference is a 180 degree flip of the conduit with respect to the bending direction. Since the final step is straightening the conduit for the removal of test specimens, it is thought that the difference in the state of the CW Reverse Bend and the CW Double Bend is negligible.

3.4 Aged condition (AG) - is accomplished by aging the (AC + CW) conduit at 650C/200h in an Argon atmosphere. This represents the final step of the Nb_3Sn reaction heat treatment that the conduit alloy must endure.

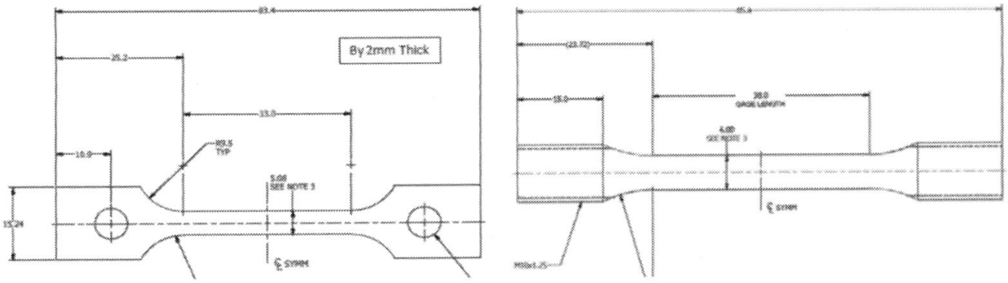

FIGURE 2. Schematic flat and round tensile specimens removed from the conduit section.

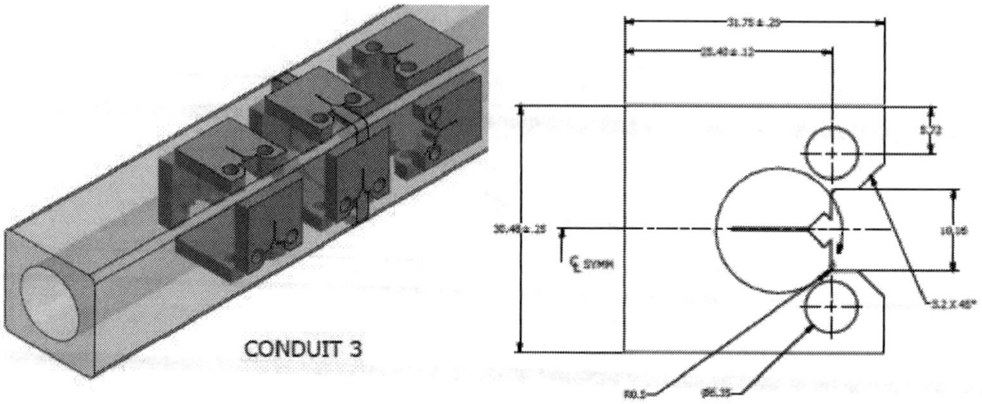

FIGURE 3. Schematic shows the location and orientation of dual purpose CT specimens.

4. Test Procedures

The mechanical properties of the base and weld metal are evaluated with tests on specimens that are electro-discharge machined (EDM) from the conduit as shown in Figures 3 and 4. All the tests are conducted on a 100 kN capacity MTS machine equipped with a cryostat to enable testing at 4 K with the specimen and fixture immersed in liquid helium. Tensile tests are conducted in displacement control at a rate = 0.5 mm/min according to procedures prescribed in ASTM E8 and E1450. Round tensile specimens were removed from conduit corners and the 2 mm thick flat tensile specimens were removed from the flat of the conduit as shown in Figures 2 and 5. Strain is measured with a 10% strain range clip-on extensometer.

For fatigue crack growth rate (FCGR) tests and J-integral (J_{Ic}) fracture toughness tests, dual-purpose compact tension (CT) specimens (Figure 3) were used. The 7 mm thick, 0.5 CT specimens are machined with a short notch (7 mm), after 2 mm crack initiation at 77 K, FCGR tests are conducted at 4 K for approximately 6 mm of crack extension (from 9 to 15 mm) or an a/W ratio = 0.6. Specimen orientation is defined by two letters, TL or LT, the 1st designates direction of applied force and the 2nd defines the crack direction. FCGR tests are conducted according to the guidelines in ASTM E647 using either the constant ΔP method or decreasing ΔK method.

The fracture toughness tests are conducted according to the guidelines provided in ASTM E1820-13. Fracture toughness CT specimens are usually oriented (TL) to generate conservative design data, with the crack direction parallel to the rolling direction. The ITER CS coil in-service stress is well defined, and a potential fatigue crack would run transverse to the conduit axis and the primary rolling (or extrusion) direction, thus one could argue that J tests should be performed on specimens with the LT orientation. Here we performed tests of the base metal with CT specimens from both orientations (4 each, TL and LT). For the butt weld material there is only one specimen orientation possible (LT). One half of the specimens (2 for each condition) were side grooved after pre-cracking as identified the results table.

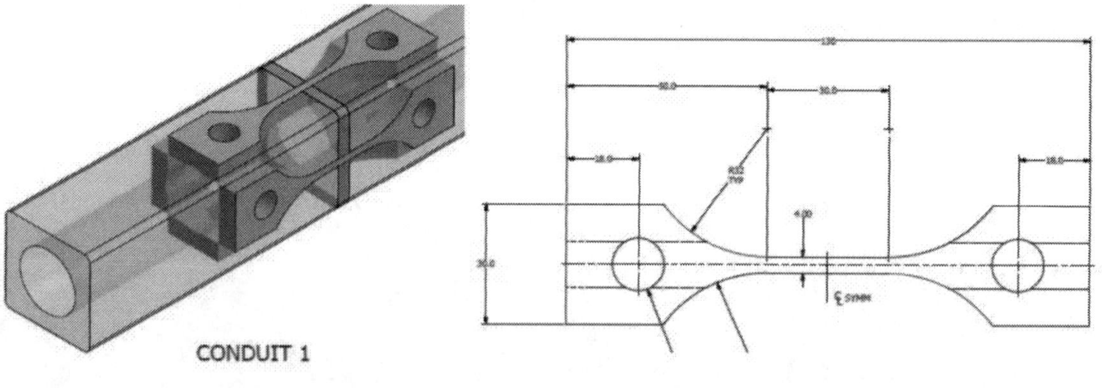

FIGURE 4. Schematic of fatigue specimens that retain the conduit inside and outside surface finish.

IOP Conf. Series: Materials Science and Engineering **102** (2015) 012002 doi:10.1088/1757-899X/102/1/012002

TABLE 2. JK2LB conduit tensile test results.

Material	Condition	Temp K	Yield Strength (MPa)	YS Std Dev (MPa)	Tensile Strength (MPa)	TS Std Dev (MPa)	Elong. in 25 mm (%)	Reduction of Area (%)	No of Tests
ITER Req. Minimum	AC+CW+AG	295	300		550		30		2
Base in Flats	AC+CW+AG	295	435	12	618	8	44	47	2
Weld in Corners	AC+CW+AG	295	466	2	619	4	34	54	2
ITER Req. Minimum	AC+CW+AG	4	850		1150		25		2
Base in Flats	AC+CW+AG	4	1085	40	1458	44	41	32	12
Base in Corners	AC+CW+AG	4	1076	5	1456	5	42	47	4
Weld in Corners	AC+CW+AG	4	1031	6	1432	5	34	51	2

Force-control axial-fatigue tests are used to generate cyclic stress vs. cycles to failure data (S-n curves). The 4 K tests are performed according to the guidelines in ASTM E466, parameters are sinusoidal tension-tension fatigue cycling, frequency f = 20 Hz, R-ratio (P_{min}/P_{max}) = 0.1. The test specimens are removed directly from the conduit as shown in the in Figure 4. The fatigue sample retains the original inside and outside surface finish of the as-manufactured conduit. Typically S-n fatigue tests are performed on polished samples but here we want to simulate the in-service conditions to get a better estimate of the in-service fatigue life.

5. Results and Discussion

5.1 Tensile Properties

Table 2 shows that the material has excellent 4 K tensile properties that easily satisfy the ITER requirements. Material variability is not very large as seen by the standard deviation of around 40 to 45 MPa for the 4 K average yield (1085 MPa) and tensile strength (1458 MPa) of the 12 base metal tests on flat specimens. The flat tensile specimens were tested to evaluate possible residual stress that may occur during processing of the conduit (Figure 5) and the test results showed negligible through thickness material variability.

The tensile properties of the welds are also excellent and only slightly lower strength than the base metal. The average 4 K tensile elongation of 34% is well above the required minimum of 25% and slightly reduced compared to the base metal's 42% elongation.

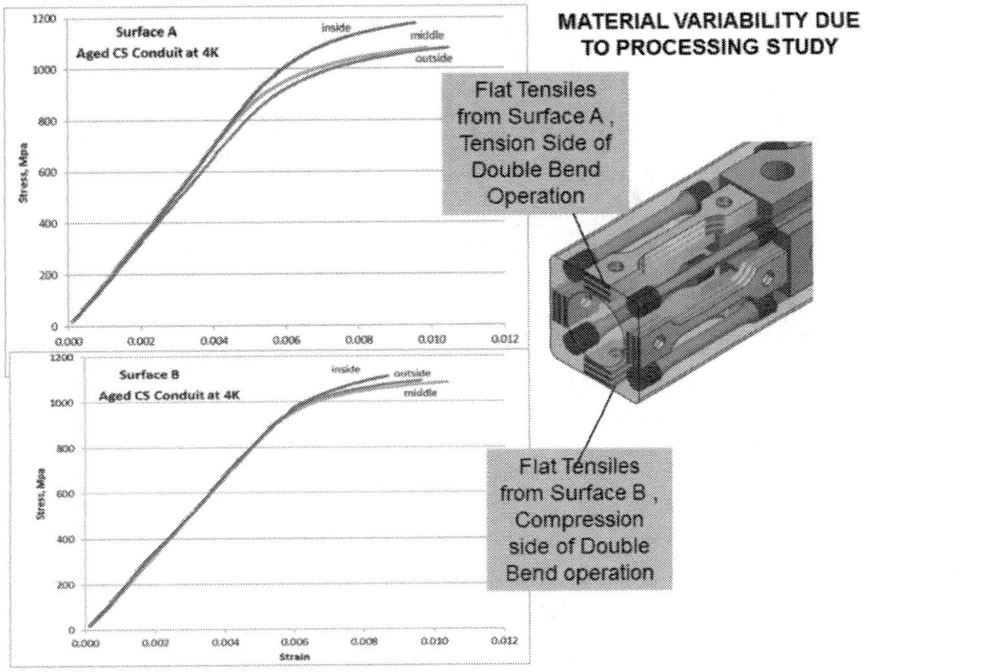

FIGURE 5. Stress-strain curves for evaluation of residual stress in conduit wall.

TABLE 3. 4 K FCGR Test Results

Material	Sample No.	Paris Law Constants (see note)		Delta K Range
		C	m	MPa√m
	CT-21	4.488E-11	4.122	23 to 48
Base	CT-22	1.677E-10	3.724	32 to 54
Metal	CT-23	5.599E-11	3.951	36 to 64
AC+CW+AG	CT-24	1.203E-10	3.789	27 to 47
	Average	2.087E-10	3.649	23 to 64
	CT-29	3.206E-10	3.676	43 to 78
Weld	CT-30	2.047E-10	3.715	29 to 50
Metal	CT-31	6.828E-12	4.558	34 to 62
AC+CW+AG	CT-32	1.277E-10	3.844	32 to 54
	Average	3.170E-11	4.212	29 to 78
Note: Relevant units for Paris Law constants				
ΔK units are MPa√m, and da/dn units are mm/cycle				

5.2 FCGR and Fracture Toughness

The fatigue crack growth rate (FCGR) test results are shown on Table 3 and the graph in Figure 6. The graph shows the analyzed data for four test each for the comparison of the average FCGR behavior of the fully processed (AC+CW+AG) base and weld metal as well as comparison with previously published comparable data for conduit alloys . The upper and lower bounds for the range of crack growth rates found in the literature [5,6] for aged 316LN intended for CICC applications are plotted in Figure 6 along with previously published FCGR measurements for JK2LB published in [2-4]. Previously published data for JK2LB nicely brackets current measurements.

The goal of fracture toughness testing is to determine the J integral at the point of fracture instability (J_Q) that satisfies the criteria to become the critical J integral (J_Ic), which can then be converted to the more commonly used critical stress intensity factor (K_{JIc}). The strict validity

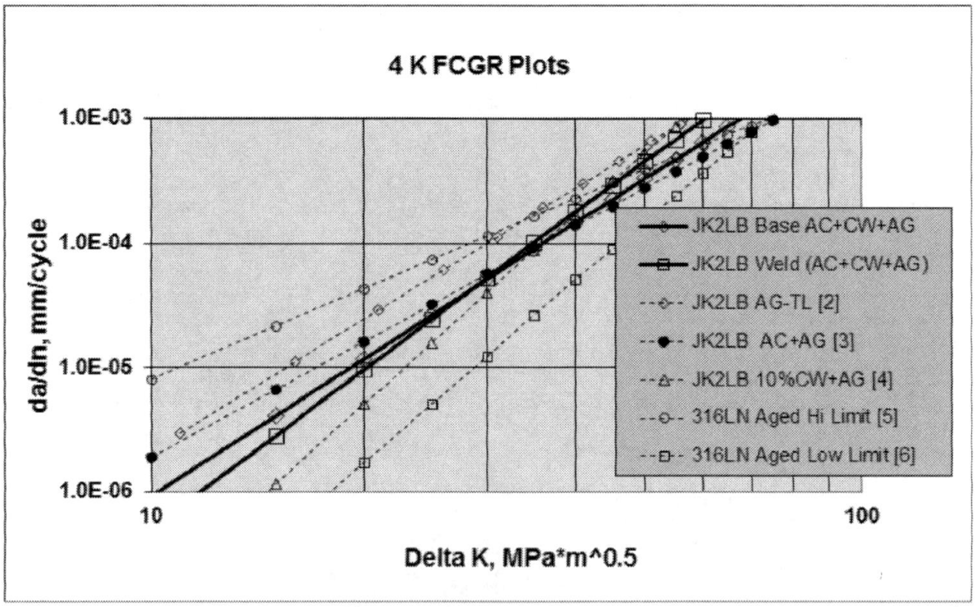

FIGURE 6. 4K Fatigue Crack Growth Rate data graph along with reference data for comparison.

TABLE 4. 4K Fracture Toughness Test Results

Specimen No.	Material	Orientation	K_{JIc} (MPa√m)	K_{JIc}, AVG (MPa√m)	STD DEV, (MPa√m)	Validity Notes
CT-9	Base	LT	360	357	14	Invalid due to J test crack tunneling
CT-10			376			Invalid due to J test crack tunneling
CT-11, side grooved			346			Invalid due to crack curvature
CT-12, side grooved			348			Invalid due to crack curvature
CT-17	Base	TL	174	177	7	Invalid due to J test crack tunneling
CT-18			188			Invalid due to J test crack tunneling
CT-19, side grooved			173			Invalid due to crack curvature
CT-20, side grooved			176			Invalid due to crack curvature
CT-13	Weld	LT	243	237	26	Invalid due to J test crack tunneling
CT-14			267			Invalid due to J test crack tunneling
CT-15, side grooved			235			Invalid due to crack curvature
CT-16, side grooved			204			Valid

requirements of the ASTM E1820 test standard are designed to ensure the resulting measurements are specimen size independent. Satisfying all the requirements is difficult, especially in cryogenic tests. The most basic qualification criteria are the geometric requirements in Equation 1;

$$B, b \geq 10 * J_Q / \sigma y \qquad (1)$$

Where: B = sample thickness, mm
b = uncracked ligament length
J_Q = J integral at the onset of significant tearing
σy = (Yield Strength + Tensile Strength)/2

Although this size criteria was met by all the specimens, with thickness (B) of 7 mm and remaining ligaments (b) of around 10 mm, almost all of these tests failed to meet other more discreet requirements related to crack front straightness as noted on the results table. Even though most of the test results are invalid, according to the strict ASTM requirements, they are considered to be good estimates of the fracture toughness and confirm the JK2LB material and welds have very good toughness at 4 K.

The fracture toughness test results are shown in Table 4. The base metal shows a high degree of anisotropy with almost a factor of 2 difference in the values measured for the two orientations (the LT toughness is greater). A similar degree of anisotropy has been seen previously in tests of JK2LB conduit material [2] with 40% higher toughness for the LT orientation compared to the TL orientation. The base metal exhibits extremely high toughness (357 MPa*m^$^{0.5}$) with the crack direction transverse to the conduit's longitudinal axis (LT orientation). The toughness of the TL orientation (177 MPa*m^$^{0.5}$) is also excellent and compares well with previously reported toughness (167 MPa*m^$^{0.5}$) [2] of aged JK2LB conduit for the TL orientation. The weld falls in between with exemplary toughness of 237 MPa*m^$^{0.5}$ as the measured average.

5.3 Axial cyclic fatigue (S-n)

The cyclic fatigue results are shown graphically in Figure 7 along with a table of the individual data pts. Previous S-n fatigue data for JK2LB [2] with slightly better fatigue performance is also plotted on the graph for comparison. The reason for the better fatigue performance of the prior data [2] from JK2LB conduit may be that the samples had a polished surface finish and the ones tested here were left with the as manufactured finish.

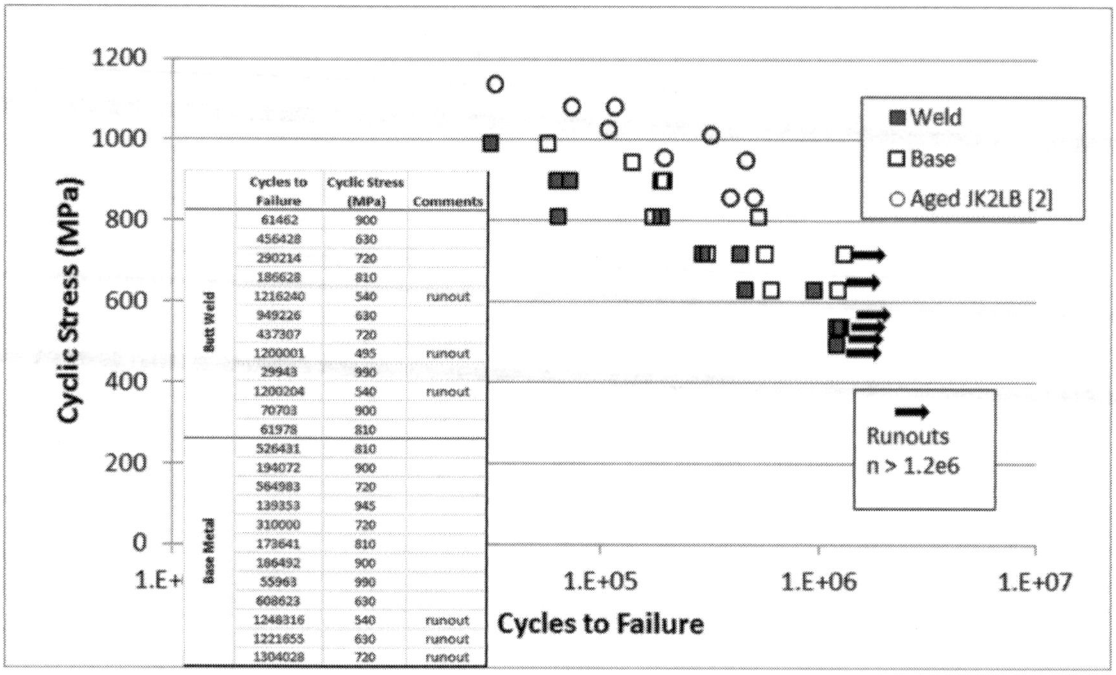

	Cycles to Failure	Cyclic Stress (MPa)	Comments
Butt Weld	61462	900	
	456428	630	
	290214	720	
	186628	810	
	1216240	540	runout
	949226	630	
	437307	720	
	1200001	495	runout
	29943	990	
	1200204	540	runout
	70703	900	
	61978	810	
Base Metal	526431	810	
	194072	900	
	564983	720	
	139353	945	
	310000	720	
	173641	810	
	186492	900	
	55963	990	
	608623	630	
	1248316	540	runout
	1221655	630	runout
	1304028	720	runout

FIGURE 7. Graph of 4K axial fatigue data for JK2LB along with [1] JK2LB data for comparison.

6. Conclusions

In summary, the production grade JK2LB conduit base metal and welds, before and after the final processing steps, are found to have excellent mechanical properties that easily satisfy the ITER 295 K and 4 K requirements. The 4 K fatigue properties of the conduit base and weld metal, have been characterized in axial fatigue test and fatigue crack growth rate tests generating data that allows designers to estimate the conduit's service lifetime and reliability.

Acknowledgements

This work was funded by US-ITER Project Office through a grant from Battelle. We would like to thank JAEA for supplying the test material.

REFERENCES

[1] Japanese Atomic Energy Agency Document No. : JADA-11002-221R3001_E

[2] Mechanical Properties of Modified JK2LB for Nb3Sn CICC Applications, R. P. Walsh, K. Han, and V. J. Toplosky, N. N. Martovetsky, T. L. Mann Jr., and J. R. Miller, *Advances in Cryogenic Engineering- Materials* **56**, (2010) pp.17-24.

[3] Hamada, Nakajima, Matsui, Kawano, Takano, Tsutsumi, Okuno, Teshima, and Soejima, " Development of Jacketing Technologies for ITER CS and TF Conductor", *Advances In Cryogenic Engineering - Materials* **54**, pp. 76-83 (2008).

[4] Feng, Kim, Ballinger and Minervini " Mechanical Properties of JK2LB at 4 K" *Advances In Cryogenic Engineering-Materials,* **52**, pp. 192-199 (2006).

[5] Nyilas, Obst and Harries, "Mechanical Investigations on Aged Stainless Steel and Incoloy 908 Materials at 4.2 K and 7 K", *Advances In Cryogenic Engineering (Materials)* **48**, pp 128-128 (2002).

[6] Kim and Feng," The 4K Mechanical Properties of Modified SS316LN for Jacket Materials in Superconducting Fusion Magnets" *Advances In Cryogenic Engineering-Materials,* **54**, pp. 92-99 (2008).

The Mechanical and material properties of 316LN austenitic stainless steel for the fusion application in cryogenic temperatures

J Sas[1], K-P Weiss[1] and A Jung[1]

[1] Karlsruhe Institute of Technology, Institute for Technical Physics, Hermann-von-Helmholtz-Platz 1, 76344 Eggenstein-Leopoldshafen, Germany

Abstract. Due to the constant increase of claims for all materials used in superconducting magnets in „magnetic fusion reactors", the article deals with the possibilities of increasing the mechanical properties of austenitic stainless steel tested at cryogenic conditions that ensure the transport of Helium to magnets. The aim of the experimental plan was to increase the mechanical properties of the steel grade 316LN tested at 4.2K from the original value Steel A: YS = 1045 MPa, UTS = 1528 MPa, A = 33 % to the value of YS = 1204 MPa, UTS = 1642 MPa, A = 34 % and Steel B: YS = 1173 MPa, UTS = 1541 MPa, A = 28 % to the value of YS = 1351 MPa, UTS = 1645 MPa, A = 17 %. The increase in mechanical properties of the steel grade under examination has been made by means of heat processing in the conditions of annealing: T_{h1} = 625 ° C / t_{h1} = 696 h. The mechanical properties of steel were evaluated using static tension tests at 4,2 K. The samples were placed in a cryostat filled with liquid helium. Except for the mechanical properties, there were also evaluated structural changes depending on the conditions of heat processing by light optical microscopy and EBSD (Electron Backscatter Diffraction). The increase of steel properties used in low temperatures was achieved by heat processing.

1. Introduction

For large-scale superconducting magnets the so-called cable in conduit conductor (CICC) allows a structural reinforcement of the bare superconducting NbTi or Nb_3Sn strands by austenitic stainless steel pipes [1]. In relation to this specific material and its characteristics, 316LN austenitic stainless steel is a prospective material for fusion reactor due to its resistance to neutron irradiation and corrosion as well as excellent mechanical properties, exemplified by high temperature strength, toughness and ductility [2].

Nitrogen addition resulted in the improvement of mechanical properties, related to short range order SRO, retardation of dynamic strain aging DSA, microstructure change from cellular dislocation to planar dislocation structure, decrease of grain size and carbide precipitation at grain boundaries. Importance of these factors is dependent on stress as well as temperature which was applied during the testing period [2].

316 LN Stainless steel was used as a jacket material for the Nb_3Sn conductors [3]. In relation to this fact, cold working effect as well as annealing on jacket's mechanical properties have to be investigated due to conductor compaction, bending and annealing, after the superconducting cable insertion into

the jacket. Moreover, since the CICC operates at about 4.5K, it is inevitable to evaluate mechanical strength at low temperatures. Mechanical properties of a stainless jacket with 316LN differs greatly from the 316L stainless steel [4], having lower elongation at failure but simultaneously higher mechanical strength. During the cryogenic test, the primary minimum requirement to be achieved on elongation for TF jacket was 30%. However, it has been proven by the experiments on different jacket materials that high elongation at failure can be reached with certain difficulties. As a result, the minimum value to be achieved was reduced to 20% and the requirement for elongation was modified as well.

However, the measurements were performed on the sub-size samples. Therefore, this study deals with the comparison of results from tensile tests as well as with the evaluation of mechanical properties and material properties in the cryogenic temperatures [5].

2. Experimental material and procedure

2.1. Experimental material

For the purposes of this study, we used the seamless tube, which is made of the modified 316LN austenitic stainless steel. The carbon content was reduced to less than 0,02 (compared to C value which is the norm - less than 0, 03%) to reduce the degradation in ductility after the reaction of the Nb_3Sn superconductor to heat treatment. The chemical composition of the modified 316LN austenitic stainless steel is mentioned in the Table 1. In order to simulate the magnet manufacturing process, the experimental tubes (original outer diameter of 48 mm) were compacted and tensile stretched 2.5% in steps at room temperature. After the magnet manufacturing process simulations, the final dimensions of the jacket were 43.5 mm outer diameter and 2.0 mm wall thickness, as mentioned in the Figure 1. After these procedures, experimental tubes were put into the furnace for aging with argon atmosphere. To specify the parameters of the heat treatment (aging), it is necessary to state that the temperature was 625°C and the duration was 696 hours. The reason for this aging procedure was to simulate the Nb_3Sn formation process.

The sub-size specimens were cut by electron discharge machining (EDM) according to ASTM E8M standard, as it can be seen in the Fig.2. The dimensions of the specimens were 12,5 mm x 2 mm in the reduced section and the gauge length was 50 mm. Every test series consisted of at least three specimens in longitudinal direction.

Chemical composition of two experimental melts (wt%)

Spec. ID	**C**	**Si**	**Mn**	**P**	**S**	**Cr**	**Ni**	**Mo**	**N**	**Fe**
B	0.013	0.152	1.668	0.019	<0.005	17.25	13.71	2.537	0.174	Bal.
A	0.011	0.332	1.79	0.006	<0.005	17.36	13.58	2.671	0.131	Bal.

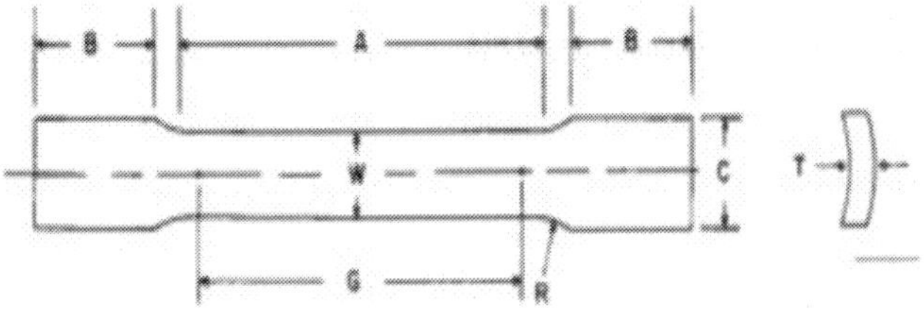

Parameter	Dimension
G-gage length	50 ± 0.1 mm
W-reduced section length	12.5 ± 0.2 mm
T-specimen thickness	Tube wall thickness
R-radius of fillet	≥12.5 mm
A-length of reduced section	≥60 mm
B-length of grip section	≥75 mm
C-width of grip section (approximate)	20 mm

Figure 1. Dimensions of tensile testing samples

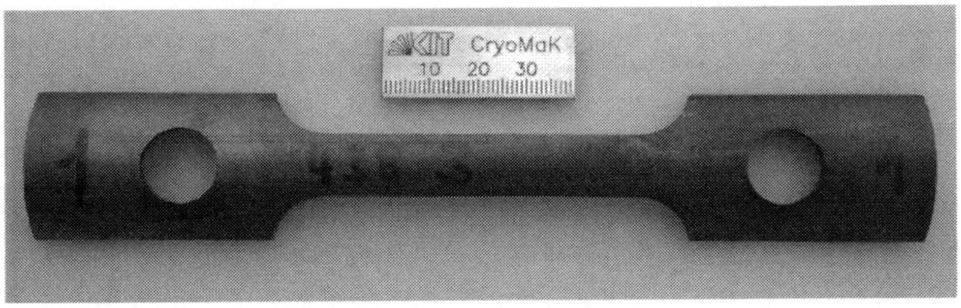

Figure 2. The sub-size sample

2.2. Test conditions
The sub-sized specimens were tested by an ATLAS universal testing machine with the load capacity of 650 kN, including the cryostat system inbuilt in the machine. Working temperature of the machine is down to 4.2 K. The sub-sized samples were immersed into the liquid helium in order to reach the given temperature 4.2 K. Tensile tests were conducted in displacement control with a strain rate of below 5 x 10^{-4} s^{-4} according to ASTM E8M. The material properties of the tested samples were examined using the scanning electron microscope with the EBSD camera (ZEISS LEO 1530) and the optical microscopy (ZEISS AXIO Vert. A1). For the analysis of chemical elements, we used the machine Bruker Q4 Tasman (optical emission spectrometer) and for the analysis of nitrogen content, we used the Bruker G8 Galileo (inert gas fusion analyser).

3. Experimental results and discussion

3.1. Mechanical properties
Mechanical properties of 316LN austenitic steel for RT measured at work [5] were compared with experimental materials A and B, which were tested at 4.2 K, shown in the FIG. 3. The measurements indicate that the value of yield and ultimate tensile strength measured at 4.2 K approximately doubled as compared to similar materials tested at RT and vice versa, total elongation demonstrated the opposite trend. By decreasing the temperature to 4.2 K, the values of total elongation were decreased for both tested materials of approximately 11 %. It was also observed that the material B has higher values of yield strength than the material A - of 128 MPa after the measurement at 4.2 K. For ultimate tensile strength and total elongation were not measured significant differences in both tested materials. Better mechanical properties (YS and UTS) of B versus A material can be attributed to the higher content of nitrogen in the chemical composition at the level of 0.17 % (material B) compared to the

material A, with the proportion of nitrogen 0.13 %. FIG. 4 shows the mechanical properties of both tested materials at 4.2 K, depending on the impact of compaction (CO) and 2.5 % tensile stretching (TS) carried out at RT (cold working), followed by aging (AG). The major influence on the value increase of yield (up to approximately 15%) and tensile strength (up to approximately 7 %) was from CO + TS. It is possible to observe the influence of aging on the values of total elongation at which the material B has its value decreased by 10 % [6].

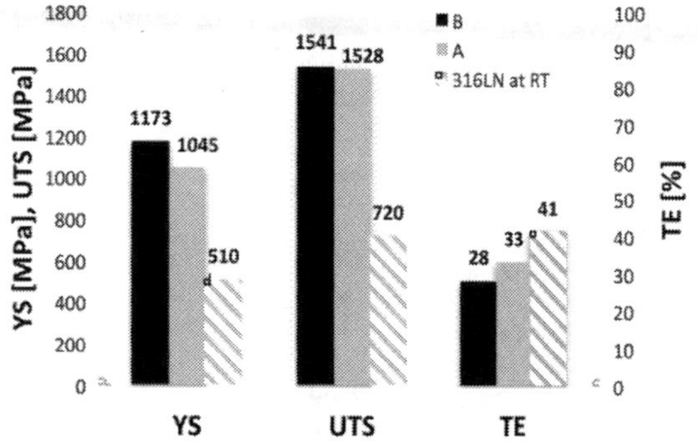

Figure 3. Mechanical properties at RT and 4.2 K of the 316 LN

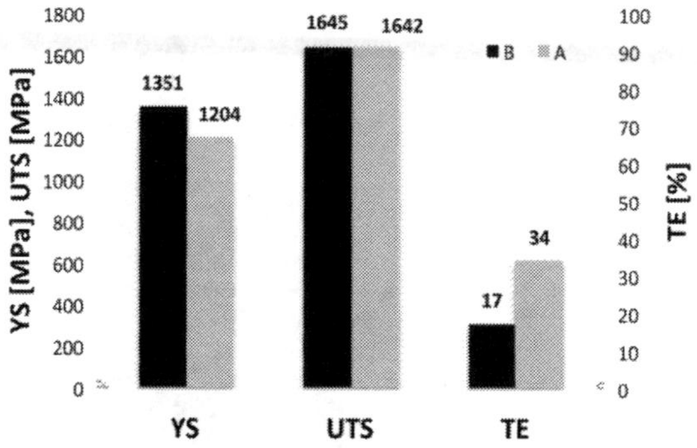

Figure 4. The Effect of compaction (RT), tensile stretching 2.5% (RT) and aging on mechanical properties at 4.2 K

3.2. Structural analysis

In Fig. 5 is shown the microstructure of tested austenitic steel for the initial testing materials (Fig 5 a,b) as well as for selected testing material compacted and tensile stretched and aged (Fig 5 c,d). The microstructure is shown in the longitudinal direction of the pipe, and it is formed by polyedric austenite grains (when twin boundaries are not considered as twin boundaries) and high fraction of Σ3 twin boundaries. The average size of austenite grains (determined from the optical microscopy) was at the initial basic material A and ranged from d_γ = 50-140 μm, wherein after CO + ST + AG, their size does not change significantly ($d_{\gamma+TS+CO+AG}$ = 60-160 μm) which means the austenitic structure was well stabilized by precipitates. The cover Size B is observed a heterogeneous particles size, exhibiting patterns of bimodal division (d_γ = 20 - 250 μm), but the material does not exhibit a change in the size of austenite grain size after the heat treatment.

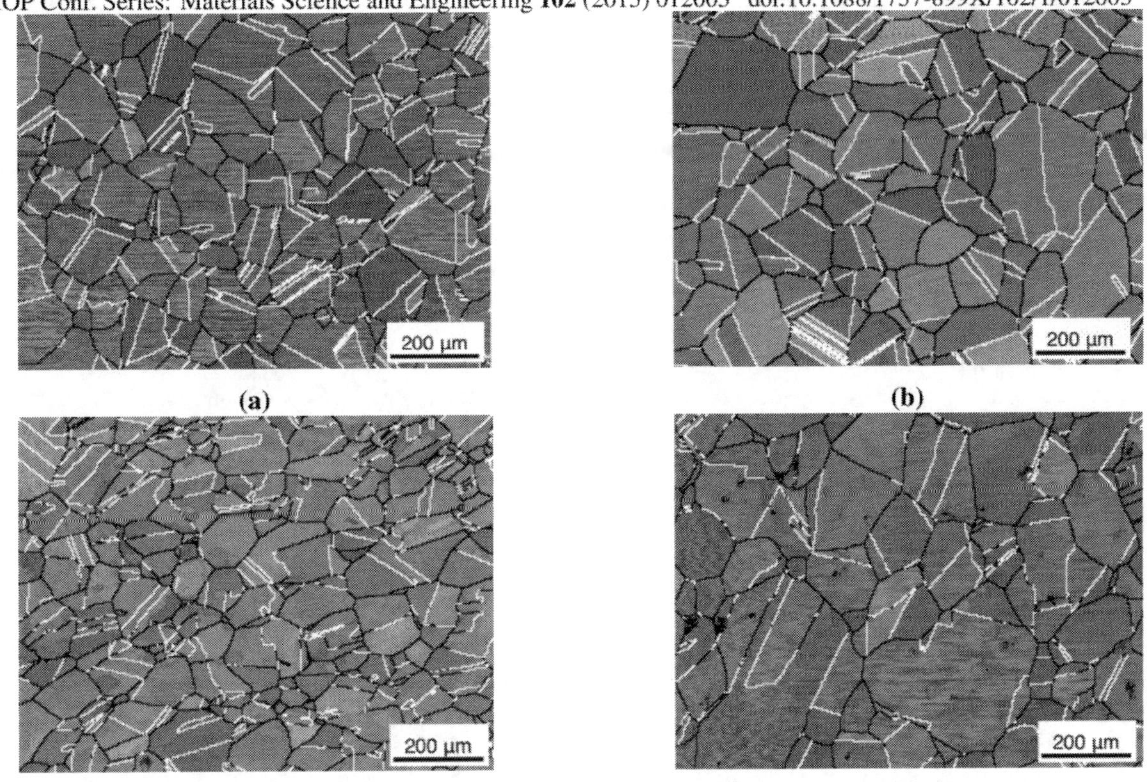

Figure 5 Band contrast map and Σ3 - Twin Boundaries (white lines), longitudinal direction of the tube = horizontal: (a) A; (b) B; (c) A after CO+TS+AG; (d) B after CO+TS+AG

The presence of the second phase was observed on both materials in the form of unidentified particles of white and black colour which were located evenly on the borders of austenite grains (Fig. 6). It can be stated (from works of various authors [6,7]) that the found particles are precipitates $M_{23}C_6$ carbide (white elements) and M_6C_6 (black elements). The reduced content of M_6C_6 in the microstructure, as opposed to $M_{23}C_6$, can be attributed to the large consumption of carbon, necessary for its creation.

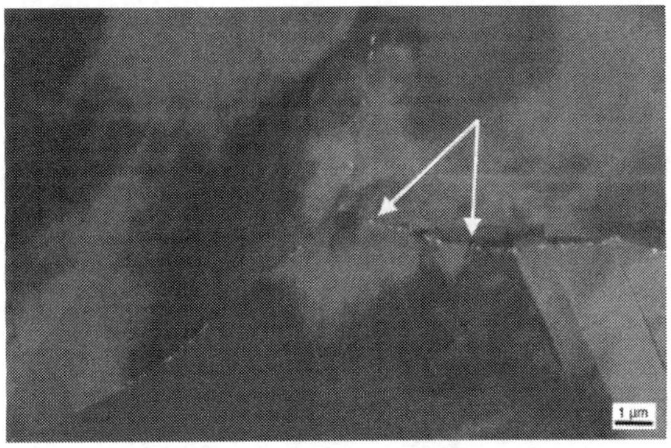

Figure 6. SEM picture of the material A after CO+TS+AG

4. Conclusion

This work dealt with the assessment of the mechanical properties of two chemical concepts 316LN austenitic stainless steel in the cryogenic temperatures; down to 4.2 K. In relation to measurements, it can be stated that the increase in yield and ultimate tensile strength (in the cryogenic temperatures) results from the impact of the increasing nitrogen content, without significant impact on the total elongation. Further increase of mechanical properties was achieved after the application of compaction, tensile stretching and aging. Testing steel microstructure was composed of polyedric austenitic grains with twins and precipitates. These materials achieved very good results, when considering the work in cryogenic temperatures, and therefore could be possibly applied in the area of fusion in the future.

Reference

[1] Mitchell N, Devred A, Libeyre P, Lim B, Savary F 2012 *IEEE TASC* **22** 4200809
[2] Kim D W, 2012 *J. of Nuc. Mat.* **420** 473-478
[3] ITER, Procurement specification for the jacketing of the cables for ITER Toroidal Field (TF) coil conductors. ITER design document. Revision V-01, April, 2005
[4] Liu H et al 2011 Mechanical tests on the ITER PF 316L jacket after compaction *Cryogenic* **51** 234–236
[5] Wu Z et al 2013 *Fusion Engineering and Design* **88** 2810-2813
[6] Sgobba S et al 2013 *Fusion Engineering and Design* **88** 2484-2487
[7] Liang Z et al 2015 *High. Temp. Mar. Proc.* aop

IOP Conf. Series: Materials Science and Engineering **102** (2015) 012004 doi:10.1088/1757-899X/102/1/012004

Permanent effect of a cryogenic spill on fracture properties of structural steels

H Keseler[1], I Westermann[1], S Y Kandukuri[2], J O Nøkleby[2], B Holmedal[1]

[1] Norwegian University of Science and Technology, Department of materials science and engineering, Alfred Getz vei 2b, N-7491 Trondheim, Norway
[2] DNV GL, Veritasveien 1, N-1363 Høvik, Norway

E-mail: ida.westermann@ntnu.no

Abstract. Fracture analysis of a standard construction steel platform deck, which had been exposed to a liquid nitrogen spill, showed that the brittle fracture started at a flaw in the weld as a consequence of low-temperature embrittlement and thermal stresses experienced by the material. In the present study, the permanent effect of a cryogenic spill on the fracture properties of carbon steels has been investigated. Charpy V-notch impact testing was carried out at 0 °C using specimens, from the platform deck material. The average impact energy appeared to be below requirements only for transverse specimens. No pre-existing damage was found when examining the fracture surfaces and cross sections in the scanning electron microscope. Specimens of the platform deck material and a DOMEX S355 MCD carbon steel were tensile tested immersed in liquid nitrogen. Both steels showed a considerable increase in yield- and fracture strength and a large increase in the Lüders strain compared to the room temperature behavior. A cryogenic spill was simulated by applying a constant tensile force to the specimens for 10 min, at -196 C. Subsequent tensile tests at room temperature showed no significant influence on the stress-strain curve of the specimens. A small amount of microcracks were found after holding a DOMEX S355 MCD specimen at a constant force below the yield point. In a platform deck material tensile tested to fracture in liquid nitrogen, cracks associated with elongated MnS inclusions were found through the whole test region. These cracks probably formed as a result of the inclusions having a higher thermal contraction rate than the steel, causing decohesion at the inclusion-matrix interface on cooling. Simultaneous deformation may have caused formation of cracks. Both the microcracks and sulphide related damage may give permanently reduced impact energy after a cryogenic exposure.

1. Introduction

Liquid nitrogen is a widely used cryogenic liquid, which has a boiling point of approximately -196 °C. Cryogenic liquids are commonly used for various cooling purposes in the industry. Storage of large quantities of gas can also be done more effectively by condensing the gas into a liquid, as this will give a large volume reduction, e.g. the volume of liquefied natural gas will reduce the volume by a factor 600 compared to the gas volume [1-3].

Cryogenic liquids need to be kept at very low temperatures, hence, their storage and transportation set specific demands for the materials used in the storage tanks. A common way of storing cryogenic liquids is in insulated metal tanks, where the inner walls are made from aluminum or austenitic stainless steel. These are materials which retain their ductility at low temperatures. Carbon steels, on the other hand, are known to experience a ductile to brittle transition when reaching a critical low temperature, normally around -50 °C. In offshore applications, it is therefore important to keep cryogenic gases out of contact with the surrounding ship or platform structure, which is usually constructed from carbon steel. Exposure of the carbon steel structure to a cryogenic liquid, may lead to embrittlement and possibly also structural failure [3].

Because it is well known that carbon steels become brittle at low temperatures, this material is normally not used for such applications. For this reason little research has been done regarding the permanent consequence of exposing carbon steel to low temperatures or, as the case during a cryogenic spill, low temperatures combined with tensile stresses. Only two previous studies have been found, which concern the low-temperature behavior of carbon steel. The earliest of these studies was done in 1957 by Owen et al. [4]. They carried out tensile tests on a carbon steel at -196 °C using various cross-head speeds. Their results showed that the fracture always occurred after gross yielding of the specimen. When the cross-head speed was increased, the fracture stress increased with the yield stress. The observation of the strain patterns in pre-polished specimens showed that the fracture always was located within the Lüders band, and also that sub-critical microcracks were formed within the Lüders band prior to fracture. Another study concerning the low-temperature tensile properties of carbon steel was made by Goodenow and Bucher [5] in 1969. They confirmed the results seen by Owen et al., that tensile fracture of carbon steel at temperatures down to -196 °C occurred after the yield point had been reached. They defined a plastic instability transition temperature, TPI, as the temperature below which inhomogeneous plastic deformation would continue after the specimen had yielded. At this temperature the lower yield point stress was equal to the tensile stress and necking would occur as a result of different temperature dependencies of the yield- and tensile stress. In other words, decreasing the temperature made the initiation of yielding increasingly difficult and reduced the materials capacity for general work hardening increasing the Lüders plateau.

In 2011 the risk management and classification company DNV received a request concerning an incident where a large amount of liquid nitrogen had been spilled on a platform deck constructed from carbon steel plates of DNV grade NV E36. The combination of low-temperature embrittlement and thermal stresses experienced by the material during the cryogenic exposure had resulted in failure of the platform deck. After such incidents, the standard procedure is to replace only the visibly failed regions of the material, after thorough visual inspection and NDT, as no other permanent damage is known to result from such an exposure. Nevertheless, a larger region of the exposed material was cut from the platform for further examination. Charpy V-notch impact tests of the exposed area showed lower values than expected, whereas metallurgical examinations and tensile tests did not show any anomalies compared to non-exposed material. The reason for the failure was found to be a flaw in a weld, which had initiated an unstable crack growth.

In the present work, investigations have been done with the purpose of finding out if any permanent changes will occur in the fracture properties of carbon steel as a consequence of a cryogenic spill. Remaining material from the exposed platform deck has been subjected to fracture analysis in addition to Charpy V-notch impact testing to see if any permanent damage or deterioration of the fracture properties could be found and to investigate the possible mechanisms behind such damage. In addition the conditions during a cryogenic spill were simulated by tensile testing carbon steels in liquid nitrogen. Tensile stresses of various magnitudes were applied to simulate the thermal stresses arising from the temperature gradients during a cryogenic spill. The material was subsequently examined to see if any permanent damage had resulted from this treatment.

2. Material and experimental procedures

2.1. Material

The carbon steel used in the first part of the experimental work was cut from the platform deck, which had been exposed to a liquid nitrogen spill. The material was a hot-rolled steel of DNV grade NV E36 with plate thickness 12 mm. For the second part a 5 mm hot-rolled carbon steel plate of grade

DOMEX S355 MCD, with a similar strength as the platform material, was investigated in addition to the NV E36. The composition of the two materials is listed in Table 1.

Table 1 Chemical composition of the carbon steel grades given in mass % of the alloying elements. The exact composition is given for the DOMEX S355 MCD, whereas the composition of the NV E36 is defined from the specification.

	C	Si	Mn	P	S	Cr	Mo	Ni	Cu	Al	Nb	V	Ti
NVE36	0.20	0.10-0.55	1.70	0.030	0.030	0.20	0.08	0.40	0.35	min 0.02	0.02-0.05	0.05-0.10	0.007-0.05
S355	0.069	0.01	0.62	0.009	0.003	0.03	0.00	0.03	0.01	0.046	0.023	0.00	0.00

Figure 1 shows an overview of the original damaged platform material that was received by DNV in 2011. The arrows indicate the extension of the secondary cracks, which were assumed to have propagated from brittle areas and arrested when reaching more ductile material. The two plates that were used in the present study are outlined with dashed lines. These plates were further divided into smaller sections as indicated with dotted lines and numbered 1 to 5. The platform material was painted, and the paint had cracked in certain areas. This was assumed to be caused by differences in the thermal contraction between the paint and the underlying material during the liquid nitrogen exposure. The areas where the paint cracking was observed were expected to have been cooled to lower temperatures. Areas indicating cryogenic exposure will further on be referred to as exposed areas, whereas those with no paint cracking will be referred to as non-exposed.

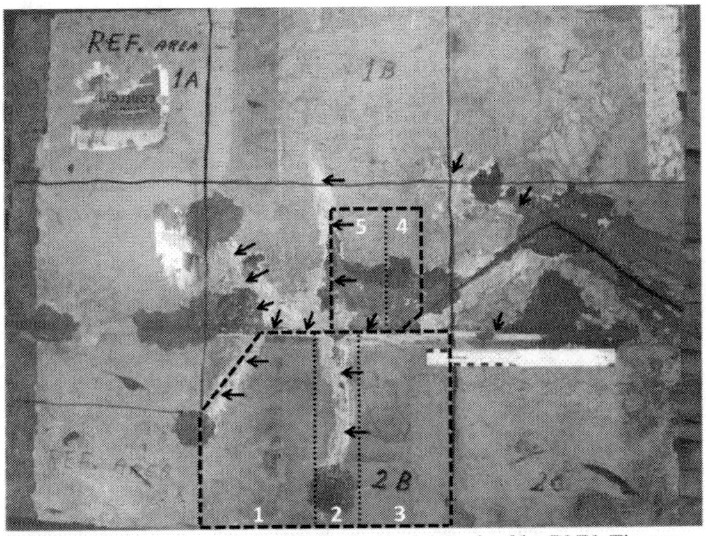

Figure 1. An overview of the platform material received by DNV. The two pieces used in this study is highlighted with dashed lines. Further subdivision is shown with dotted lines and the pieces are marked with numbers 1-5. The arrows indicate the extension of the crack formation.

2.2. Examination of the platform material by Charpy V-notch impact testing

Charpy V-notch impact testing was carried out using a Zwick/Roell RKP 450 apparatus with a theoretical impact velocity of 5.235 m/s, total pendulum mass of 32.85 kg and work capacity of 450 J. The specimens were cooled to 0 °C in an ethanol bath before testing.

Standard Charpy-V notch specimens were prepared as close to the plate surface as possible. The specimens were made with a square cross section of 10 mm × 10 mm, length of 55 mm and a 2 mm deep V-notch with angle 45° and 0.25 mm root radius, located at mid-length. 46 specimens aligned in the transverse direction were made from plate section 1, cf. Figure 1. From section 3 of the plate, 39 longitudinal Charpy specimens were made. Each of the sections contained both exposed and non-exposed areas.

2.3. Simulation of a cryogenic spill by tensile testing in liquid nitrogen

For the low temperature tests, a tensile test apparatus of the type MTS 880 was applied, with a tensile speed of 1 mm/min. An isolated plastic container was provided around the specimen holders to contain the liquid nitrogen during the immersed tests. No extensometer could be used in the cryogenic set-up. Therefore, the specimen strain was calculated as the difference in displacement taking into account the machine stiffness.

Small tensile specimens with a gauge diameter and length of Ø3 mm and 13 mm, respectively, were made from the DOMEX S355 MCD steel parallel to the rolling direction, and a few from section 2 of the NV E36 platform material (6 specimens). The liquid nitrogen was filled into the isolated container after the specimen was firmly tightened to the manual grips in the test rig. A holding time of 10-15 min was necessary to ensure that the specimen and the equipment had reached -196 °C.

Tensile curves at -196 °C were obtained as a reference for both the materials. To simulate tensile stresses from a cryogenic spill, samples were then held at different constant tensile force levels for 10 min at -196 °C. For the DOMEX S355 MCD three or four parallel samples were held at 1000 N, 4000 N and 6500 N. For the NV E36 material only two parallel samples were held at 5000 N, due to the low number of available samples. After 10 min the force was released, the nitrogen was removed, and the specimens were left in air at room temperature.

Subsequently all the prestressed specimens were further tensile tested to fracture at room temperature to see if the previous cryogenic treatment had affected the tensile properties. Reference tensile curves at room temperature were also conducted for comparison. These were tested in an MTS 810 test machine equipped with an extensometer allowing measuring the strains up to approximately 16%. The remaining strain to fracture was calculated by extrapolation of strain versus displacement data.

2.4. Metallographic examination

Two small samples were cut from each of the sections 2 and 4, from a non-exposed and an exposed area, respectively. A Leica Axiovert 25 optical microscope was used to compare the microstructure in the exposed and non-exposed material both parallel and perpendicular to a reference weld. This investigation was also necessary to determine the rolling direction of the investigated material.

Specimen preparation was carried out by standard metallographic preparation, i.e. mechanical grinding on SiC-paper with decreasing roughness followed by mechanical polishing with diamond spray of 3 µm and 1 µm. Finally, the specimens were etched for 10 sec in a 2 % Nital solution.

2.5. SEM examinations

The fractographic examination of both Charpy impact tested specimens and the tensile tests were carried out in a Zeiss Supra 55 VP LVFE SEM using the secondary electron signal. The SEM was operated at 10-20 kV and a working distance of 11-17 mm.

The Charpy impact specimens were investigated edge on, but also in a cross section perpendicular to the fracture surface, as the specimens were cut and polished in the length direction. The tensile specimens were similarly investigated edge on as well as in the cross sectional length direction. The latter was done for both specimens that only had been strained at cryogenic temperatures as well as specimens strained to fracture.

For the edge on investigations, the specimens were cleaned in an ultrasonic bath with acetone, and for the mid-section investigations, the samples were prepared similarly as for the metallographic examination.

3. Results and discussion

3.1. Charpy V-notch impact test of the platform material

The light optical images in Figure 2 show the microstructure of the platform material. In (a) the microstructure from a non-exposed area in a plane perpendicular to the rolling direction, and (b) for a cryogenic exposed material in a plane parallel to the rolling direction. In both cases the microstructure consists of ferrite and elongated pearlite and contains long manganese sulphide inclusions elongated in the rolling direction; cf. Figure 2(b). No obvious differences between the two microstructures could be observed of the exposed and non-exposed areas of the plate.

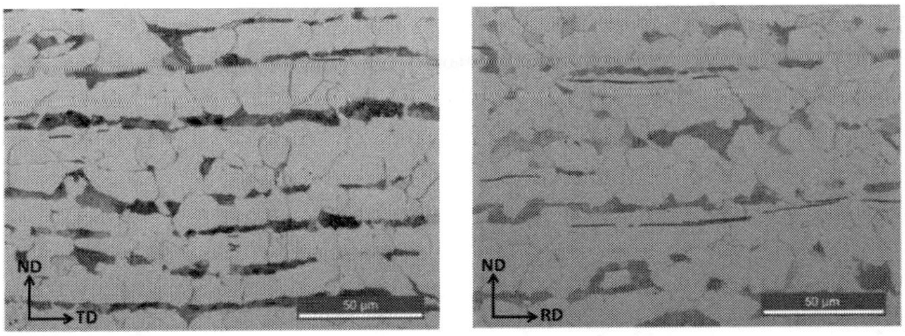

(a) (b)

Figure 2. The microstructure of (a) material from a non-exposed area in a plane perpendicular to the rolling direction, and (b) from cryogenic exposed material in a plane parallel to the rolling direction.

Figure 3 shows the numbering and position of the transverse Charpy specimens taken from section 1 of the plate in Figure 1. The Charpy impact energy for the given specimen is marked in the white boxes. The average value of the measured impact energies was 38.68 J. However, large variations are observed within the plate, with a maximum value of 76.49 J and a minimum of 24.47 J.

Similarly, Figure 4 shows the position and numbering of the longitudinal Charpy specimens taken from section 3, where the Charpy impact energy is marked for each specimen. In the longitudinal direction, the Charpy impact values are in general higher with an average value of 102.83 J, but also here large variations are found in the energy, ranging between 52.20 J and 144.71 J

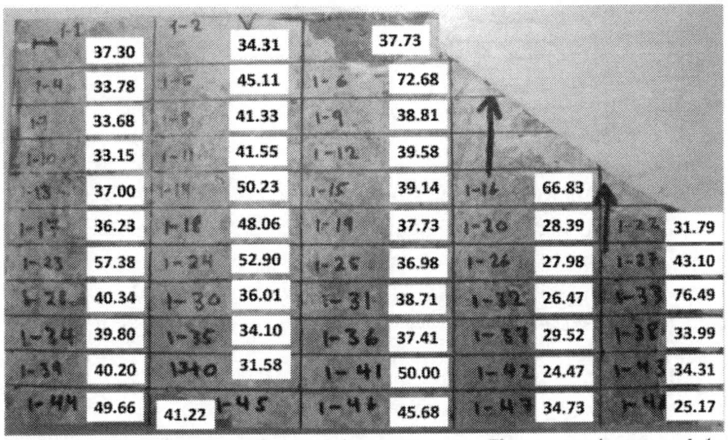

Figure 3. The sectioning of plate 1 into transverse Charpy specimens and their corresponding impact toughness values, given in Joules [J], at 0 °C.

The specified Charpy V-notch impact energy of NV E36 steel at -40 °C is 34 J in the longitudinal direction and 24 J in the transverse direction [6]. These values are given as minimum values and the real impact values of the material are expected to be significantly higher. In this work the impact test were performed at 0 °C. As a rule of thumb, the impact energy increases by 1 Joule for 1 °C increase in temperature. At 0 °C this leads to required impact energies of approximately 74 J and 64 J in the longitudinal and transverse direction, respectively. According to the standard [7] the minimum requirement should be met by the average of three test pieces. It can be accepted if one individual value is below the specified value, provided that it is not less than 70 % of this value.

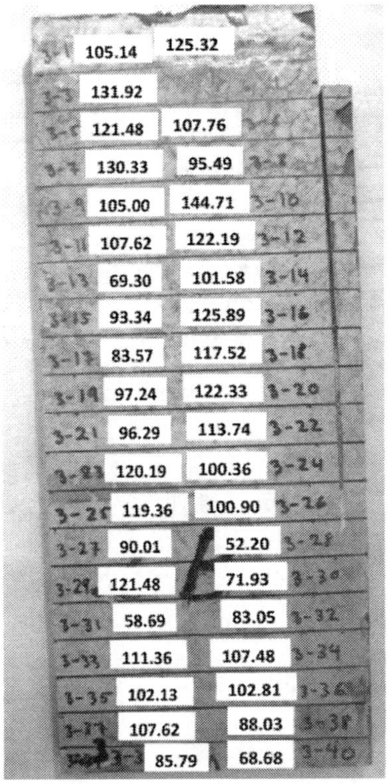

Figure 4. The sectioning of plate 3 into longitudinal Charpy specimens and their corresponding impact toughness values, given in Joules [J], at 0 °C.

IOP Conf. Series: Materials Science and Engineering **102** (2015) 012004 doi:10.1088/1757-899X/102/1/012004

The longitudinal Charpy specimens are well within the requirement, whereas the transverse specimens are far too low in impact energy. A rather large variation in impact energy is observed. The reason for this might be random variations caused by microstructural inhomogeneities through the material. However, it could also be a consequence of stress relief occurring in the region close to the main crack.

3.2. Simulation of a cryogenic spill by tensile testing in liquid nitrogen

In Figure 5 the reference curves obtained at -196 °C for the NV E36 and the DOMEX S355 MCD are presented. The noise is a side effect from the test machine used and should be neglected. For the DOMEX S355 MCD a clear upper yield point is observed. Then the stress drops and slowly decreases until fracture at around a strain of 15 %. The NV E36 show a less clear upper yield point. The stress-strain curve then enters a stage where the stress is almost constant, before fracture occurs at strains around 26 %. Compared to the reference curves obtained at room temperature shown in Figure 6 the yield strength significantly increases with decreasing temperature. However, the appearance of the stress-strain curves is quite different. The long horizontal or decreasing region after the upper yield point is believed to be caused by propagation of Lüders bands across the length of the specimen. The large Lüders strains seen at -196 °C can be explained by the low work hardening rate at this temperature. This is in accordance with the observations by Goodenow and Bucher [5], who observed a reduced work hardening rate and increased upper yield strength with decreasing temperature. Instead of a ductile-to-brittle transition, they observed a so-called plastic instability transition when the temperature decreased below a certain point. At this plastic instability transition temperature, the tensile stress is equal to the lower yield stress and necking occurs within the Lüders strain. In the case of the NV E36 steel tested here, the work hardening rate was probably just large enough for the Lüders band to spread over the length of the test region.

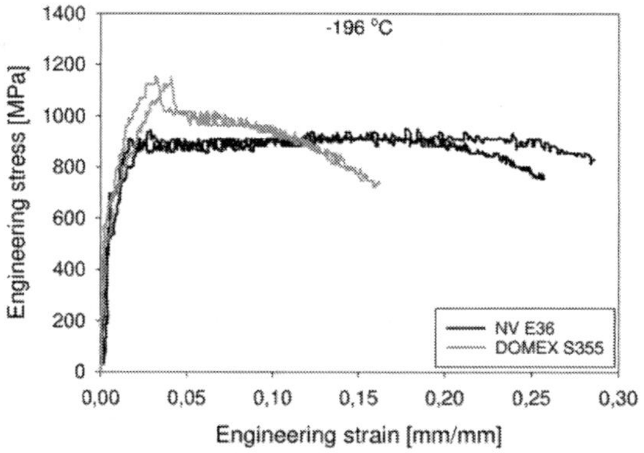

Figure 5. The engineering stress-strain reference curves for the two materials tested at -196 °C.

Figure 6 shows the room temperature engineering stress-curves of reference material that has not been exposed to liquid nitrogen, along with material that has been held at a force below the yield point in liquid nitrogen for 10 min. Only the stress-strain curve for the material held at the maximum constant force (6500 N) is shown here for the DOMEX S355 MCD.

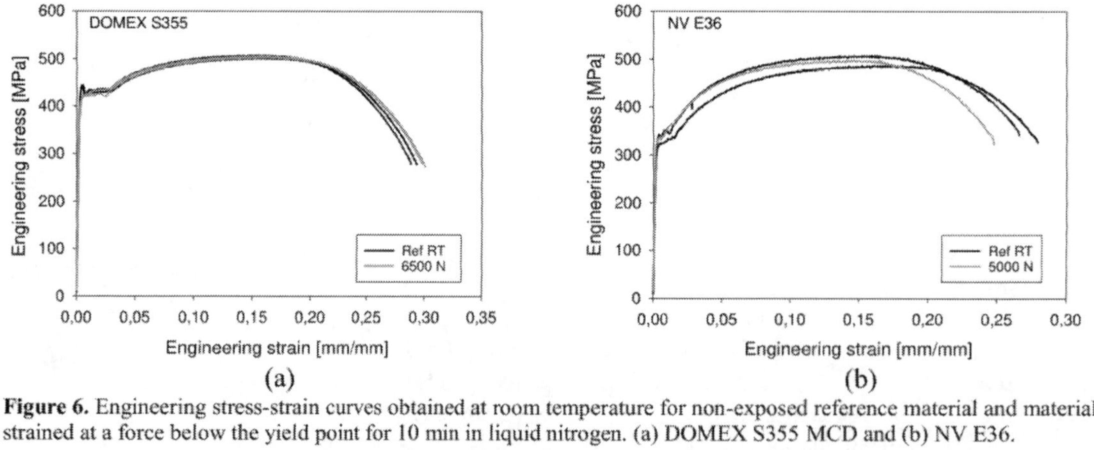

Figure 6. Engineering stress-strain curves obtained at room temperature for non-exposed reference material and material strained at a force below the yield point for 10 min in liquid nitrogen. (a) DOMEX S355 MCD and (b) NV E36.

No particular difference is observed between the stress-strain curves, except that the upper yield point gradually disappeared for the DOMEX S355 MCD material when the tensile force applied under cryogenic conditions was increased. The fracture surface of the pre-strained material also appeared ductile after tensile testing at room temperature.

SEM investigations of the longitudinal cross section of a DOMEX S355 MCD sample that only had been exposed to liquid nitrogen at a force of 6500 N revealed a few microcracks as shown in Figure 7. Some of the cracks were located at grain boundaries, but mostly they appeared in the grain interior. For comparison the longitudinal cross section of an untreated specimen was investigated, and no microcracks were found in this material.

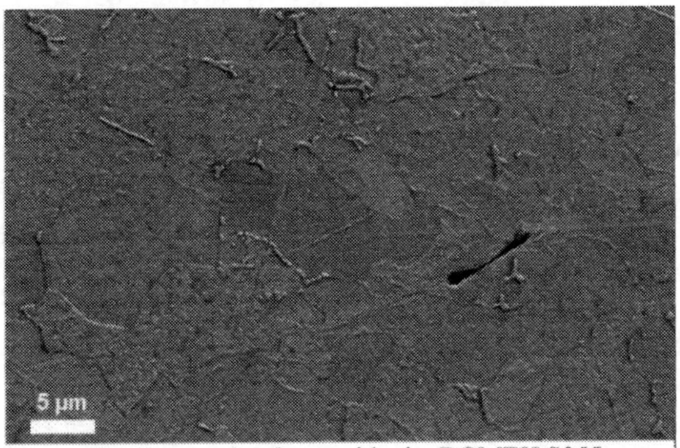

Figure 7. Microcracks observed in the DOMEX S355 subjected to 6500 N under cryogenic conditions.

Figure 8 shows two types of microcracks that were found in the NV E36 material tested at -196 °C. Figure 8(a) shows a type of microcrack which was only found at small distances from the fracture surface. These cracks were mainly oriented parallel to fracture surface. Another type of microcrack found in the longitudinal cross section is shown in Figure 8(b). These cracks were always directed along the rolling direction and are associated with elongated manganese sulphides. Such cracks were observed to occure through the entire test region of the tensile specimen, and the extent of these cracks could be through several ferrite grain boundaries. The reason for the formation of the sulphide related cracks is believed to be a combination of the low temperature and the high stresses in the material. Brooksbank and Andrews [8] showed that MnS inclusions have a linear thermal expansion coefficient that is higher than for steel (18×10^{-6} °C^{-1} vs. 12×10^{-6} °C^{-1}). This means that when the material is cooled, the MnS inclusions will contract at a higher rate than the steel. The result can be decohesion, and possibly also void formation at the inclusion-matrix interface.

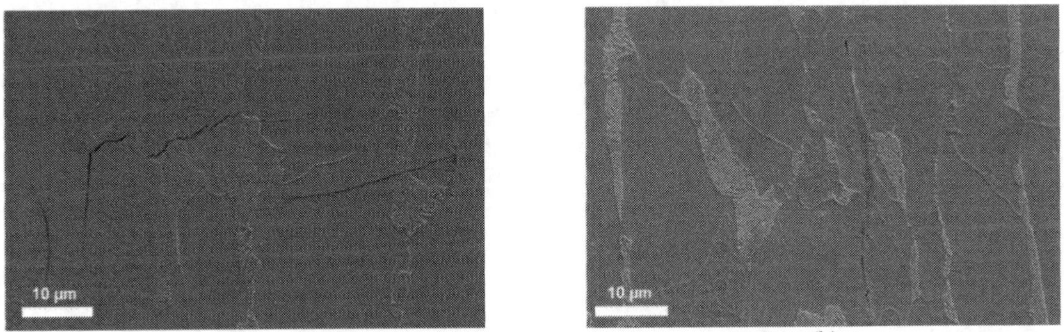

(a) (b)

Figure 8. Two types of microcracks observed in the RD-ND plane in the NV E36. (a) microcracks near the fracture surface and (b) cracks associated with MnS.

The latter type of cracks may also explain the observed direction dependency observed in the Charpy impact tests, where the poorest impact energy is observed when the notch is located parallel to the rolling direction as sketched in Figure 9. Cracks aligned with the direction of the notch will have a detrimental effect on the properties.

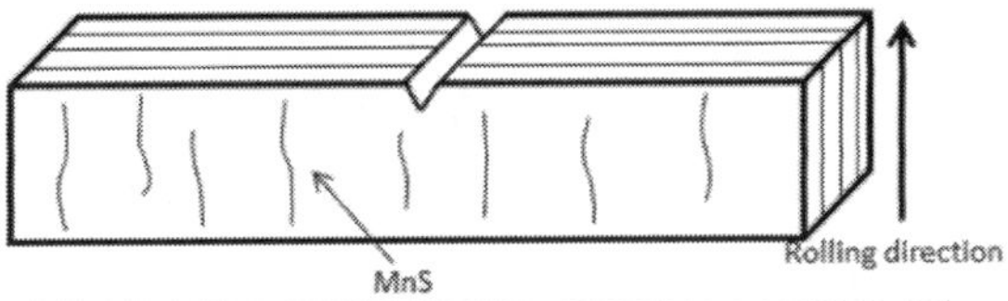

Figure 9. Sketch of the transverse Charpy specimens showing the alignment of the microstructure in the rolling direction.

4. Conclusion

Material from the platform deck that had been exposed to a liquid nitrogen spill was subjected to Charpy V-notch impact testing. The impact energy was below the minimum requirement for the transverse specimens. The impact energy of the longitudinal specimens was above the requirement. This is believed to be either due to differences in the temperature and stress history of the regions the specimens were taken from, or due to an orientation dependent damage mechanism. No evidence of pre-existing damage was found in the Charpy specimens. However, from examination of specimens tensile tested in liquid nitrogen, two possible mechanisms for reduced impact energy after a cryogenic spill were found.

The first mechanism is believed to be the formation of microcracks before fracture, occurring when carbon steel is cooled to low temperatures and simultaneously subjected to high thermal tensile stresses. A small number of microcracks were seen in a specimen of DOMEX S355 MCD carbon steel after holding 10 min at a force below the yield point in liquid nitrogen. The presence of microcracks is expected to reduce the impact energy in both the transverse and longitudinal orientations.

The second possible mechanism which was found is the formation of damage associated with elongated manganese sulphide inclusions. It is believed that cracks were formed due to manganese sulphides having a higher thermal contraction rate than the steel. When the material was cooled this caused decohesion and possibly also void formation at the inclusion-matrix interface. Due to the orientation of the manganese sulphides in the rolling direction, the reduction in impact energy is expected to be larger for in the transverse direction than the longitudinal direction in accordance with what is measured.

The tensile behaviour of the platform material and a DOMEX S355 MCD carbon steel at -196 °C showed largely increased yield- and fracture strength compared to at room temperature. A large increase in the Lüders strain occurs for both the materials due to a low work hardening rate at this temperature. Both steels showed plastic instability, as the fracture occurred within the Lüders strain region. The DOMEX S355 MCD steel showed the most pronounced plastic instability, as necking started directly after the upper yield point. For the platform material, necking would first occur when the Lüders band had spread through approximately the whole test region of the tensile specimen.

A cryogenic spill was simulated by cooling specimens of the two materials to a temperature of -196 °C in liquid nitrogen, and applying a constant force below the yield point for a period of 10 min. This showed no significant effect on the subsequent room temperature tensile behavior, indicating that a cryogenic spill will not permanently affect the tensile behavior of these steels.

From the results of this work, it is believed that a cryogenic spill can permanently damage the fracture properties of carbon steel. Therefore, the standard repair procedure after such incidents, which is to only remove the visibly damaged material, may be insufficient. Microcracks or sulphide related damage may be present in apparently undamaged material and can reduce the impact energy in these regions.

References

[1] Tomsic J 2000 "Dictionary of materials and testing" 2nd ed. (Warrendale, PA: Society of Automotive Eng.)
[2] Mokhatab S, Mak J Y, Valappil J V, Wood D A 2013 *Handbook of Liquefied Natural Gas* (Oxford: Gulf Professional Publishing)
[3] Kumas S, Kwon H T, Choi K H, Lim W, Cho J H, Tak K, Moon I 2011 *Appl. Energy* **88** 4264
[4] Owen W S, Averbach B L, Cohen M "Brittle fracture of mild steel in tension at -196C" 1957 Technical report: SSC-109 Prog Rpt. National Technical Information Service
[5] Gooden R H and Bucher J H 1969 *J. Fluids Eng.* **91** 603
[6] Det Norske Veritas 2011 "Rules for classification of ships/high speed, light craft and naval surface craft" (Part 2) (Høvik: Det Norske Veritas)
[7] Det Norske Veritas 2012 "Metallic materials – offshore standard" (Høvik: Det Norske Veritas)
[8] Brooksbank D and Andrews K W 1968 *J Iron Steel Inst* **206** 595

IOP Conf. Series: Materials Science and Engineering **102** (2015) 012005 doi:10.1088/1757-899X/102/1/012005

Hydrogen Environment Embrittlement on Austenitic Stainless Steels from Room Temperature to Low Temperatures

Toshio Ogata

Environment & Energy Materials Div., National Institute for Materials Science
1-2-1 Sengen, Tsukuba, Ibaraki, 305-0047, Japan

E-mail: OGATA.Toshio@nims.go.jp

Abstract. Hydrogen environment embrittlement (HEE) on austenitic stainless steels SUS304, 304L, and 316L in the high pressure hydrogen gas was evaluated from ambient temperature to 20 K using a very simple mechanical properties testing procedure. In the method, the high-pressure hydrogen environment is produced just inside the hole in the specimen and the specimen is cooled in a cooled-alcohol dewar and a cryostat with a GM refrigerator. The effect of HEE was observed in tensile properties, especially at lower temperatures, and fatigue properties at higher stress level but almost no effect around the stress level of yield strength where almost no strain-induced martensite was produced. So, no effect of HEE on austenitic stainless steels unless the amount of the ferrite phase is small.

1. Introduction

Stainless steels are also used in fuel-cell vehicle applications for storage tanks, mobile containers, and pipes due to their ductility at low temperatures and in hydrogen environments. Recently, it was also discussed that at a hydrogen gas station for the vehicle, high pressure hydrogen gas will be produced by evaporating liquid hydrogen quickly. The hydrogen embrittlement in meta-stable stainless steels has been evaluated so far [1-3] using thermally or electochemically hydrogen-charged materials. It has been shown that HEE in the temperature range from 200 K to 250 K is caused by strain induced martensite, while little or no HEE occurs at 77 K.

A hydrogen environment causes embrittlement at ambient temperature [4] and low temperatures [5-7] down to 80 K under 1.1 MPa gaseous hydrogen. However, no tensile or fatigue data was available neither for the higher pressure hydrogen and low temperature environments nor fatigue properties. A simple testing method for mechanical properties evaluation under high pressure hydrogen gas was developed [8] without using a high pressure gas chamber and was easily applied to low temperatures [9, 10]. The effects of HEE on fatigue properties of SUS304L at 10 MPa and 70 MPa and cryogenic temperatures were evaluated using the simple method [11, 12]. Stainless steels with higher strength would be used in the instruments of high pressure hydrogen system and the HEE of those steels was a major concern.

In this work, the HEE on the tensile and fatigue properties of meta-stable austenitic stainless steels were examined with behaviour of strain-induced martensitic transformation to study an available limit in high pressure hydrogen environment.

2. Experimental detail

2.1. Material and Specimens, and Testing

The materials used in this study are the commercial austenitic stainless steels, SUS 304, 304L and SUS 316L, hot-rolled and solution-treated round bars. The diameter of the bars was 13 mm. The typical chemical compositions of the materials are listed in TABLE 1. Round tensile specimens were machined from the bars and their diameter in the gage length was 6.25 mm and the length of parallel part was 34 mm. Hour-glass type round fatigue specimens were also machined and their diameter was 6 mm. The hole in the specimen was machined by a wire-cut and its diameter was less than 2 mm.

FIGURE 1 shows an illustration of this method. High pressure gas was filled into a small hole in the specimen from a gas cylinder or a portable compressor for 70 MPa test. This system has no compressor up to the cylinder pressure of 10-15 MPa and requires no extra cost for the testing. The temperature of the specimen with the high pressure gas was controlled by surrounding environment temperatures.

2.2. Testing

Tensile tests were carried out at from 298 K (RT) to 25 K. In the test at 25 K and 77 K, the specimen was in the cryostat with refrigerator and in liquid nitrogen dewar, respectively. In the tests from 100 K to 170 K and from 190 K to 280 K, the specimen was cooled by temperature–controlled nitrogen gas and temperature–controlled alcohol, respectively. The specimen temperature was measured by a thermo-couple attached on the specimen surface at middle of gage length. The crosshead-speed during tensile test was 3.6 mm/h. The effect of HEE is evaluated by the ratio of reduction of area (relative reduction of area) obtained in tensile tests using the specimen filled in hydrogen gas to that obtained in helium gas environment [6-10]; the ratio of 1.0 means no effect of hydrogen and the ratio becomes smaller when the effect of hydrogen increases. The tests were started one hour later after the inside pressure was set.

Fatigue tests were carried out at a stress ratio of 0.01 (tensile-tensile) and a frequency of 10 Hz at 298 K and 190 K.

TABLE 1. Chemical compositions of stainless steels used in this study (wt. %).

	C	Si	Mn	P	S	Ni	Cr	Mo
SUS304	0.06	0.50	1.53	0.029	0.026	8.19	18.41	-
SUS304L	0.016	0.53	1.03	0.029	0.003	9.42	18.01	-
SUS316L	0.025	0.48	1.52	0.032	0.018	12.04	16.84	2.04

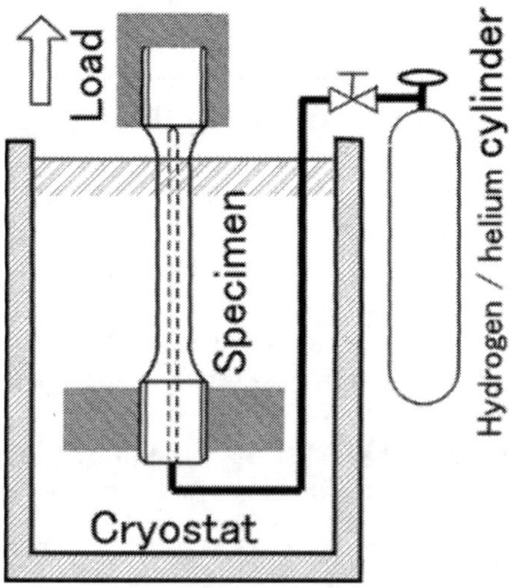

Figure 1. An illustration of simple testing method for high pressure gas environments.

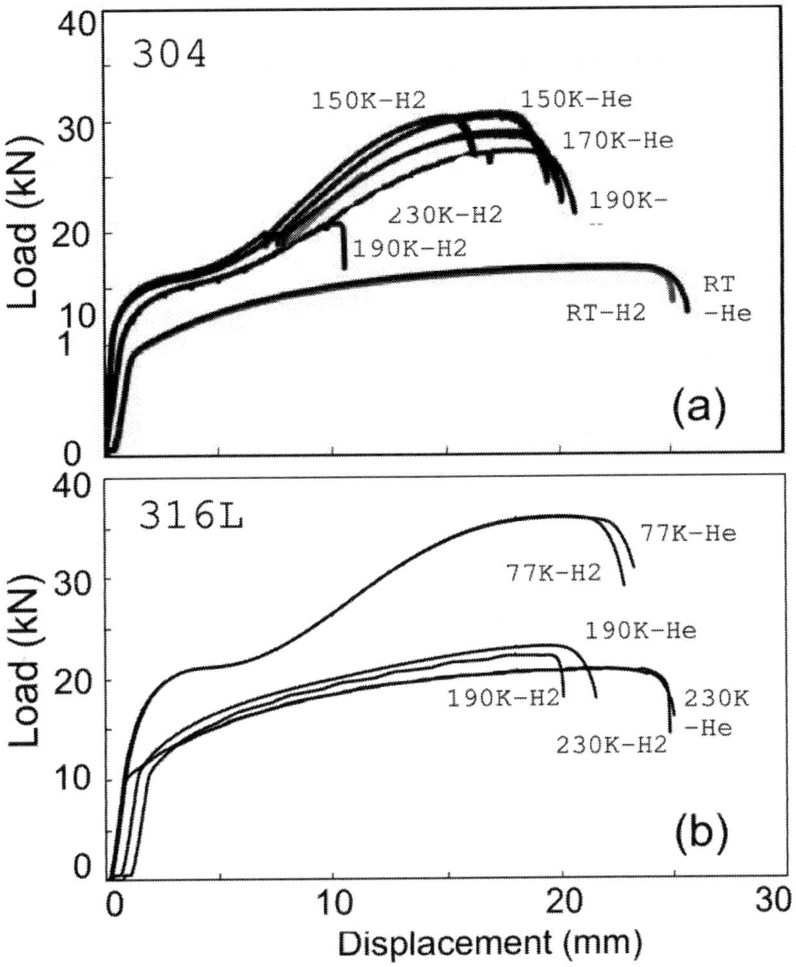

FIGURE 2 Effects of hydrogen and specimen temperature on load-displacement curves for SUS304L (a) and SUS316L (b) [9].

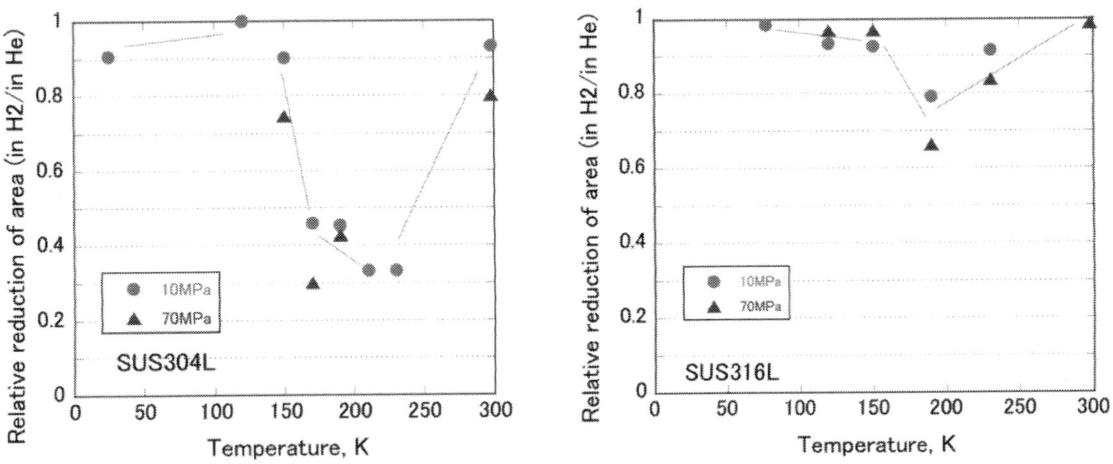

Figure 3. Effect of temperature on the relative reduction of area of SUS304L and 316L in 10 and 70 MPa hydrogen environment [10].

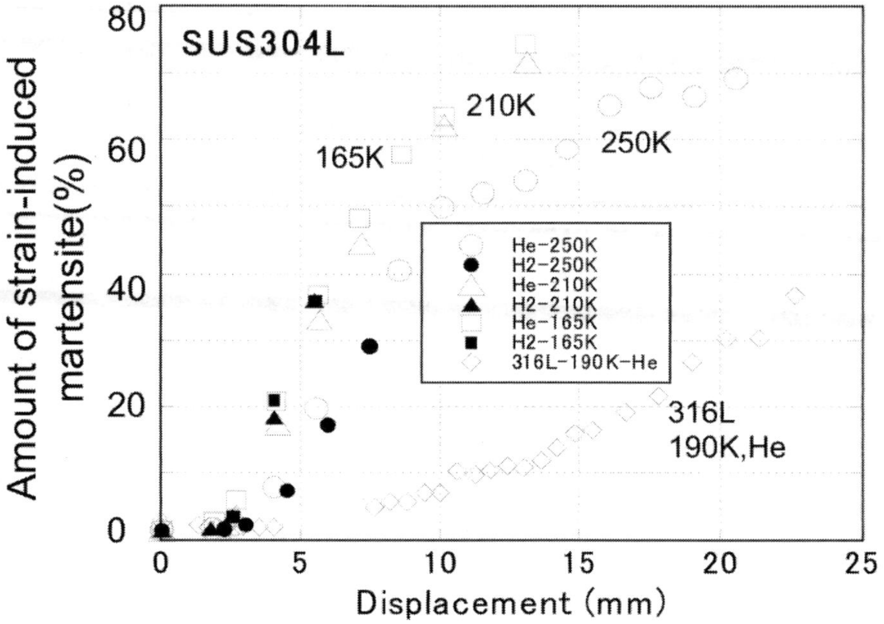

FIGURE 4 Amount of strain-induced martensite during tensile tests at low temperatures in hydrogen and helium environment.

3. Results and discussion

3.1. Tensile properties

Figure 2 shows Load-displacement curves during tensile tests for SUS304L and 316L in 13 MPa hydrogen gas and 12 MPa helium gas at various temperatures [9]. The hydrogen environment tests for SUS304L show premature failure compared to the helium environment tests, which results in lower ultimate tensile strength, elongation, and reduction of area. In SUS316L, almost no HEE and small effect at 190 K were observed. Large HEE was observed in SUS304 at RT and 190 K and in SUS304L at 190 K, but almost no HEE below 100 K for those steels.

FIGURE 3 shows HEE as a relative reduction of area for SUS 304L and 316L from RT to 25 K [10]. At room temperature, 298 K, SUS 304L showed small HEE, however, with decreasing temperature, the HEE increased and showed maximum around 200 K but the HEE decreased below 200 K and almost disappeared below 120 K.

Meta-stable austenitic stainless steel, such as SUS304 and 304L, showed remarkable HEE at low temperature, which is due to the increase of strain-induced martensitic phase at the temperatures. The author reported the amount of the martensitic phase during tensile tests at each strain level for SUS304L and 316L [13, 14]. The increase of the amount of the martensitic phase behaviour clearly explains the HEE of those steels from room temperature to 200 K. The diffusion of hydrogen in fcc materials is very slow below room temperature and almost no diffusion occurs at low temperature. So, the amount of the strain-induced martensitic transformation and the bcc martensitic phase are essential to the embrittlement in meta-stable austenitic stainless steels and diffusion of hydrogen play an important role in the embrittlement, which means that the effect of HEE increases in work-hardened meta-stable stainless steels, that is, there is no HEE until the materials deformed at a certain strain.

FIGURE 4 shows change of amount of strain-induced martensite during tensile tests at low temperatures in hydrogen and helium environment. The amount of strain-induced martensite during tensile tests was measured by Ferrite Scope and its probe was pointed on the specimen during tensile tests. The measured value was converted to the amount of martensite using the value obtained in the

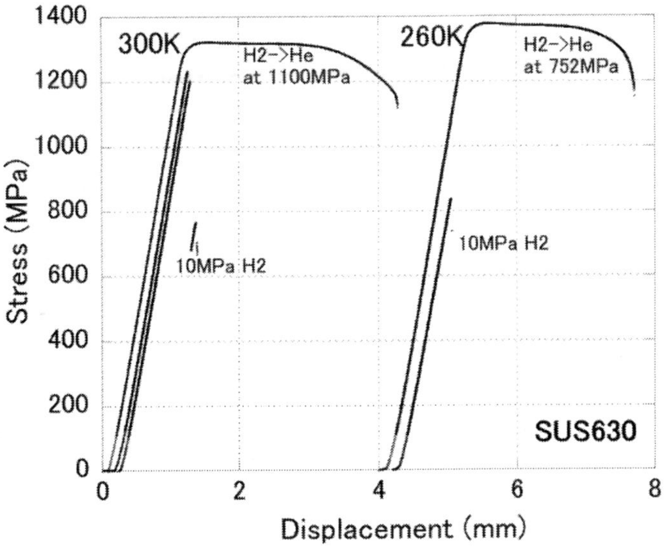

Figure 5. Gas change experiments for SUS630.

previous works [13, 14].

First, no effect of hydrogen on the amount of martensite was obserbed. In hydrogen environment, specimens fractured over 30% of martensite phase, which might be a critical amount of martensite for HEE in static tests. So, if hydrogen gas was removed before the critical point, there is no HEE until specimen fracture.

Figure 5 shows the results of experiments that hydrogen gas was changed to helium gas during tensile test for commercial SUS630. The temperature dependency of HEE on SUS630 was reported in previous work [12]. At 300K in 10 MPa hydrogen, this steel failed at 1236 MPa, .then the change gas at 1200 MPa was tried, but specimen failed in several seconds during stop loading for exchange gas. The specimen changed gas at 1100 MPa was deformed as if no hydrogen from the beginning.

At 260K and in 10 MPa hydrogen, this steel failed at 832 to 1006 MPa, the specimen changed gas at 752 MPa, 90 % of 832 MPa, was deformed as if no hydrogen from the beginning.

These tests proved that no HEE until at a certain stress level or amount of bcc phase.

3.2. Fatigue properties

Stress vs. Number of cycles to failure curves for SUS304, 304L, and SUS316L obtained in hydrogen and helium environment at RT and 190K were shown in FIGURE 6a), b), and c) [11,12]. In SUS304 and 304L, fatigue lives in hydrogen environment decreased at higher stress level, 200 MPa or 250 MPa, at RT and 190 K. But no clear influence at a stress level of 150 MPa. This stress level is corresponding to the yield strength at RT. In SUS316L, influence of hydrogen is small even at a stress level of 250 MPa.

FIGURE 7 shows relation between amount of strain-induced martensite and stress-strain curves at the early-stage of tensile tests for SUS304L at RT and 190K. At RT and/or below 0.2 % proof stress, almost no martensite was produced even at low temperatures. At 190 K, the amount of martensite was about 8 % at 5 % strain. The HEE on fatigue properties can be discussed as follows:

Meta-stable austenitic stainless steel, SUS304 and 304L, showed remarkable HEE in relative reduction of area at low temperature around 190 K [5-7, 9], which is due to the increase of strain-induced martensitic phase during plastic deformation at the temperatures. A remarkable HEE can be seen in the evaluation of relative reduction of area at low temperature around 190 K, however, fatigue properties at 190 K were better than those of at RT, which is mainly because that the strength at 190 K is higher than at RT. The amount of martensite might increase during testing above the stress level of 0.2 % proof stress, but bellow the proof stress almost no strain-induced martensitic transformation might occur, so no effect of HEE can be seen.

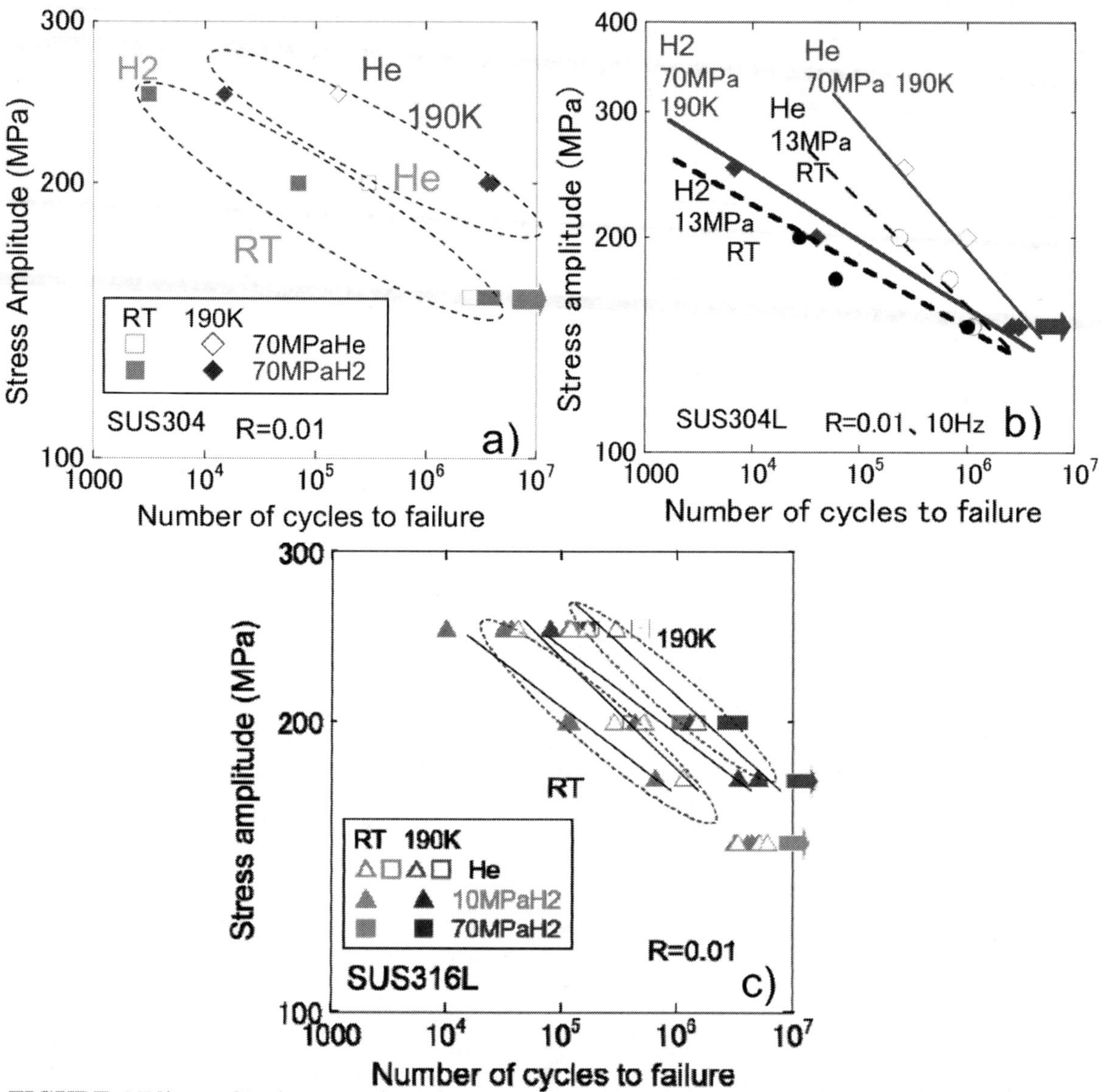

FIGURE 6 Effects of hydrogen and specimen temperature on fatigue properties of SUS304, 304L, and SUS316L at RT and 190K [11,12].

In SUS316L, the austenitic phase is much more stable than that of SUS304L and the effect of HEE is less also in S-N curves

4. Conclusions

- No effect of hydrogen on the amount of martensite was observed during tensile tests.
- In the meta-stable stainless steels, there is no HEE until the certain amount of deformation in tensile tests..
- No HEE until at a certain stress level or amount of bcc phase.
- Fatigue life of SUS 304 and 304L in hydrogen environment decreased at higher stress level, 200 MPa or 250 MPa, at RT and 190K. No influence at a stress level of 150 MPa. In SUS316L, influence of hydrogen is small even at a stress level of 250 MPa.

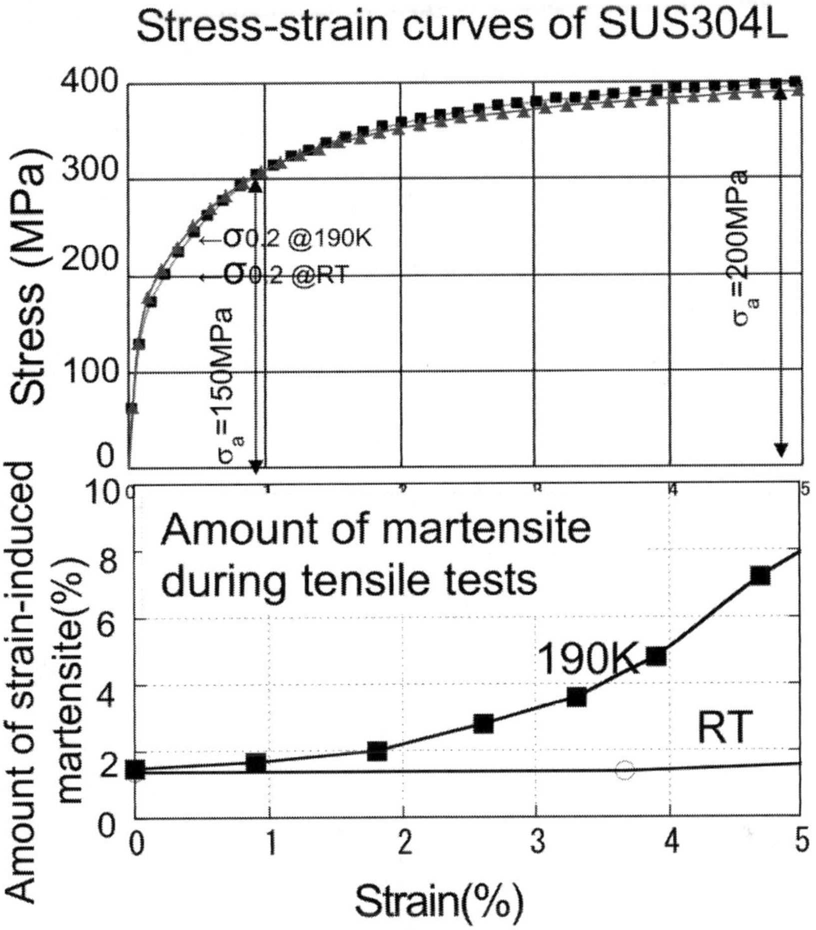

FIGURE 7 Relation between amount of strain-induced martensite and stress-strain curves at the early-stage of tensile tests for SUS304L at RT and 190K.

Acknowledgments
This work was carried out through "Establishment of Codes & Standards for Hydrogen Economy Society" program administrated by New Energy and Industrial Technology Development Organization (NEDO) in Japan

References

[1] Walter R J and Chandler W T 1973, NASA CR-124410
[2] Thompson A W 1974, Hydrogen in Metals, edited by I. M. Bernstein and A. W. Thompson, ASM, 91
[3] Huang J H and Altstetter C J 1995, *Metall. Trans. A*, **26A**, 845-849
[4] Gray H R 1974, *ASTM STP* **543**, 133-151
[5] Han G, Fukuyama S and Yokogawa K. 1998, *Acta mater.*, **46**(13), 4559-4570
[6] Sun D, Han G, Vaodee S, Fukuyama S, and Yokogawa K 2001, *Mater. Sci. and Technology*, **17**, 302-308
[7] Fukuyama S, Zhang L, Wen M, and Yokogawa K 2003, *J. Japan Inst. Metals*, **67**(4), 161-164.

[8] Ogata T 2008, *J. Japan Inst. Metals*, **72**, 125-131
[9] Ogata T 2008, *Advances in Cryogenic Engineering Materials* vol **54**, 124-131
[10] Ogata T 2009, *Proceedings of ICEC 22-ICMC 2008*, 1061-1065
[11] Ogata T 2010, *Advances in Cryogenic Engineering Materials* vol **56**, 25-32
[12] Ogata T 2012, *Advances in Cryogenic Engineering Materials* vol **58**, 39-46
[13] Ogata T, Yuri T, and Ono Y 2006, *Advances in Cryogenic Engineering Materials* vol **52A**, 122-129
[14] Ogata T, Yuri T, and Ono Y 2007, *Cryogenic Engineering in Japan* **42**, 10-17

IOP Conf. Series: Materials Science and Engineering **102** (2015) 012006 doi:10.1088/1757-899X/102/1/012006

Tensile properties and conductivities of a precipitation hardened and cold-rolled Cu-0.3Cr-0.1Zr alloy at cryogenic temperature

O Umezawa[1]
Faculty of Engineering, Yokohama National University, 79-5 Tokiwadai, Hodogaya, Yokohama, 240-8501, Japan

E-mail: umezawa@ynu.ac.jp

Abstract. Cu-0.3Cr-0.1Zr alloy (in mass%) is one of typical precipitation hardened copper-based alloys, and exhibited an excellent combination of high strength and high conductivities at the temperature range of 4 K to 300 K. The tensile properties, electrical resistivity, thermal conductivity and magnetization of precipitation hardened and cold-rolled Cu-0.3Cr-0.1Zr alloy were measures in comparison with oxygen free copper at cryogenic temperatures. The Cu-0.3Cr-0.1Zr showed higher yield ratio (yield strength / tensile strength) and lower the ratio of yield strength to electrical resistivity at cryogenic temperature than oxygen free copper. It exhibited high electrical and thermal conductivities, excellent non-magnetic stability and very low magnetic permeability at 4.2 K.

1. Introduction

The National Research Institute for Metals (NRIM, now National Institute for Materials Science, NIMS) has accumulated physical and mechanical properties at the temperature range of 4 K to 300 K for structural metallic materials since 1980s. Their electrical resistivity, magnetization, thermal conductivity and tensile properties were determined as a part of the work.[1,2] Precipitation hardened copper-based alloys as well as oxygen free copper in the cold-rolled condition were the candidate of a high strength and high conductivity structural material at cryogenic temperature. Since few cryogenic data for those alloys were available [3-5], precipitation hardened copper-based alloys developed as the lead frame in device [6-8] were chosen to evaluate for their cryogenic properties. Figure 1 represents a combination of their tensile strength and electrical conductivity (IACS) in the heavily cold-rolled condition (99% reduction). Cu-0.3Cr-0.1Zr alloy (in mass%) exhibits an excellent combination of high strength (600 MPa) and high electrical conductivity (80% IACS) at room temperature (Figure 1(a)). Furthermore it keeps higher tensile strength among the precipitation hardened copper-based alloys at the both temperatures of 4 K and 77 K (Figure 1(b)), and Cu-Cr-Zr system has no ferrous elements. In the present study, therefore, tensile properties and conductivities of precipitation hardened and cold-rolled Cu-0.3Cr-0.1Zr alloy plates were evaluated in comparison with oxygen free copper at cryogenic temperature, i.e. from 4 K to 77 K.

2. Experimental detail

[1] To whom any correspondence should be addressed.

2.1. Test materials

Test materials are Cu-0.3Cr-0.1Zr alloy plates and oxygen free copper (OFHC Cu) rods. The nominal composition of Cu-0.3Cr-0.1Zr alloy was Cr: 0.3, Zr: 0.1, Mg: 0.05, Si: 0.02 and Cu: balance (in mass%). The Cu-0.3Cr-0.1Zr plate which ingot was obtained by a vacuum melting was forged, hot-rolled (heated at 1093 K), and solution-treated (at 1223 K, water-quenched). The plate was cold-rolled, aged (at 773 K for 3.6 ks, water-quenched), and finally cold-rolled. The samples with final reduction ratios of 20%, 40% and 60% were provided. Figure 2 represents TEM (transmission electron microscopy) bright field image of Cu-0.3Cr- 0.1Zr alloy in the cold-rolled reduction of 99%. The materials were under over aged condition because of grown intermetallic particles, and dislocation cell structure was developed in the matrix as cold-rolling.

The OFHC Cu rod was as received (no annealed) with 4-nine grade, and its specification was JIS H3510 (article: C1011BD-H). Cold-rolling was done with various reduction ratios between 6% and 75%. As a reference, some precipitation hardened copper-based alloy sheets in commercial which were 99% cold-rolled of Cu-0.3Cr-0.1Zr (OMCL1), Cu-0.10Fe-0.034P (KFC-SH), Cu-0.15Fe-0.05P-0.1Sn-0.1Ag-0.1Zn (KLF201), Cu-3.2Ni-0.7Si-1.25Sn-0.3Zn (KLF125) and Cu-1.5Ni-0.3Si-0.03P (PMC102) (in mass%) [6-8] were examined.

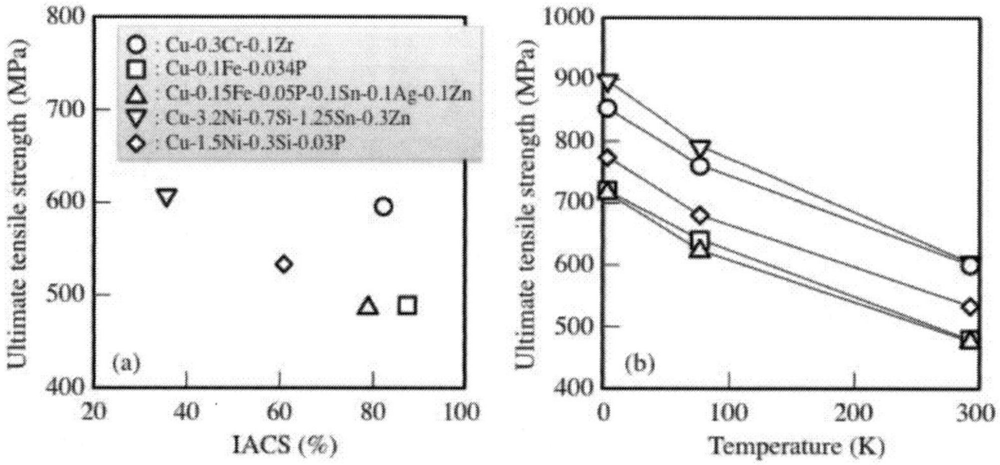

Figure 1. High strength and high conductivity copper-based alloys developed for the lead frame (a) and their tensile strength at 4 K and 77 K (b).

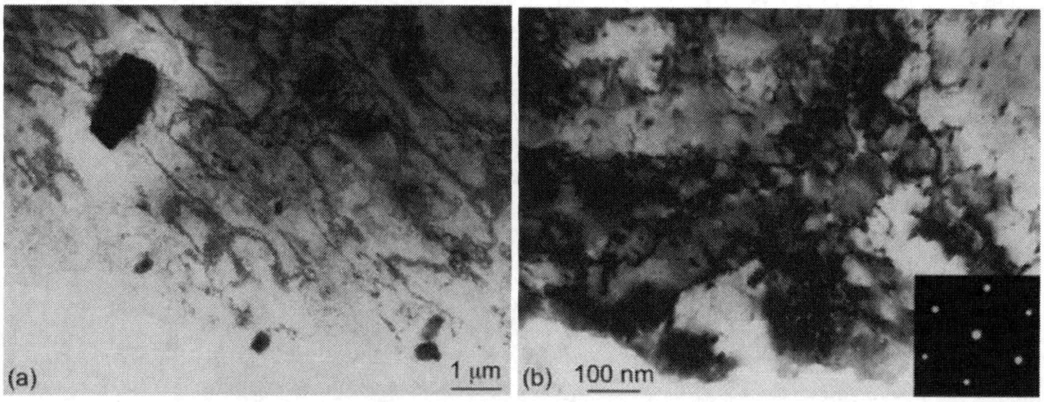

Figure 2. TEM bright field images of Cu-0.3Cr-0.1Zr alloy with the cold-rolled reduction of 99%: (a) precipitates of intermetallic phases and dislocation structure, and (b) dislocations in the matrix. Beam direction is near [011].

2.2. Measurements

Tensile test specimen was cut from the plates or rods along the rolling direction. Tensile test (displacement-controlled) was carried out at a strain rate of about 4 x 10-4 sec-1 using a screw-driven machine. The specimens were immersed in liquid helium (4 K), liquid nitrogen (77 K), and ambient air (293 K), respectively. Yield strength was evaluated by the off set value of 0.2% proof stress using MTS extensometer.

The electrical resistivity was measured using bar-shaped specimens by four-point probe technique between liquid helium temperature (4.2 K) and room temperature (293 K).

The thermal conductivities were measured by the steady state heat flow method from 3.5 K to 60 K. Samples were 10 mm x 4 mm x 50 mm strips abrasively polished. The samples were held in a chamber evacuated to 10-5 Pa. The temperature gradients along the samples were measured using two calibrated carbon-glass resistors fixed on copper holders. The distance between the thermometers were 20 mm. The sample temperature was defined as the mean value of the two. The sample heater was made of a manganin wire, and the power applied to the sample was measured using the four-probe method.

The magnetization measurements at 4.2 K were made using a computerized PAR-EG&G vibrating-sample magnetometer. Point-by-point measurements were taken in stepped fields between -3 and 3 MA/m provided by a superconducting solenoid. The typical sample was a 3 mm-diameter cylinder. The shearing correction by demagnetizing factor, N=0.27, was made.[9]

3. Results and discussion

3.1. Tensile properties

Figure 3 shows the load displacement curves of the 20% cold-rolled Cu-0.3Cr-0.1Zr material at low temperatures. As the test temperature decreased from 293 K to 4 K, both strength and elongation increased. Especially, the higher homogeneous elongation and work hardening rate were, the lower the temperature was. Both ultimate tensile strength and 0.2% proof stress were higher as the increase of

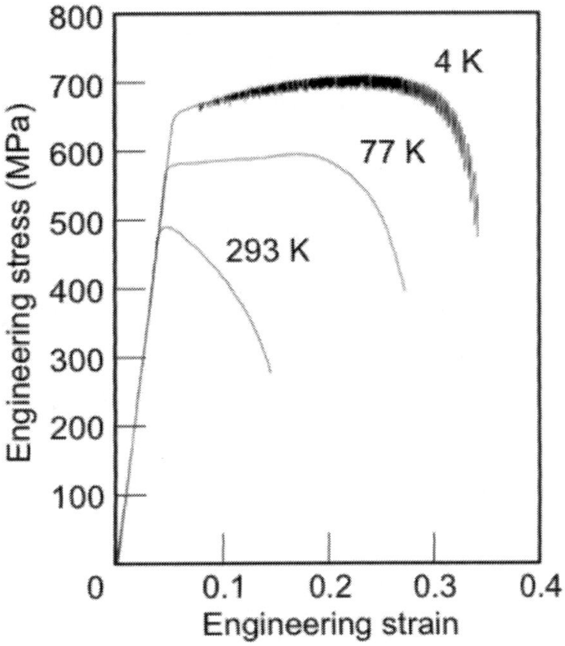

Figure 3. Stress-strain curves of Cu-0.3Cr-0.1Zr 20% cold-rolled material at cryogenic temperatures.

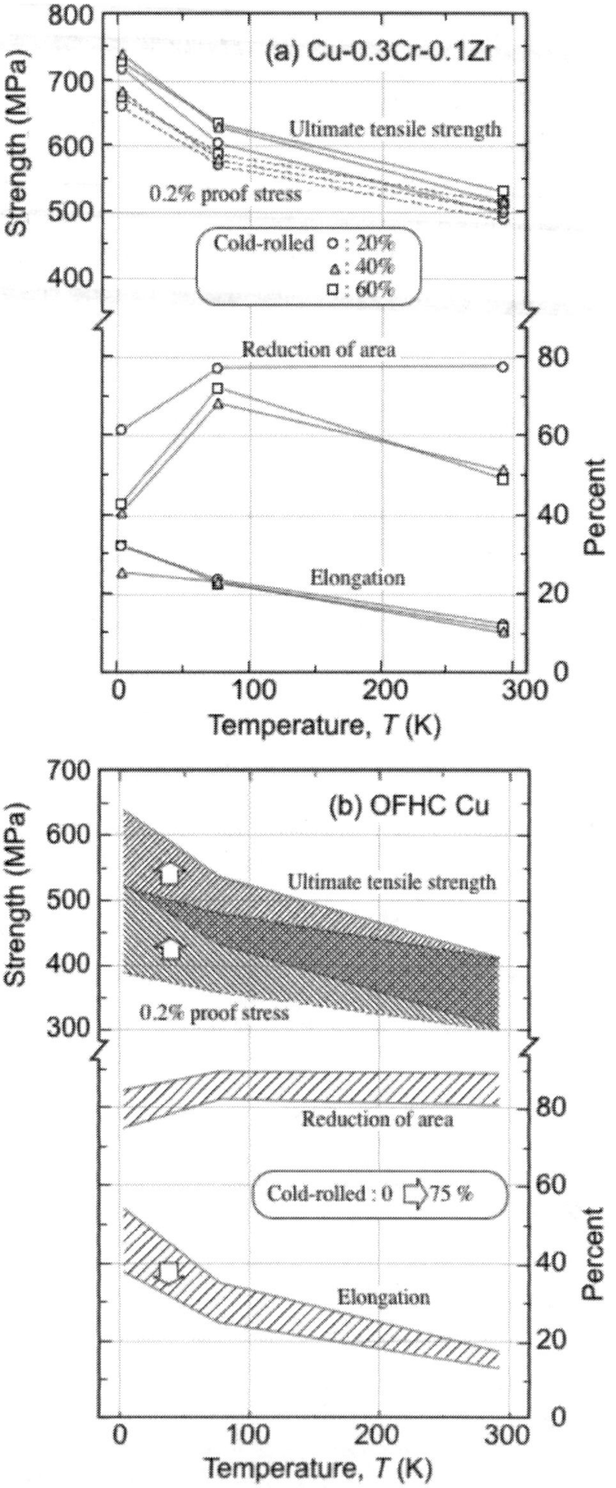

Figure 4. Effects of cold-rolling on tensile properties of (a) Cu-0.3Cr-0.1Zr alloy and (b) OFHC Cu at cryogenic temperatures.

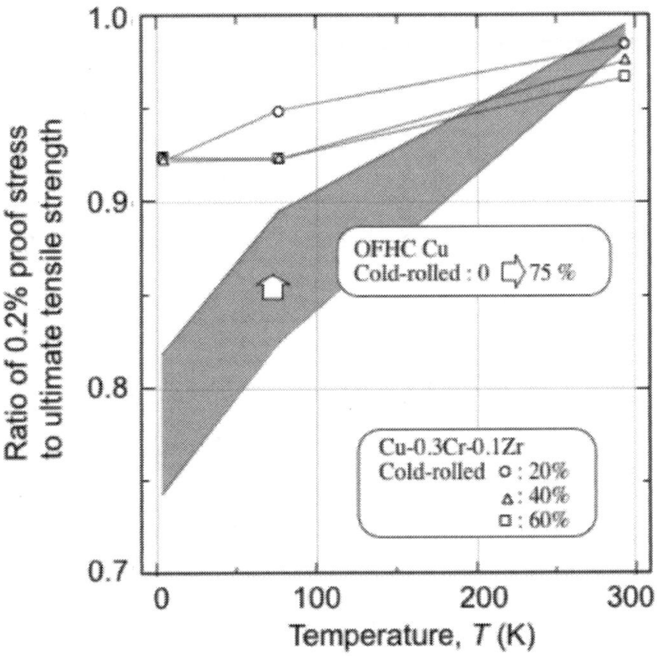

Figure 5. Ratio of 0.2% proof stress to ultimate tensile strength of Cu-0.3Cr-0.1Zr alloy and OFHC Cu at cryogenic temperatures.

cold-rolled ratio from 20% to 60%, although the elongation was almost the same without the cold-rolled ratios at each temperature (Figure 4(a)). These tensile properties' dependence on test temperature was similar to that of OFHC copper. Figure 4(b) summarized the tensile properties of cold-rolled OFHC Cu with the reduction from 0% to 75%. Both ultimate tensile strength and 0.2% proof stress were higher as the increase of cold-rolled ratio, although the elongation was lowered. The yield ratio (yield strength / tensile strength) of Cu-0.3Cr-0.1Zr, however, was higher than that of OFHC Cu at lower temperature as shown in Figure 5, since both precipitation strengthen and dislocation strengthen may introduce high yield strength with less temperature dependence. The precipitation behavior and the dislocation structure developed on Cu-Cr-Zr and Cu-Fe-P alloys have been evaluated and discussed with their mechanical property in the references [10-12].

3.2. Conductivities

The electrical resistivity of Cu-0.3Cr-0.1Zr alloy and OFHC Cu was summarized in Figure 6(a). The residual resistivity ratio (RRR) of each material was also interpolated in the figure. The resistivity of materials is increased as the increase of cold-rolled ratio for each alloy. The RRR in Cu-0.3Cr-0.1Zr alloy is roughly from 4 to 5, although the RRR in OFHC Cu is much higher than that in Cu-0.3Cr-0.1Zr alloy and strongly depend on the cold-rolled ratio. Slight solid solubility in copper may affect the high resistivity in the precipitate strengthen alloys,[12] although the resistivity of OFHC Cu is rapidly decreased at around several ten degrees [1,4]. Then the ratio of yield strength to electrical resistivity was plotted at test temperatures in Figure 6(b). The ratio in OFHC Cu is much higher than that in Cu-0.3Cr-0.1Zr alloy especially at lower temperature.

The thermal conductivity of Cu-0.3Cr-0.1Zr alloy is about one tenth of heavily cold-rolled OFHC Cu at cryogenic temperature as shown in Figure 7. However, it was twice as high as that of phosphorous-deoxidized copper [1,4]. Thus Cu-0.3Cr-0.1Zr alloy exhibited high electrical and thermal conductivities as well as the excellent combination with high strength in the cooper alloys at cryogenic temperature.

IOP Conf. Series: Materials Science and Engineering **102** (2015) 012006 doi:10.1088/1757-899X/102/1/012006

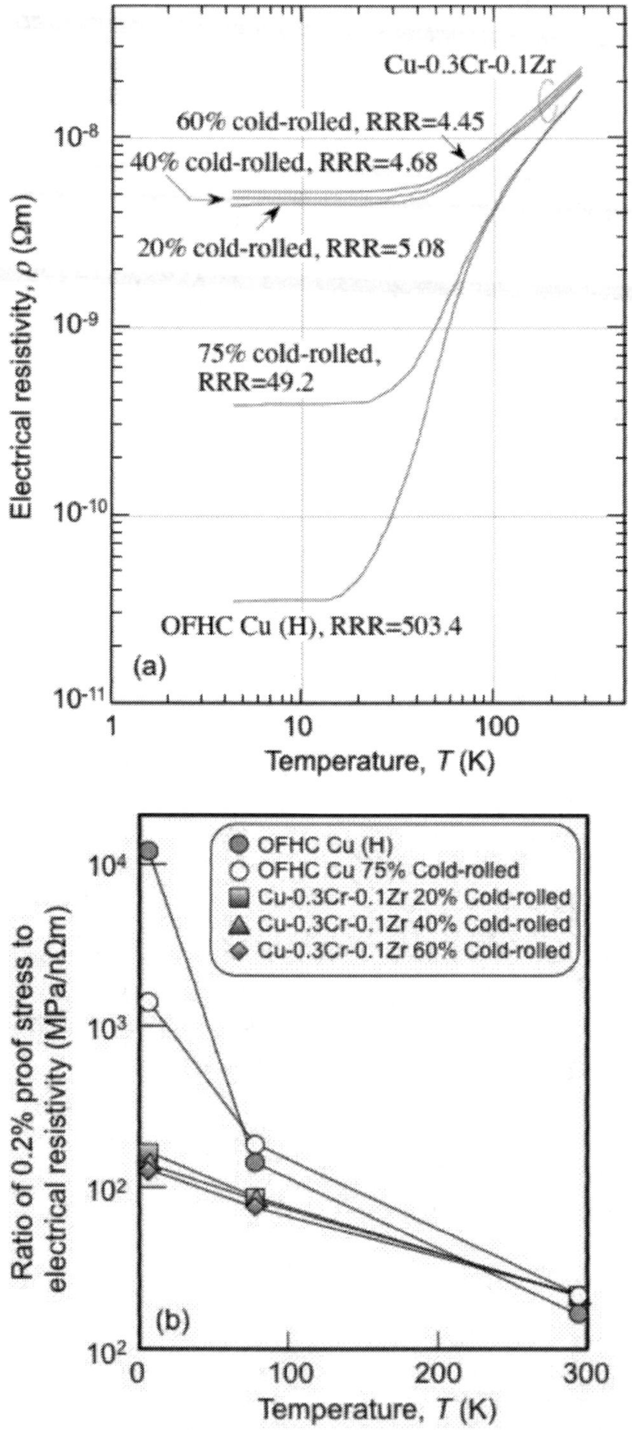

Figure 6. Electrical resistivity (a) and the ratio of yield strength to resistivity (b) for Cu-0.3Cr-0.1Zr alloy and OFHC Cu at cryogenic temperatures.

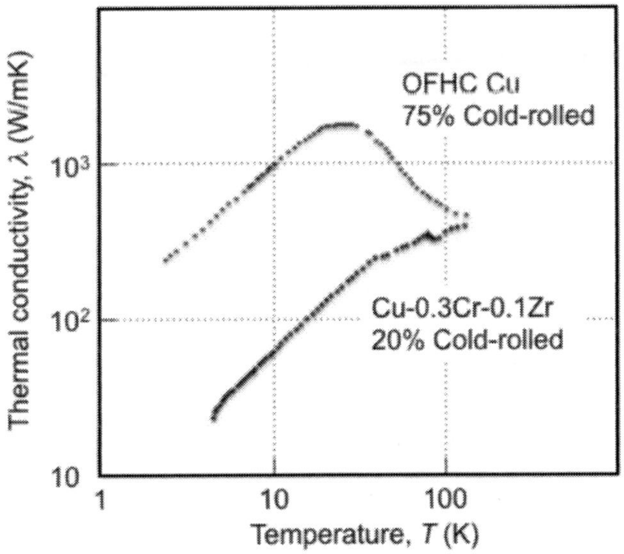

Figure 7. Thermal conductivity of cold-rolled Cu-0.3Cr-0.1Zr and OFHC Cu at cryogenic temperatures.

3.3. Magnetization

Figure 8 represents the variation of d.c. magnetization with applied field at 4.2 K for Cu-0.3Cr-0.1Zr and Cu-0.15Fe-0.05P-0.1Sn-0.1Ag-0.1Zn materials cold-rolled with 99%. The Cu-0.3Cr-0.1Zr exhibits paramagnetism at 4.2 K, since its magnetization is directly proportional to the applied field in Figure 8(a). On the other hand, the plots of Cu-0.15Fe-0.05P-0.1Sn-0.1Ag-0.1Zn develop a saturation shown in Figure 8(b). But the magnetization of the alloy at 3 MA/m is about 0.25 kA/m. Thus the alloy exhibits very low magnetization and susceptibility under high fields, although it has higher permeability in the tested alloys. All tested alloys exhibited excellent non-magnetic stability up to high fields and very low magnetic permeability.

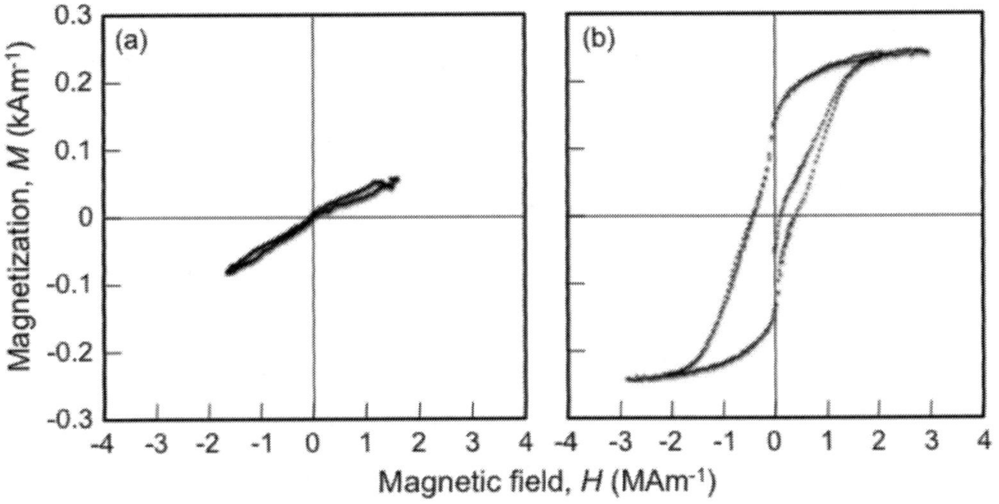

Figure 8. Magnetization curve of heavily cold-rolled sheets of (a) Cu-0.3Cr-0.1Zr and (b) Cu-0.15Fe-0.05P-0.1Sn-0.1Ag-0.1Zn at 4.2 K.

4. Conclusions

The tensile properties, electrical resistivity, thermal conductivity and magnetization were determined at cryogenic temperature for precipitation hardened and cold-rolled Cu-0.3Cr-0.1Zr alloy in comparison with OFHC Cu. The Cu-0.3Cr-0.1Zr alloy exhibited an excellent combination of high strength and high conductivities at the temperature range of 4 K to 300 K. Major results are summarized as follows:

The Cu-0.3Cr-0.1Zr showed higher yield ratio and lower the ratio of yield strength to electrical resistivity at cryogenic temperature than OFHC Cu. The Cu-0.3Cr-0.1Zr alloy exhibited high electrical and thermal conductivities, excellent non-magnetic stability and very low magnetic permeability at 4.2 K, although its slight solid solubility in copper resulted in the higher resistivity and lower thermal conductivity than OFHC Cu.

Acknowledgments

The author wishes to thank Drs. T. Numazawa, M. Uehara and K. Nagai, NRIM, for their technical help on the measurements of thermal conductivity and magnetization, respectively.

References

[1] Umezawa O 1993 *Handbook of Superconductivity and Cryogenics* ed Cryogenics and Superconductivity Society of Japan (Tokyo: Ohmsha) pp 88-93, 1084-1089
[2] Umezawa O and Ishikawa K 1992 *Cryogenics* **32** 873
[3] Clark A F, Childs G E and Wallace G H 1970 *Cryogenics* **10** 295
[4] Metals and Ceramics Information Center 1977 *Handbook on Materials for Superconducting Machinery* (Battele: Columbus Laboratory)
[5] Reed R P and Clark A F 1983 *Materials at Low Temperatures* (Ohio: American Society for Metals)
[6] Kobayashi M, Iwamoto T and Izumida M 1988 *Bull. Japan Inst. Metals: Materia Japan* **27** 284
[7] Miyamoto M 2000 *Kobe Steel Engineering Reports* **50-3** 62
[8] Kim Y, Lim C, Ryu I, Ryu C, Han J K and Reed R P 1988 *Advances in Cryogenic Engineering Materials* vol 34 (New York: Plenum Press) pp 693-702
[9] Bozorth R M 1951 *Ferromagnetism* (New York: Van Nostrand)
[10] Batra I S, Dey G K, Kulkarni U D and Banerjee S 2001 *J. Nuclear Materials* **299** 91
[11] Hatakeyama M, Toyama T, Nagai Y, Hasegawa M, Eldrup M and Singh B N 2008 *Materials Transactions* **49** 518
[12] Aruga Y, Ozaki R and Kajihara K 2009 *J. Japan Inst. Metals* **73** 33

IOP Conf. Series: Materials Science and Engineering **102** (2015) 012007 doi:10.1088/1757-899X/102/1/012007

Design and fabrication of a cryostat for low temperature mechanical testing for the Mechanical and Materials Engineering group at CERN

I Aviles Santillana[1, 2], R Betemps[1], A Gerardin[1], M Guinchard[1], S A E Langeslag[1] and S Sgobba[1]

[1] CERN, CH-1211 Genève, Switzerland
[2] University Carlos III of Madrid, Campus de Leganes, Av. Universidad 30, 28911 Madrid, Spain

E-mail: Ignacio.Aviles.Santillana@cern.ch

Abstract. Mechanical testing of materials at low temperatures is one of the cornerstones of the Mechanical and Materials Engineering (MME) group at CERN. A long tradition of more than 20 years and a unique know - how of such tests has been developed with an 18 kN double-walled cryostat. Large campaigns of material qualification have been carried out and the mechanical behaviour of materials at 4 K has been vastly studied in sub – size samples for projects like LEP, LHC and its experiments. With the aim of assessing the mechanical properties of materials of higher strength and/or issued from heavy gauge products for which testing standardized specimens of larger cross section might be more adapted, a new 100 kN cryostat capable of hosting different shapes of normalized samples has been carefully designed and fabricated in-house together with the associated tooling and measurement instrumentation. It has been conceived to be able to adapt to different test frames both dynamic and static, which will be of paramount importance for future studies of fracture mechanics at low temperatures.

The cryostat features a double-walled vessel consisting of a central cylindrical section with a convex lower end and a flat top end closure. The transmission of the load is guaranteed by a 4 column system and its precise monitoring is assured by an internal load cell positioned next to the sample in the load train. This innovative approach will be discussed together with other non-conventional instrumentation solutions.

A validation of the whole system has been carried out, where bending efforts on instrumented samples have been measured. Additionally, dedicated tooling has been fabricated for the device's optimization. The preliminary results obtained confirm an excellent performance of the system and enhance the analysis of materials under extreme conditions with state of the art instrumentation.

1. Introduction

At the European Organization for Nuclear Research (CERN), mechanical testing of materials under extreme conditions has always been of great importance for the construction of state of the art equipment for the biggest particle accelerator complex worldwide. The use of cryogenics for experiments at CERN dates back to the 1960's with the track chambers, where liquid hydrogen and neon were used for working in the 30 K range at a pressure of around 5 atmospheres [1]. From the 1970's to the 1990's, the development of superconducting magnets and accelerating cavities for numerous projects (ISR, ALEPH, LEP) generalized the use of cryogenics at CERN at temperatures down to 4.5 K, and from the 1990's until today, the use of cryogenics has increased in the whole accelerator complex (superconducting accelerator and detector components in the Large Hadron Collider (LHC)). There are even centralized CERN-wide distribution lines of liquid nitrogen and liquid helium to small scale cryogenic experiments such as the Antiproton Decelerator (AD) or the Isotope Separator On Line DEvice (ISOLDE) [1].

It stands to reason that mechanical characterization of materials at low temperatures has always been one of the main pillars of the Mechanical and Materials Engineering (MME) group at CERN. In the accelerator complex, high energy magnets, superconducting radiofrequency cavities, particle detectors and all the associated instrumentation is working at a temperature below helium's boiling point (4.5 K), hence a detailed characterization and a comprehensive study of the mechanical behavior of materials in such an extreme condition becomes essential. Within this scope, a total of four reduced size cryostats like the one shown in figure 1 were designed and fabricated at CERN to be able to respond to the high demand of mechanical characterization at cryogenic temperatures. With these 18 kN cryostats, a huge number of test campaigns have been executed, not only for material characterization [2] but also for more advanced studies such as discontinuous plastic flow and local heating during tensile testing at cryogenic temperatures [3]. Despite the high reproducibility and exploitability of the results, they were obtained with use of sub-size samples according to standard ASTM E8. In the frame of new particle accelerators (Future Circular Collider), improvement of the existing ones (High Luminosity LHC) and amongst others, the push forward in high field superconducting magnets (11 T), an increasing demand of tensile tests close to the operating temperature of these High Energy Physics (HEP) devices exists. To be able to face this increasing need, the MME group decided to design and fabricate in-house a 100 kN cryostat capable of accommodating standard size tensile specimens.

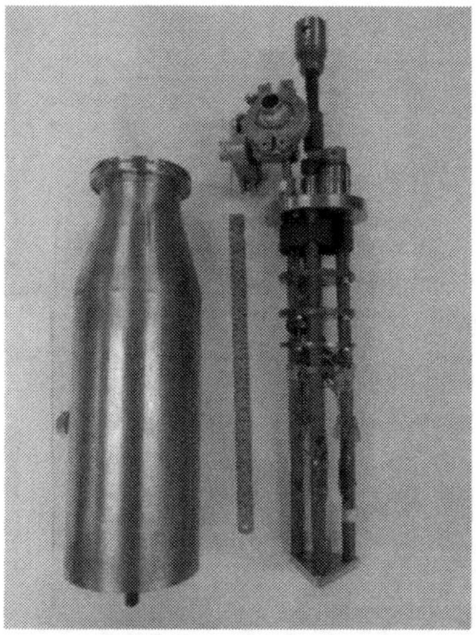

Figure 1. 18kN cryostat. On the left, the female part consisting of a dewar; on the right the male part consisting of inner columns, thermal screens, sample holder and pulling rod.

2. Conception and design features

When the need of a 100 kN cryostat was acknowledged, it was also decided that the latter would have been designed, fabricated and assembled in house, profiting of the available experience of designing and fabricating cryogenic equipment and of the flexibility of a fully tailored solution. In this section, the different technical choices adopted with respect to others found in other laboratories will be presented and discussed. They are gathered in four different categories: thermal, structural, load train - sample holders and instrumentation solutions (see figure 2).

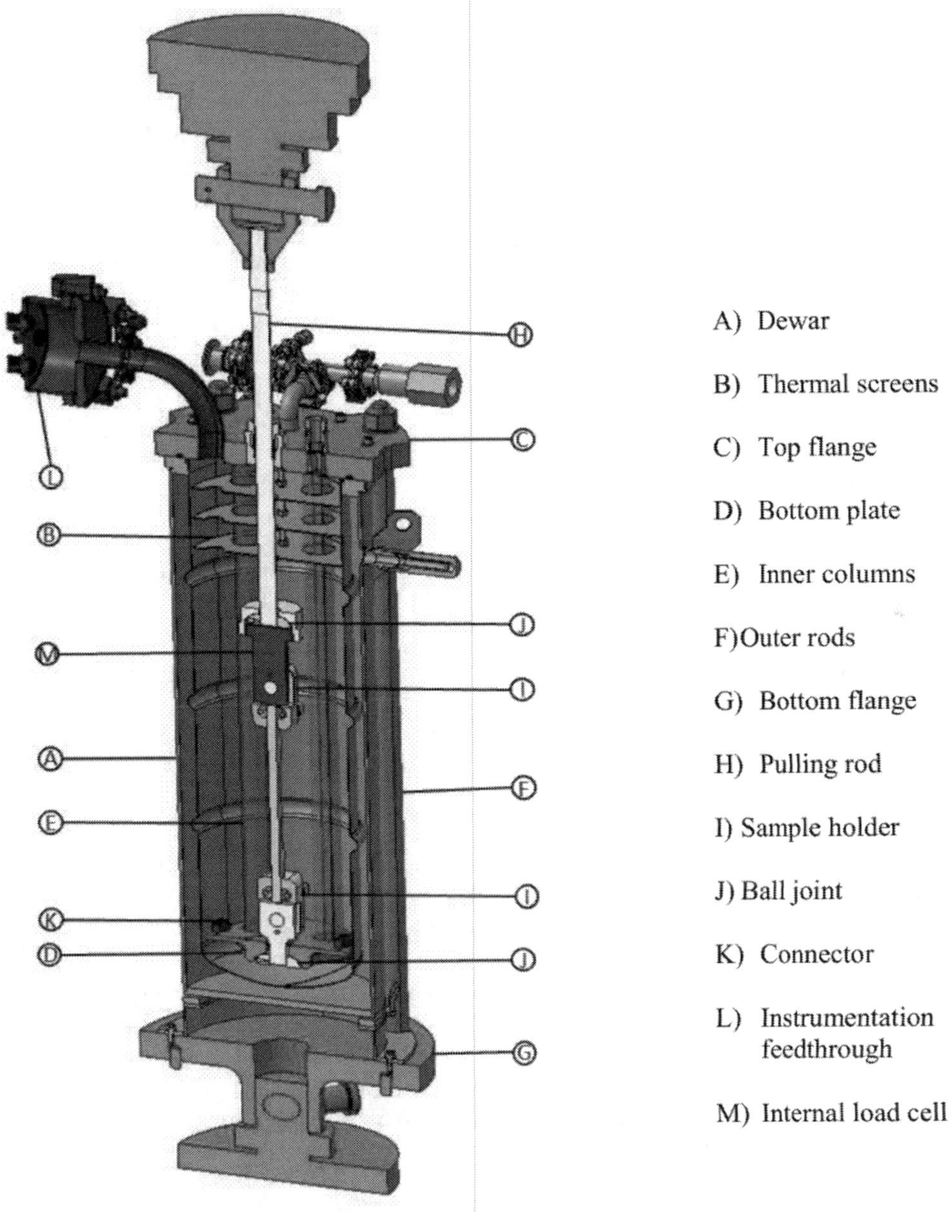

Figure 2. Assembly drawing of the cryostat tensile system

2.1. *Thermal*

The thermal performance of a cryostat with such a big volume becomes very important in order to minimize cryogen consumption. The main part which ensures the thermal performance of the system is a stainless steel double-wall vessel (dewar) (A) with a 0.8 mm inner wall thickness. The convolutions shown in Figure 2 avoid buckling during the purging which needs to be completed prior to cryogen filling. Between the two walls, a thermally isolating volume pumped down to a vacuum of 10^{-4} mbar and 20 layers of a multi-layer insulation (MLI). A single dewar was used instead of the double one sometimes employed when working with liquid helium mainly to avoid complexity and to ease handling. Time between tests consumption spent in disassembling and reassembling the system for the single dewar solution greatly compensates the diminution in thermal efficiency when compared to the more complex double dewar.

Three ETP copper thermal screens (B) each 1.5 mm thick are used in the top part in order to minimize the heat exchange with the massive top plate. Copper is chosen due to its excellent thermal conductivity. The three thermal screens are pierced to provide access to the filling pipe.

For the insertion of the cryogens in the system, an inlet and outlet are located in the top flange, the former being aligned with the pierced surfaces of the thermal screens. Their leak tightness is guaranteed by a sliding rubber o – ring seal within a bronze threaded tube to prevent the gas to escape. Bronze is used for this kind of connections due to its low friction coefficient and its increased thermal contraction with decreasing temperature with respect to steel (which avoids galling) [4].

When designing new tooling, the main approach is minimization of their mass together with an increase in the contact surface with the coolant. This has to be ensured without jeopardizing the mechanical stability of the part.

All in all, the cool down time between two tests stays in the order of 20 minutes to 25 minutes if the filling is done at 1.2 bar – 1.3 bar absolute. Helium consumption for a single test is calculated to be between 25 liters and 30 liters, depending on sample ductility. These two figures prove the excellent thermal performance of the cryostat, considering its internal volume is approximately 15 liters.

2.2. *Structural*

The structural part of the system consists of four internal columns, four outer rods, a top flange, an inner bottom plate and an outer bottom plate. Their main purpose is to provide mechanical stability to the system without introducing undesired efforts. They also contain instrumentation elements such as connectors and cabling.

2.2.1. Top flange. The material for the top flange (C) is austenitic AISI 316 LN (1.4429) stainless steel conform to the CERN technical specification, which requires electroslag remelting and multi directional forging, resulting in a homogeneous product, completely austenitic and featuring a low inclusion content. The top flange has a thickness of 27 mm and contains the inlet and outlet for cryogenic fluids, a connection for the instrumentation feedthrough and a connection for the overpressure security systems (a 1.5 bar valve and 2 bar burst disk), respectively.

Three elements aim at ensuring leak tightness: a groove which is machined in the lower surface in which a rubber o – ring is placed, eight silver coated screws which pull the dewar towards the flange and four outer rods, which are tightened with a precise 50 N•m torque to secure the parallelism of the top flange and the outer bottom flange.

In order to give to the pulling rod access to the test volume, an o – ring system similar to the one used for the inlet and outlet previously described is used. This solution was chosen due to its simplicity with respect to more complicated ones like bellows. However its implementation is more delicate as an incorrect positioning of the hole in the flange would induce a misalignment and a deviation from uniaxial stress in the material studied. Because of this, the deviation from the ideal position has been measured to be lower or equal to 22 μm.

2.2.2. Inner bottom plate. The inner bottom plate (D) hosts the lower sample holder including a ball joint, the four instrumentation connectors and the four internal columns. Its design has been optimized to reduce mass without endangering its mechanical stability (figure 3). A titanium base alloy (Ti6Al4V) was chosen with the goal of maximising the yield strength while at the same time minimising the heat input into the cryostat.

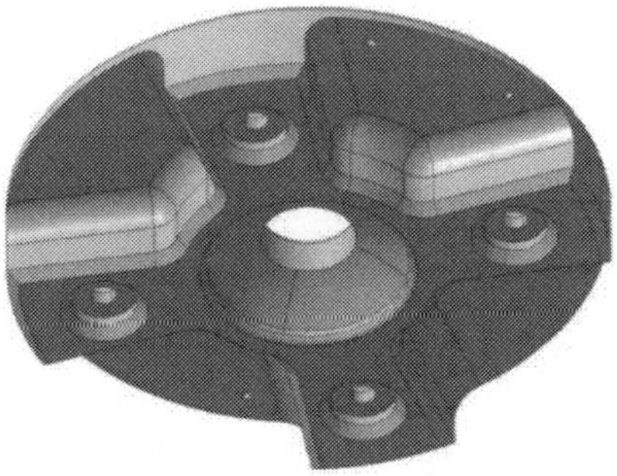

Figure 3. 3D model of the inner bottom plate

2.2.3. Internal columns. The top flange and the bottom plate are connected through four austenitic stainless steel (AISI 316L) hollow columns (E). As these elements work under compressive loads, steel was preferred to titanium due to its higher rigidity in order to avoid buckling. The wall thickness (2.6 mm) and outer diameter (30 mm) were again optimised to minimise heat input guaranteeing mechanical stability for a given permissible load. On the top part of each column, a ring was welded to ensure a correct perpendicularity of the columns with respect to the top flange. In order to have a better access to the sample and more room for instrumentation, a four columns system was chosen instead of a three columns one.

2.2.4. Outer rods. The outer rods (F) connect the top flange with the outer bottom plate. They are solid bars of 36NiCrMo16 carbon steel (1.6773), typically used for tooling subjected to high stresses. With a diameter of 28 mm, they are connected to the top and bottom flanges with a threaded M20 connection. Their length after taking into account the thread length has to be accurate to a tolerance of 100 µm (± 50 µm) to safeguard the parallelism of the two flanges and thereby avoid deviations from uniaxial stress regime.

2.2.5. Bottom flange. For the connection of the lower part of the cryostat to the tensile machine, EN S355J0 steel (1.0553) flange (G) is used. As it is shown in Figure 2, the attachment is performed with a standard pin connection, available for most universal testing machines.

2.3. *Load train and sample holders*

Experience has shown that in order to obtain satisfactory results during tensile testing and especially with highly brittle materials, it is essential to achieve and maintain a precise specimen alignment during the test. On this basis, the load train of the cryostat has been conceived to provide an accurate specimen alignment mainly allowed by a shoulder – headed sample holders and ball joints at both extremities, respectively.

2.3.1. Pulling rod. The pulling rod, in its lower extremity, is screwed to the upper ball joint which is in turn connected to the sample holder. Its upper extremity extends beyond the cryostat through the sliding rubber o – ring seal as described in §1.1. It is then fixed to an adaptor which is connected to the machine through a pin connection to allow free axial movement. The roughness of the rod is specified to be Ra < 1.6 μm in order to avoid high frictional forces at the o – ring seal. The material used is grade 5 titanium, which among commercial metals shows the required ductility in uniaxial tension at low temperatures and features a favourable combination of high yield strength and fair thermal conductivity, hence minimizing heat conduction into the cryostat along the rod [5].

2.3.2. Sample holders. Shoulder–headed sample holders (I) were preferred over pin connections to minimise stress concentration in the heads of the tensile specimen. A very tight shape tolerance of 0.12 μm is imposed in the shoulder region to guarantee a correct positioning of the sample and a homogeneous application of the load. The fabrication of the remainder of the tensile specimen obviously imposes the same shape accuracy. The first set of sample holders were fabricated in AISI 316L stainless steel, but in order to be able to test highly resistant materials without introducing deformation at the level of the tooling, a new set will soon be fabricated. The material properties demanded are high resistance and reasonable ductility at cryogenic temperatures, leaving a shortlist of two candidates: Ti6Al4V and Inconel 718.

The sample holders are fabricated in two identical halves which are placed symmetrically at both sides of each sample head. They are connected through a large pin to the load cell at the top and a connector at the bottom. This connection allows rotation around the tensile axis. A set of collars is placed around the two halves of the sample holders to keep them (and the specimen) in position and avoid their separation during testing or cool down.

When testing flat samples, the sample holders are designed to be able to hold samples of different thicknesses by adding a set of adaptors which prevent the movement of the specimen in the direction perpendicular to the application of the load. These adaptors, the sample holders and the collars are shown in figure 4.

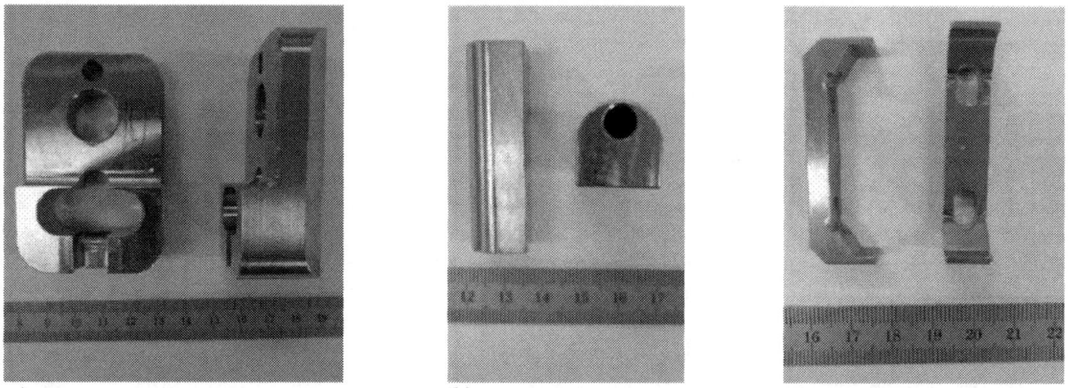

a) b) c)

Figure 4. a) Flat sample holders. b) Adaptors for different thicknesses. c) Collars for sample holders

2.3.3. Ball joints. The idea of placing the tensile specimen between two ball joints comes from beam theory, which states that no bending moments can be induced in any beam section found between the ball joints (or hinges) [6]. The system is designed in such a way that one ball joint (J) and its counterpart have exactly the same geometry, while individually maximising the contact area. In this way, a better distribution of the load is achieved and hence, a lower frictional force is implied between the parts. A potential beneficial effect on the performance of the ball joints induced by solid lubricants is currently being studied.

2.4. *Instrumentation*

A key aspect during the conception of a cryogenic tensile system is the instrumentation. It is extremely important not only to guarantee an accurate monitoring of the signals but also achieving flexibility of the system, meaning creating the possibility of adding new sensors according to the test campaign to be carried out.

2.4.1. Connectors. Four sets of 3M® connectors (K) are attached to the inner bottom plate. One of them is dedicated exclusively to the internal load cell. Another one is used either for two extensometers or possibly for two LVDT sensors for measuring displacement. The last two are available for any additional sensors which could be required (temperature sensors, strain gages, capacitive sensors), each of them counting 12 pins. Additionally, there are four optical connections where optical fibres can be coupled, which can be used as sensors. The presence of these four optical connections opens the possibility of implementing a great variety of sensors based on glass fibre, which are unaffected by magnetic field. All the cabling associated to the connectors is sent to the outside through the instrumentation feedthrough (L) which offers a total of four Fischer® connections to read out all the different signals.

2.4.2. Internal load cell. One of the most innovative solutions which the system features is the implementation of an internal load cell (M) placed close to the top part of the specimen to be tested. The material chosen was Ti5Al2.5Sn ELI due to the combination of its high tensile properties, toughness and reasonably low fatigue crack growth rates at cryogenic temperature. The load cell measuring function is based on a Wheatstone bridge configuration formed with four strain gages acting as the four active arms of the bridge. The strain gages used for this load cell are HBM's 1-XC11-3/350, 0 and 90° T rosettes with 350 Ω resistance and a k factor of 2.22.

The design of the piece to which the Wheatstone bridge is glued to be implemented as a load cell has been designed in such a way that the top part of the sample holder can be directly mounted through a pin and the pulling rod can be easily screwed to it via a specially designed adaptor. A finite element analysis has been carried out to verify that the deformation in the flat surfaces, at the position of the strain gages, is homogeneous and it remains below 25 % of the elastic limit.

Following the instrumentation of the load cell, a calibration was performed according to the instructions of the international standard ISO 376:2011. Overall 10 tensile loading steps have been performed to up to 80 kN at room temperature. These loading steps have been repeated for six cycles overall in order to validate the measurement uncertainty of the results including factors such as repeatability, reproducibility and creep errors. According to this calibration, the sensor shows a combined standard uncertainty of **uc (F) = 1.9000e-03 + 1.8390e-02**, what makes the load cell a class 1 force proving instrument, as required by ISO 6892.

2.4.3. Extensometers. The extensometers used for low-temperature tensile testing generally utilize strain gages mounted to a bending beam element. In our case, we have two different types available: the C–shape and the W–shape (figure 5), which are used depending on the gage length of the specimen. They are both a development of the Cryogenic Engineering & Materials Expertise (CEME). The C–shape one is very suitable for standard size flat samples, with a gage length of 50 mm and a maximum elongation at break of 50 %. The W–shape is more adapted to small size standard samples, with a gage length of 25 mm. They have both been calibrated by the supplier and also in house at 4 K, 77 K and room temperature.

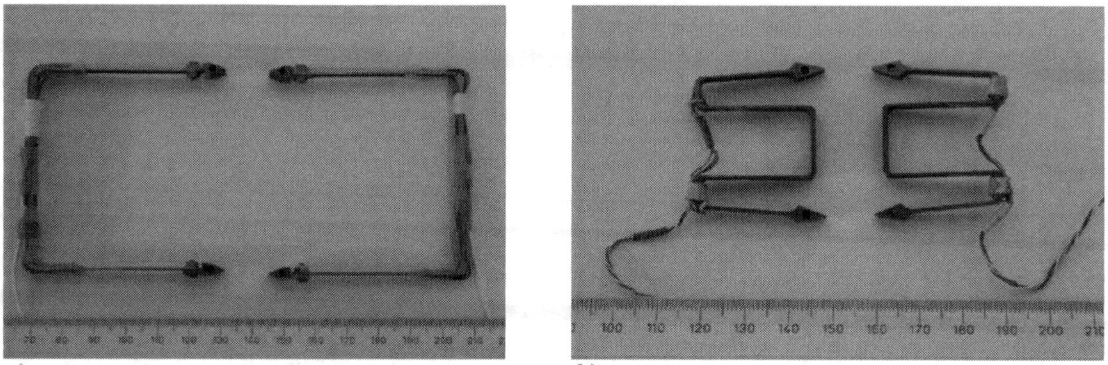

a) b)

Figure 5. Extensometers for uniaxial tensile tests. a) C–Shape extensometers. b) W–shape extensometers

3. Validation and results

3.1. *Validation*

For the validation of the cryostat, ASTM E1450 was used as a reference. This international standard states that the maximum bending strain should not exceed 10 % of the axial strain. The procedures for measuring specimen alignment are inspired on ASTM E1012. Flat and round (buttonhead) specimens were utilized for verification of the alignment. The loading scheme was also inspired by ASTM E1012, which states the percentage of bending has to be measured in at least three points in the elastic region, preferably equispaced between them. In our case, five points were measured, being the highest at approximately 75% of the expected yield strength of the material to avoid plastic deformation. Stainless steel AISI 304L is used for the validation.

3.1.1. Flat specimens. The specimens are fabricated following the requirements of both ASTM E8 and ISO 6892-1. Special attention was put in the shape accuracy as described in §2.3.2 and the flatness, with a tolerance of ± 15 μm. A single strain gage is glued to the parallel surfaces in the middle of the calibrated length. They are then connected in a half–bridge configuration which gives as an output the difference of strain measured over the strain gages. The tensile strain is calculated with the force measured by the internal load cell, the cross section and the Young's modulus of the material. Percentage of bending subsequently follows from the fraction of bending strain over the tensile strain. The values for ten room temperature tests in figure 6. Between every test, the specimen was demounted and remounted in the sample holders.

3.1.2. Buttonhead specimens. The specimens are fabricated following the requirements of both ASTM E8 and ISO 6892-1 for tensile tests at room temperature. In this occasion, 8 strain gages are positioned as described in ASTM E1012 in two rows: one on the top and one on the bottom. As an independent reading of the 8 signals was impossible at cryogenic temperature, the strain gages of the same row situated diametrically opposed one to the other were connected and measured as described in §3.1.1. The validation procedure was the same as for the flat specimen. figure 6 summarizes the results, which show consistently a percentage of bending below 10 % for all signals or slightly above for some isolated positions.

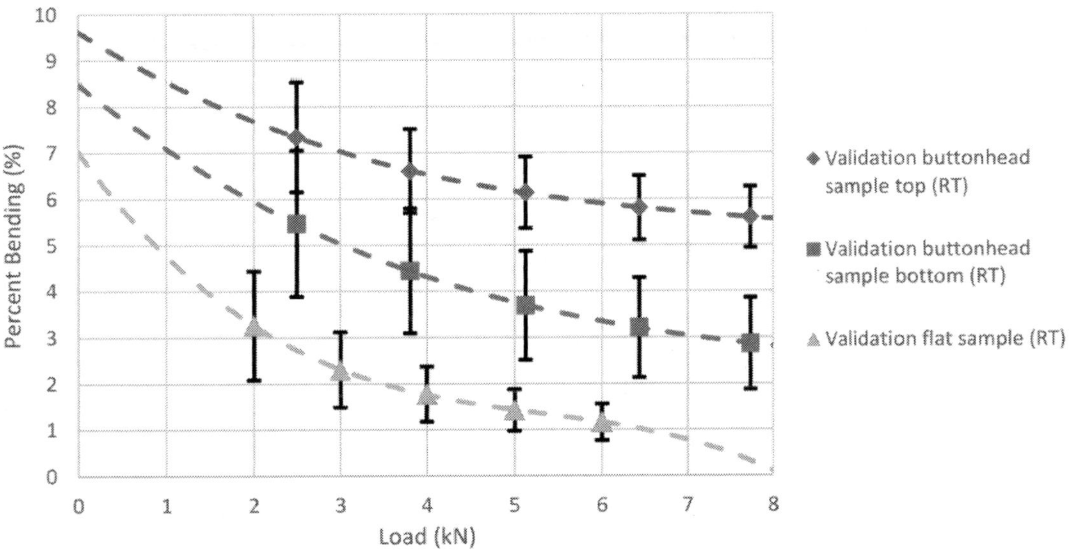

Figure 6. Summary of the results for uniaxial tensile validation tests of flat and buttonhead samples at room temperature. The trend lines correspond to a grade 3 polynomial

3.2. Results

Preliminary results for uniaxial tensile tests of two different materials are summarised in table 1. For INERMET®180 flat tensile specimens were tested at a constant stroke displacement of 0.5 mm/min. For the OFE – copper, buttonhead round specimen geometry was used. It is in a ¼ hard temper state. The stroke displacement was set to 1 mm / min. Both testing speeds are fulfilling ASTM E1450, where it is stated that strain rate should not exceed 10^{-3} s^{-1}.

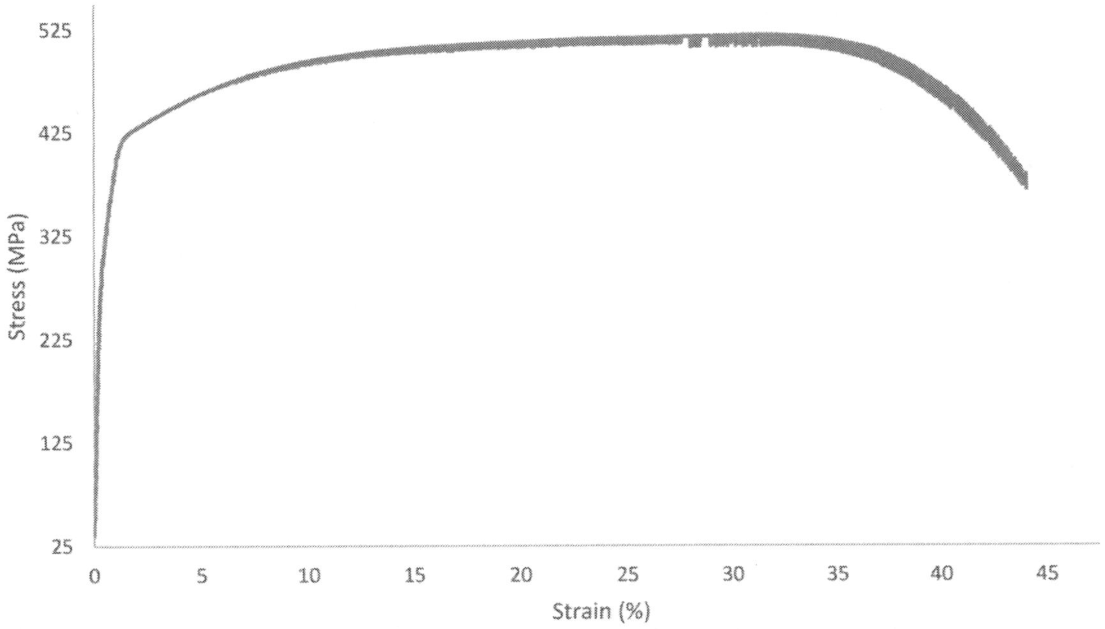

Figure 7. Stress vs strain curve for a representative sample of OFE – copper. ¼ hard temper state. Buttonhead specimen. Test speed: 1 mm/min. Serrated yielding is observed in the plastic region

The results show a very brittle behaviour for the INERMET®180, not being possible to calculate its yield strength as rupture occurs in the elastic region. For the OFE – copper, the results correspond to a cold worked temper state, with the associated increase in ductility typical at cryogenic temperatures for this material. It can be seen in figure 7 that with the current setup, discontinuous plastic flow can be observed for the three OFE – copper sample which were tested, becoming more pronounced after necking occurs.

Table1. Summary results of the tensile tests at 4 K for different materials

Material	*Rp0.2*	UTS	*A*
OFE - copper	427 MPa	525 MPa	43.4 %
INERMET®180	Broke in elastic region	1048 MPa	0.4 %

4. Conclusions

A 100 kN cryostat has been designed, fabricated, assembled and commissioned to perform uniaxial tensile testing at temperatures ranging from 300 K to 4 K. Its dimensions have been optimized to be able to work in the current UTS machine available in the EN/MME group at CERN and would also fit in a standard size servo-hydraulic machine. It is able to test different standard size specimens, which give the opportunity to assess material properties at cryogenic temperatures of flat and round products thanks to the tooling and ancillaries which have also been designed and manufactured in house.

All the different solutions which have been implemented have been comprehensively analysed and are herein presented and discussed in detail, including geometries, tolerances and materials. The instrumentation of the device, which is a key aspect of the design, has been successfully installed and calibrated. The sensors and connectors which are chosen are thoroughly described and discussed.

A validation of the cryostat has been carried out according to the international standards in use. The deviation from uniaxial stress is confined to less than 10 % even when extrapolating the trend lines to low loads. Additionally, some preliminary test results at 4 K are shown, where a tungsten alloy show a very brittle behaviour at 4 K while OFE – copper exhibits typical values obtained for a cold worked temper state, including a very high ductility. Discontinuous plastic flow is observed for all the three OFE – copper samples which were tested, confirming this phenomenon can be measured with the current setup.

5. References

[1] Barth K 2006 Operation of cryogenics for CERN experiments and LHC test facilities *Workshop on Cryogenics Operations* (California).

[2] Sgobba S and D'Urzo C, Fabbricatore P, Sequeira Tavares S 2004 Mechanical performance at cryogenic temperature of the modules of the external cylinder of CMS and quality controls applied during their fabrication, *IEEE Trans. Appl. Supercon.*, **14** (2), 556-559.

[3] Barisone E et al. 2007 Development and Use of Advanced Equipment for Measuring Serrated Flow and Local Heating during Tensile Testing of Materials at Cryogenic Temperatures EUROMAT (Germany).

[4] Ekin J W 2006 Experimental techniques for low – temperature measurements (New York: Oxford University press)

[5] McClintock R M and Warren K A 1960 Tensile cryostat for the temperature range 4 ° to 300 °K *Advances in Cryogenic Engineering, Proceedings of the 1960 Cryogenic Engineering Conference University of Colorado and National Bureau of Standards Boulder*, 372 – 378 (Colorado: Springer).

[6] Beer F P and Johnston E R Jr 2012 Mechanics of Materials (McGraw-Hill)

IOP Conf. Series: Materials Science and Engineering **102** (2015) 012008 doi:10.1088/1757-899X/102/1/012008

Apparatus and method for low-temperature training of shape memory alloys

A M Swanger[1], J E Fesmire[2], S Trigwell[3], T L Gibson[4], M K Williams[5] and O Benafan[6]

[1] NASA Kennedy Space Center, Cryogenics Test Laboratory, NE-M5, KSC, FL 32899 USA
[2] NASA Kennedy Space Center, Cryogenics Test Laboratory, UB-R1, KSC, FL 32899 USA
[3] Sierra Lobo ESC, Kennedy Space Center, ESC-850, KSC, FL 32899 USA
[4] Vencore ESC, Kennedy Space Center, ESC-872, KSC, FL 32899 USA
[5] NASA Kennedy Space Center, Exploration Research & Technology, UB-R3, KSC, FL 32899 USA
[6] NASA Glenn Research Center, Cleveland, OH 44135 USA

Email: adam.m.swanger@nasa.gov

Abstract. An apparatus and method for the low-temperature thermo-mechanical training of shape memory alloys (SMA) has been developed. The experimental SMA materials are being evaluated as prototypes for applicability in novel thermal management systems for future cryogenic applications. Alloys providing two-way actuation at cryogenic temperatures are the chief target. The mechanical training regimen was focused on the controlled movement of rectangular strips, with S-bend configurations, at temperatures as low as 30 K. The custom holding fixture included temperature sensors and a low heat-leak linear actuator with a magnetic coupling. The fixture was mounted to a Gifford-McMahon cryocooler providing up to 25 W of cooling power at 20 K and housed within a custom vacuum chamber. Operations included both training cycles and verification of shape memory movement. The system design and operation are discussed. Results of the training for select prototype alloys are presented.

1. Introduction

Thermal management for space launch systems, hardware, spacecraft, and habitats entails the need to move heat from one region to another for energy supply or to manage the transmission of heat to protect a given component, habitat surface, or person. At NASA Kennedy Space Center (KSC) and NASA Glenn Research Center (GRC) research work is being conducted to combine novel SMA material systems with novel approaches for the management of heat flow in the range of 4 K to 400 K (cryogenic to moderate temperatures). Although research into potential applications of low temperature shape memory alloy (SMA) materials has been explored, the science and understanding of SMA phenomena in the cryogenic realm is still in its infancy [1-4]. Envisioned are new systems incorporating SMA materials and a plurality of moving elements that can respond to certain temperatures to transmit the desired heat flux between two broad area surfaces. The broad area surfaces could be cryogenic tanks, habitat walls, space radiators, or any number of different systems, and be comprised of metals, composites, or hybrid materials.

In the design and development of these prototype adaptive thermal management systems, there exists the need for thermo-mechanical training (or shape training) at very low temperatures. A custom Apparatus for Low-Temperature Training of Materials (ALTM) was devised, designed, and built to support this effort, as well as other applied research in low-temperature thermal management systems. Figure 1 shows the low-temperature material training apparatus in operation at the KSC Cryogenics Test Laboratory (CTL).

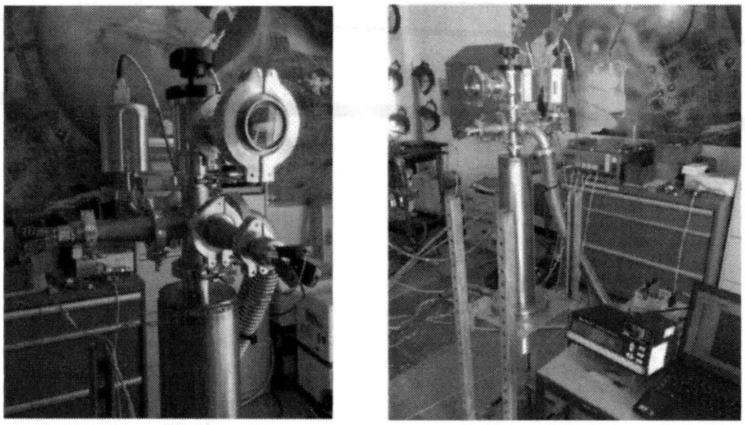

Figure 1. Apparatus for Low-Temperature Training of Materials (ALTM): actuator and upper feed-through (left); overall view with cryocooler in background, and data acquisition in foreground (right).

Novel features of the training apparatus include: the capability to remove and replace different specimens; video camera feed; and a temperature range from ambient down to near 20 K (liquid hydrogen temperature) using a cryocooler. A miniature video camera was adapted for use inside the vacuum chamber to understand the movement of the material test specimen in response to the temperature condition imposed by the cryocooler. The live video feed also provides clear insight on the movement of the actuator assembly within the apparatus.

The SMA materials used were experimental alloys being developed by GRC and evaluated for incorporation into novel designs for adaptive, multifunctional thermal management systems in future cryogenic applications; alloys providing two-way actuation at cryogenic temperatures were the chief targets. The mechanical training regimen for these low-temperature SMA materials was focused on the controlled movement of rectangular strips, with S-bend configurations, at temperatures around 30 K. Operations included both training cycles and verification of shape memory movement.

2. Shape memory alloy test specimens

The term "Shape Memory Alloy" or SMA refers to a group of materials that demonstrate an ability to return to some previously defined shape when subjected to the appropriate thermal and/or mechanical procedure. Typically, these materials can be deformed at some relative lower temperature, and then, upon exposure to some higher temperature, will return to their original, pre-deformed shape— materials that exhibit this property are referred to as one-way shape memory. However, some materials also undergo shape change upon re-cooling without an externally applied load, hence, are referred to as two-way shape memory; these are the types studied during the effort presented here.

Generally, SMA undergoes a martensitic transformation from the high symmetry, austenitic phase to a lower symmetry, martensitic phase in response to thermal and/or mechanical stimuli. A wide range of these materials exhibit the shape memory property, and in some cases, they can also display the two-way behavior after processing or training methodologies [3, 5, 6]. However, most of the commercially available alloys exhibit actuation properties at temperatures above 0 °C, which hinders their use in sub-zero or cryogenic temperatures. Alloy developments for the cold regime is particularly challenging when dealing with composition control, multiple phases (e.g., intermediate R-phase), and other processing parameters. Additionally, testing of cold SMAs requires specialized equipment to operate at extremely low (i.e. cryogenic) transition temperatures.

Thus, the focus of the current work is the thermo-mechanical training of rectangular, S-shaped SMA specimens considered to exhibit two-way behavior at low temperatures for use in a prototype design for demonstrating adaptive thermal behavior. The test specimens measure approximately 40 mm long by 10 mm wide by 1 mm thick with an out-of-plane S-bend of 10 mm. Figure 2 shows a typical SMA test specimen used during two-way training with the ALTM. The strips were fabricated from NiTi- based ternary alloys with elemental addition including Fe, Cr and Co (1-5 at. %).

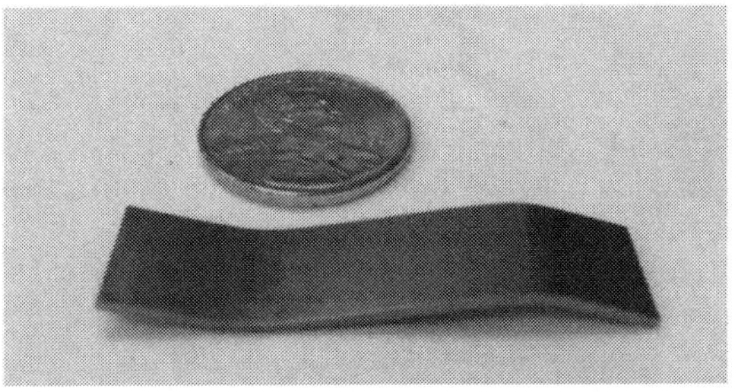

Figure 2. Typical S-shaped SMA specimen

3. Training apparatus requirements and design

Treatment and shape training of SMAs has been done extensively at ambient to elevated temperatures; however, few efforts have addressed these methods at low temperatures (i.e. cryogenic) [3, 6, 7]. With the transformation temperatures of some specimen alloys used during testing thought to be below that of liquid nitrogen (77 K)—and use of a cryogen with a lower boiling point such as helium, hydrogen or neon was deemed either too dangerous or impractical—training options were limited. The solution came in the form of a commercial Gifford-McMahon cryocooler system that was already available at the CTL.

This unit is a Cryomech AL230 model capable of 25 W of refrigeration power at 20 K. The cold-head dimensions are 340 mm in length with a copper heat head diameter of 83 mm. A custom vacuum enclosure with a single 40 mm KF-style vacuum port at the closed end (visible in figure 1) was integrated with the cryocooler to provide the working environment necessary to reach extremely low temperatures.

Procedurally, the specimens needed to be subjected to high temperature annealing in the S-shape, followed by numerous rounds of cool-down, inelastic deformation via compression until flat while cold, and then controlled warm-up unloaded in order to achieve two-way actuation in the desired direction. Figure 3 illustrates this process in three stages.

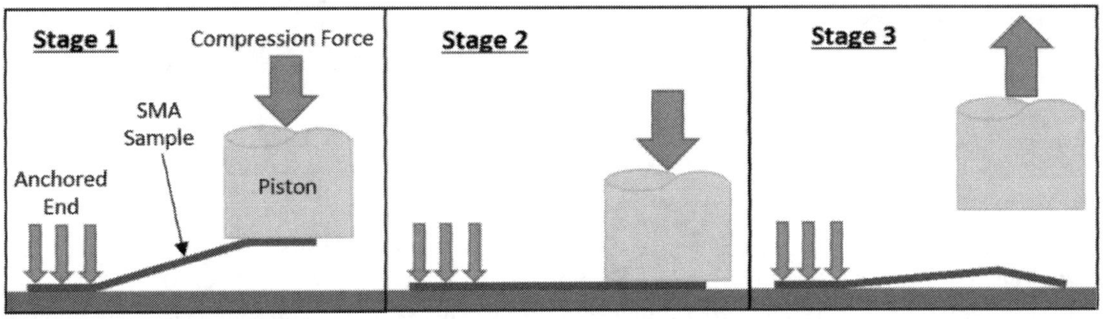

Figure 3. Simplified SMA training sequence. Stage 1: SMA and piston have thermalized and compression commences; Stage 2: specimen is compressed until flat; Stage 3: specimen is unloaded, piston is drawn a distance away, and the SMA starts to return to its original shape during warm-up back to ambient temperature.

With only 25 W of refrigeration power available, a very low heat-leak design was required to maintain the SMA test specimens at the cryogenic training temperatures. This requirement was also a challenge as a substantial piston was necessary to impart the force needed to completely flatten the specimens. The piston represents the potential for a large conductive heat leak from the ambient vacuum chamber to the specimen. Also, components coming into contact during compression—namely the SMA specimen and the piston face—needed to be as close to thermal equilibrium as possible to avoid unwanted specimen heat loads.

Another important design driver was that a single 40-mm port was the only way to connect to the vacuum pumping system, piston actuator, and pressure transducers; as well as the only path of entry into the vacuum chamber for the piston and thermocouple wires.

The design solutions were to use a six-way vacuum fitting to provide more access points at the chamber interface, and to construct the piston in three pieces. A low-conductivity G10 cylinder (tube)—that transfers force to the specimen and provides a path for chamber evacuation and instrumentation lead wires through carefully placed radial holes—bolted to an aluminum cold-head, employing a press-fit G10 face that contacted the specimen during compression. One end of the G10 cylinder was magnetically coupled to, and actuated by, a 40 mm vacuum valve outside the chamber (roughly 0.5 m away from the cryocooler cold-head). Coupling was necessary to draw the piston from SMA specimen prior to warm-up; magnets were employed to ease operation and assembly of the system. At the bottom end an aluminum piston head acted to maintain axial piston position (via a slip-fit into a machined guide), and provided a cold thermal anchor for the G10 face. This piston head was machined from a piece of 6061 aluminum hexagonal bar stock to provide flat surfaces for attach points in case thermal straps were found to be necessary.

Composite G10 was used for the piston face instead of aluminum for two reasons: 1) lower thermal conductivity for greater thermal stability when compression was initiated and 2) removability so thermocouples could be directly embedded and the entire plug pressed in using a shop vise. Figure 4 shows a 3-D rendered model of the piston assembly and the aluminum piston head with G10 face and embedded thermocouples.

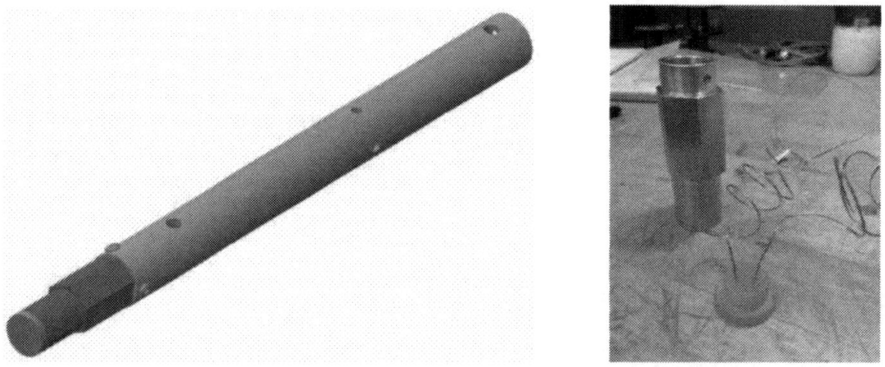

Figure 4. Compression piston assembly (left); and piston head, G10 face, and embedded thermocouples (right).

To house the SMA specimen and guide the piston, a test specimen holing fixture assembly was designed that integrated with the cryocooler cold head. This unit had to securely fix one end of the S-shaped specimen in order to provide good thermal contact for cool-down, and act as an anchor point during compression. The assembly was comprised of four main components: a copper foundation block directly affixed to the cryocooler cold-head, a cylindrical aluminum piston guide, a brass SMA fixture block, and a stainless steel bearing block.

The guide was bored-out to accommodate the aluminum piston head shown in Figure 5 with a slip-fit tolerance. This design feature allowed for full piston travel during compression (roughly 10 mm), while maintaining an accurate axial position. The fit between the two components was tight enough to promote reasonable heat transfer yet not so tight as to cause binding. Additional tolerances were also given to account for the slight imperfections in axial alignment between the vacuum chamber port and the fixture assembly. At the base of the guide a square notch was machined out to allow the free end of the SMA specimen to protrude into the cylinder; positioning it directly underneath the piston face.

Anchoring the other end of the specimen was achieved by sandwiching it between a square brass bar with an accurately sized relief cut into the bottom, and the copper foundation block. This fixture block was then secured to the copper via an 8-32 threaded rod and nut. This configuration afforded good thermal contact between the copper cold-mass and the specimen, and provided a strong anchor-point during compression.

Due to the shape of the specimen it was realized that during compression the bottom edge of the free end would be first to contact the flat, copper foundation block. Because the SMA material was much harder than copper, a stainless steel bearing block was employed at the interface to prevent gouging. Figure 5 shows the test specimen holding fixture and compression piston assembly integrated with the cryocooler cold head.

Figure 5. Test specimen holding fixture assembly integrated with the cryocooler (left and middle); compression piston mounted to holding fixture (right).

In this configuration, the solid conduction heat leak through the piston was calculated to be 130 mW (between 300 K and 20 K). Considering solid conduction to be the dominate source of heat leak—since the other cold components would be super-insulated and in high vacuum—the estimated demand for less than 0.1% of the total available refrigeration power effectively satisfied the design requirement.

4. Training apparatus set-up and operation

Untrained, flat SMA specimens were first placed on a stainless steel training jig that bent them to the correct S-shape. They were then annealed in an oven at 450 °C for one hour, and immediately quenched in a water bath. A single test specimen was then positioned in the training fixture followed by mounting of the compression piston assembly. Six type-E thermocouples were strategically placed within the system: one each on the cryocooler cold head and aluminum piston guide and redundant pairs on the specimen itself and G10 piston face. These wires fed through the middle of the G10 piston cylinder inside the vacuum chamber volume via radial holes, ran up through the 40 mm port, back out through holes drilled higher up on the cylinder, and eventually interfaced at two thermocouple feed-throughs (visible in figure 1). Figure 6 shows a system cut-away of the overall hardware configuration.

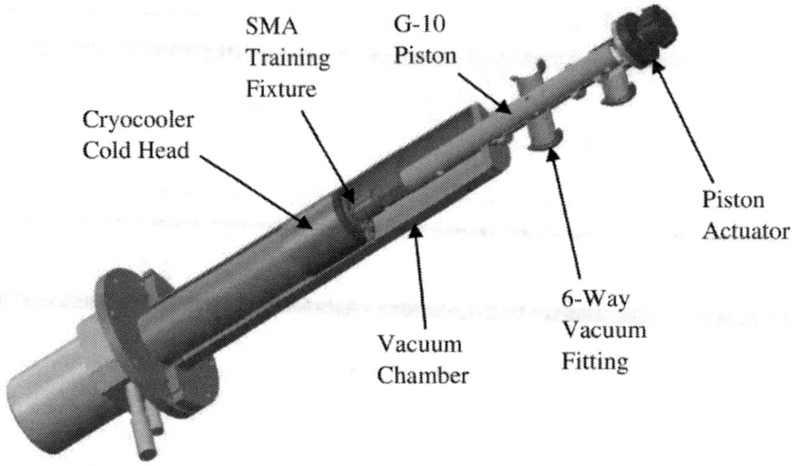

Figure 6. Cut-away of ALTM hardware configuration.

With the test specimen and instrumentation in place the cold head, holding fixture, and aluminum piston were wrapped with 20 layers of Cryolam multi-layer insulation (MLI). The vacuum chamber was then carefully slid down over the G10 cylinder. Thermocouples were interfaced to their respective feed-throughs at the 6-way vacuum cross, along with two Baratron capacitance manometers (100 and 0.1 torr ranges), and a vacuum pumping system. Finally, the piston actuator (i.e. a 40-mm vacuum angle valve) was interfaced to the top of the assembly, and a sight glass was employed in order to verify piston movement (visible in figure 1). All instrumentation was fed into Lab View data acquisition software, and displayed/recorded on a laptop.

After a one to two hour vacuum pumping period, the cryocooler was turned on to begin system cool-down. The test specimen and piston face were allowed to thermalize for approximately two hours. Compression was then applied using the actuator until the SMA specimen was pressed completely flat. The specimen was held in this position for one hour, after which time the piston was raised and the system was allowed to warm to room temperature under vacuum. This process was repeated four additional times, producing a total of five training cycles per SMA specimen. Figure 7 shows a typical temperature plot for an SMA training cycle with the ALTM.

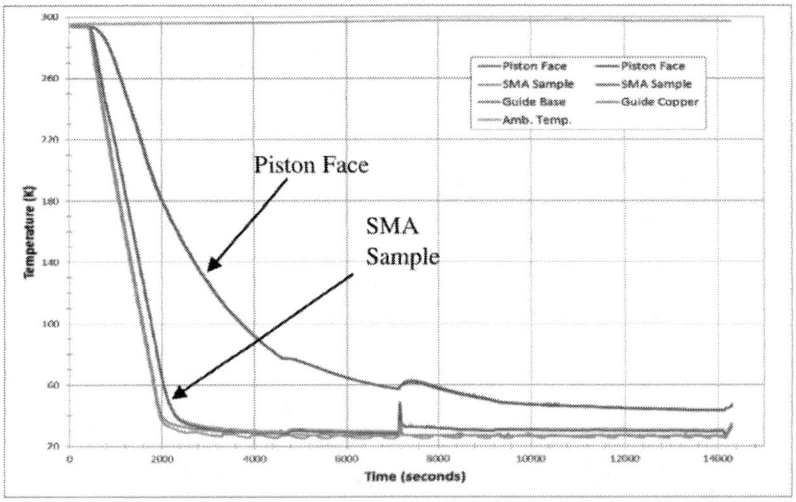

Figure 7. Example plot of temperature versus time for one training cycle of an SMA test specimen. (Note, the spike seen at ≈7200s is due to initiation of SMA sample compression)

5. Verification of two-way SMA actuation

After the SMA specimens had been subjected to the prescribed training regimen it was necessary to determine if cryogenic two-way actuation had been achieved. Numerous methods were explored in an attempt to accomplish this important but difficult task. Ultimately, the most successful and definitive method was to visually verify movement using a small camera routed through the vacuum chamber. This visual method was done using an inexpensive USB-style camera with integrated lighting. The head of the camera (10-mm diameter by measured 51-mm length) had four white LEDs whose brightness could be varied at the USB plug. To penetrate the vacuum chamber, the cord was cut and each of the five wires fitted with female, barrel-style pins; these pins then interfaced with an 8-pin, 25 mm KF-style vacuum feed-through. Due to space constraints, the camera had to be mounted vertically inside the chamber, and aimed downward at a 45° mirror in order to see the edge of the S-shaped specimen. This configuration proved highly successful, producing high-resolution videos of each specimen during cool-down down to 30 K, and verifying that two-way actuation was in fact achieved in some cases. Figure 8 shows the setup for visual verification as well as a before-and-after of an actuated SMA test specimen.

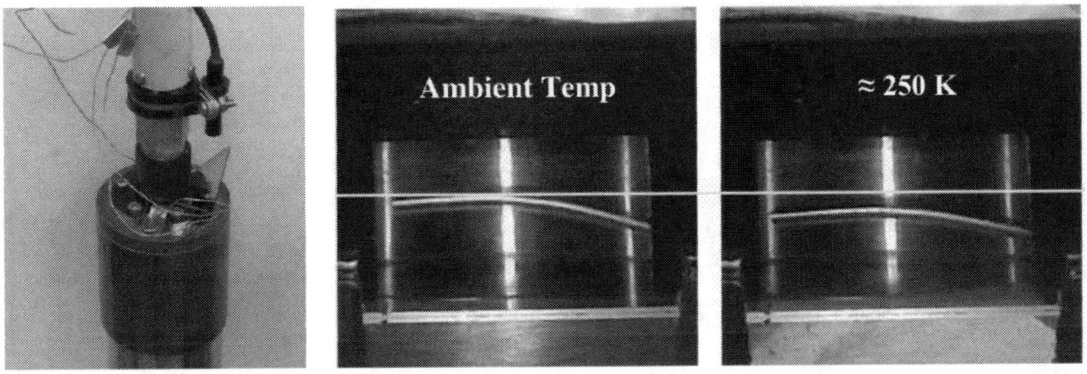

Figure 8. Visual verification setup showing USB camera and mirror (left); before-and-after screenshots showing SMA test specimen actuation (middle is warm, right is cold).

6. Conclusion

For the purposes of training shape memory alloys at cryogenic temperatures, the Apparatus for Low-Temperature Training of Materials (ALTM) was designed, built, and tested at the Cryogenics Test Laboratory of NASA Kennedy Space Center. Various S-shaped experimental SMAs were trained using the new apparatus at temperatures as low as 30 K, and resulted in successful two-way actuation in several cases. This result demonstrates the ALTM design and validates the developed thermo-mechanical training regimen for SMA prototype materials. A small video camera was also incorporated to provide verification of SMA actuation and proved to be extremely effective.

Major design challenges for the training apparatus included the need to keep heat-leak into the system to a minimum due to the limited refrigeration power available (25 W at 20 K); incorporation of a compression piston that could be actuated outside the vacuum chamber; and routing of all the required instrumentation and functional hardware through a single 40-mm vacuum port interface. These challenges were all successfully addressed resulting in an efficiently packaged, fully functional system that satisfied all the project goals as well as opening up new possibilities for materials research.

Research into the development, training and integration of low-temperature SMA materials shows great promise for many future applications in adaptive, multi-functional, and autonomous thermal management systems for in-space propulsion, cryogenic storage and transfer for ground and space systems, space habitation systems, and in-situ resource utilization. It is expected that along with the great possibilities come an infinite array of potential alloy compositions and inherent challenges as part of the development process. The motivation for the new ALTM was aimed at SMA training and proved very successful. It is fully expected that its capability will be extendable to a wide variety of other materials and situations where exercising material specimens at cryogenic temperatures is needed.

Acknowledgments
This work was funded by the NASA Space Technology Mission Directorate under Center Innovation Fund program and is part of the project *Low Temperature Two-Way Shape Memory Alloys for Thermal Management Systems*. Alloy development at GRC is supported by ARMD Advanced Air Transport Technology (AATT) and Transformational Tools & Technologies (TTT) Projects. Also, the authors would like to acknowledge Cryogenics Test Laboratory engineer Gino Carro, and technicians Johnny Kerce and Mark Velasco for their outstanding design, fabrication and testing support. O.B. thanks R. D. Noebe at NASA GRC for help in alloy development.

References

[1] Benafan O, Notardonato W U, Meneghelli B J and Vaidyanathan R 2013, Design and development of a shape memory alloy activated heat pipe-based thermal switch *Smart Materials and Structures* **22** 105017.

[2] Lemanski J, Cryogenic shape memory alloy actuators for spaceport technologies: materials characterization and prototype testing, Masters [thesis], Orlando (FL): University of Central Florida; 2005

[3] Malarria J, Lovey F and Sade M 2009 Two way shape memory effect in CuZnAl single crystals after pseudoelastic cycling at low temperatures *Materials Science and Engineering A* **517** 118-124.

[4] Abramov V Ya, Aleksandrova N M, Borovkov D V, Khmelevskaya I Yu, Korotitskiy A V, Makushev S Yu, Polyakova N A, Popov N N and Prokoshkin S D 2006 Martensitic transformations and functional properties of thermally and thermomechanically treated Ti-Ni-Nb based alloys *Materials Science and Engineering A* **438-440** 553-557.

[5] Meng X, Chen F, Cai W, Wang L and Zhao L 2006 Two-way shape memory effect and its stability in a Ti-Ni-Nb wide hysteresis shape memory alloy *Materials Transactions* **47** 724-727

[6] Flores Zuniga H, Rios-Jara D, Belkahla S, Nika V and Guenin G 1996 The training and re-training procedures for the two way memory effect and its degradation in a Cu-Al-Be alloy *Scripta Materialia* **34** 1899-1904

[7] Lopez-Ferreno I, Breczewski T, Ruis-Larrea I, Lopez-Echarri A, No M L, and San Juan J 2013 Thermal treatments and transformation behavior of Cu-Al-Be shape memory alloys *Journal of Alloys and Compounds* **5775** S463-S467.

IOP Conf. Series: Materials Science and Engineering **102** (2015) 012009 doi:10.1088/1757-899X/102/1/012009

Design of load-to-failure tests of high-voltage insulation breaks for ITER's cryogenic network

S.A.E.Langeslag[1], **E.Rodriguez Castro**[1,2], **I.Aviles Santillana**[1,3], **S.Sgobba**[1], **A.Foussat**[4]

[1] CERN, CH-1211 Genève - Switzerland
[2] University of Vigo, Campus de As Lagoas-Marcosende, 36310 Vigo - Spain
[3] University Carlos III of Madrid, Campus de Leganes, 28911 Madrid - Spain
[4] ITER Organization, Route de Vinon sur Verdon, CS 90 046, 13067 St. Paul lez Durance Cedex - France

E-mail: `Stefanie.Langeslag@cern.ch`

Abstract. The development of new generation superconducting magnets for fusion research, such as the ITER experiment, is largely based on coils wound with so-called cable-in-conduit conductors. The concept of the cable-in-conduit conductor is based on a direct cooling principle, by supercritical helium, flowing through the central region of the conductor, in close contact with the superconducting strands. Consequently, a direct connection exists between the electrically grounded helium coolant supply line and the highly energised magnet windings. Various insulated regions, constructed out of high-voltage insulation breaks, are put in place to isolate sectors with different electrical potential. In addition to high voltages and significant internal helium pressure, the insulation breaks will experience various mechanical forces resulting from differential thermal contraction phenomena and electro-magnetic loads.

Special test equipment was designed, prepared and employed to assess the mechanical reliability of the insulation breaks. A binary test setup is proposed, where mechanical failure is assumed when leak rate of gaseous helium exceeds $10^{-9} Pa \cdot m^3/s$. The test consists of a load-to-failure insulation break charging, in tension, while immersed in liquid nitrogen at the temperature of 77 K. Leak tightness during the test is monitored by measuring the leak rate of the gaseous helium, directly surrounding the insulation break, with respect to the existing vacuum inside the insulation break. The experimental setup is proven effective, and various insulation breaks performed beyond expectations.

1. Introduction

Magnet systems for plasma confinement, in fusion energy experiments such as ITER, include a wide range of coils composed of cable-in-conduit conductors (CICCs) [1]. These CICCs consist of various types of stainless steel jackets, densely filled with superconducting strands based on either Nb_3Sn or NbTi, and subsequently compacted [2]. Cooling, necessary for the operation of these superconducting cables, is realised by direct passage of supercritical helium through the conduits, thereby directly mitigating heat at the superconducting strands. This concept provides a highly efficient cooling scheme by a concentrated liquid helium supply from the cryogenic circuit to the superconducting strands [3], however it intrinsically connects the high-current pulsed coils to the helium supply pipes of the cryogenic distribution system.

High-voltage insulation breaks (IBs) provide the required electrical isolation between the CICCs and the helium supply pipes, to prevent a direct electrical connection between high-current coils and the electrically grounded cryogenic supply system [4]. The IBs will be situated at the helium inlet and outlet channels of the wide range of parallel cryogenic circuits foreseen in the ITER magnet system. A large number, over 1000, of these electrical separation units will therefore be indispensable for a successful operation of the tokamak. These axial insulation breaks are build up from stainless steel end-fittings, hermetically connected via a glass-reinforced resin composite body of sufficient length, to prevent electrical discharge during magnet operation (figure 1).

The IBs do not only have to comply with electrical requirements, but also sustain demanding conditions resulting from elevated mechanical stress originating from electro-magnetic loads, dilatation effects and high internal pressure during magnet operation [4]. Therefore the cryogenic insulation system does not only rely on the electrical, but also the structural properties of the individual IBs.

To validate the state of the art IBs and check their ability to withstand the highly demanding operating conditions, a binary test setup is proposed. An experimental setup was designed, manufactured and commissioned, allowing for an extensive assessment of the structural reliability of currently produced IBs. Additionally the results are presented of measurements obtained for certain IBs tested according to this procedure.

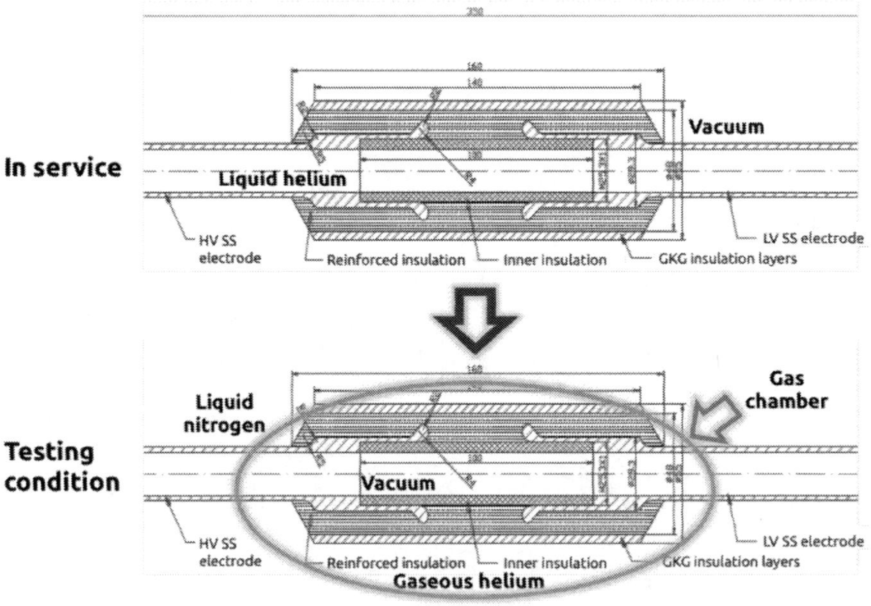

Figure 1: Technical drawing of a 30 kV insulation break, demonstrating the environmental conditions during operation and testing, respectively. Note that the cooling scheme in testing condition is reversed with respect to the operating condition.

2. Materials

Commissioning tests of the experimental setup involved a smaller type 4 kV break type, while subsequent tests, presented in § 4, were performed on 30 kV IBs with dimensions described in figure 1. All priorly described IBs, denoted as IB4kV, IB30kV1 and IB30kV2, respectively, were formerly tested up to very low loads, max. 2 kN, in a single traction-compression test [5]. These test items can therefore not be considered to be in a fully pristine state.

3. Experimental setup

A test system has been developed for reliability testing of individual IBs during magnet operation. The IBs structural integrity is challenged by applying increasing load while monitoring its leak-tightness to gaseous helium. Mechanical loads in the IBs, originating from dilatation effects and electro-magnetic forces, include tensile, bending and torsional components. As a result of the large share of the tensile component derived from the alternating hoop forces in the pulsed coils, increased by possible thermal contraction effects, the loading scenario during IB testing has been chosen uniaxial in tension.

Some of the mechanical loads will be introduced during magnet commissioning, in essence cool-down between 293 K and 4 K, however some of them will also occur while at the 4 K operating temperature. To take into account temperature effects the measurements are performed at cryogenic temperature, here liquid nitrogen temperature (i.e. 77 K) to avoid closed-circuit complications to be encountered when testing at liquid helium temperature (i.e. 4 K). In the measurements mechanical failure is assumed when the leak rate exceeds $10^{-9} Pa \cdot m^3/s$.

3.1. Sample adjustment

To adapt the IBs in order to apply the testing load scenario, modifications are made to both extremities of the as-received IBs. On one extremity a small flange Gyrolok® coupling is TIG welded, with a Böhler ASN 5-IG (material number 1.4453) filler wire, to the IB helium pipe via a U-shaped tube to allow easy access for pumping. The connection between this vacuum tube (U-tube) and the IB's stainless steel end fitting is realised via an increased diameter thick ring which acts as a load bearing surface during tensile testing.

On the other extremity a lap joint TIG weld was realised with the same filler metal between the IB end fitting and a threaded rod for top-end IB fixation to the universal testing machine in order to allow for load transmission. These adaptations have been performed in a manner to guarantee their mechanical integrity and leak tightness to avoid any influence on the outcome of the tests. Computed radiographic inspections have been performed on the welded extensions with a Philips MG161L 160 kV + MCN 166 X-ray tube, according to examination standard EN 14784-2 class B. All welds were established to conform to EN ISO standard 5817, level B. Additionally, two PT100 temperature sensors are placed on each IB, on the stainless steel end-fittings adjacent to the glass-resin composite body, in order to monitor real-time temperature at the sample.

3.2. Setup design features

The objective of the measurement is to test the IBs in traction at 77 K while monitoring leak tightness, and identifying the load at which the leak rate exceeds the imposed value of $10^{-9} Pa \cdot m^3/s$.

To this aim, the cooling scheme of the insulator break is inverted. Where in service condition a vacuum exists around, and liquid helium flows through the IB, in testing condition a vacuum is created inside the IB and it is cooled externally. This concept, shown in figure 1, is adopted for a better vacuum achievement and a leak rate reading on the vacuum pump solely influenced by the IB's structural integrity, hence avoiding environmental impact.

The setup consists of a G10 open liquid nitrogen cryostat to realise a measurement process at 77 K. Low density gaseous helium is employed as flow agent for leak detection, for which a gas dome is employed directly surrounding the to be measured IB. The gas dome is supported from the IB via a small fraction of the top end sample thread. The weight of the dome is minimised by constructing it out of a low density aluminium alloy, as to limit external factors on the measurement results. Moreover, the cross-section of the dome is limited, to realise an as

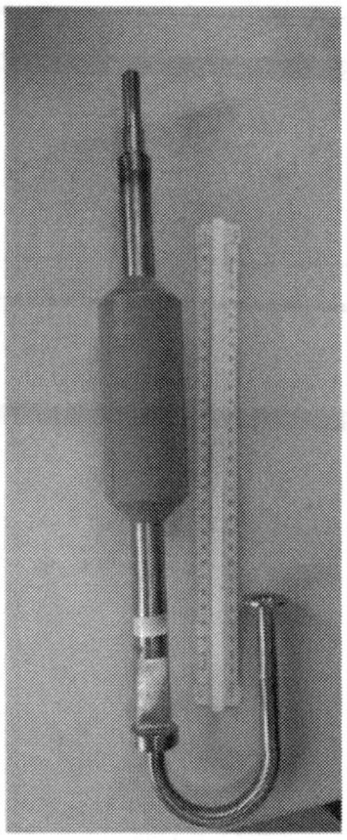

Figure 2: Insulation break sample, adjusted for load-to-failure testing.

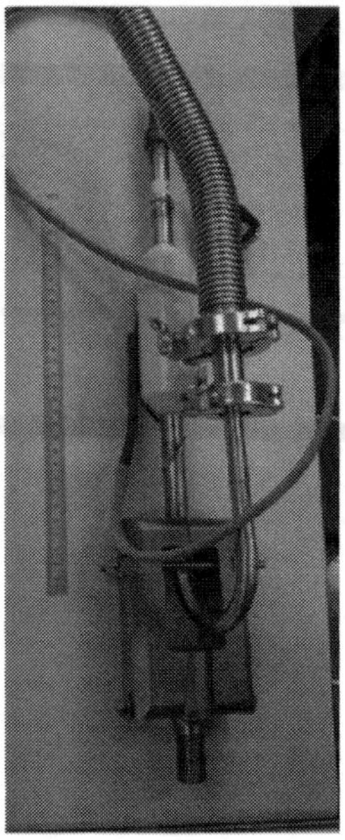

Figure 3: Adjusted insulation break sample including attached tooling for load and vacuum application.

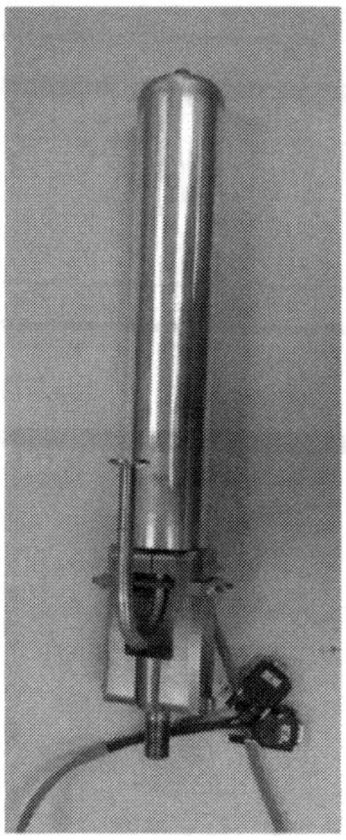

Figure 4: Adjusted insulation break sample including attached tooling for load application and gas dome.

direct as achievable cooling at the position of the IB, i.e. to reduce the gaseous column between the IB and its coolant, for which the highly thermally conductive Al-alloy is also beneficial.

The bottom tooling is designed to grip around the load bearing ring on the bottom end of the sample, while simultaneously allowing for the passage of the vacuum tube (U-tube). Vacuum inside the IB is obtained with use of a Leybold, PhoeniXL 300, mass spectrometer pump, on which the leak rate during testing is consistently monitored.

Measurements are performed on a UTS200 universal tensile/compression machine, from UTS Testsysteme GmbH. The machine can reach a maximum force of 200 kN with a position accuracy (stroke displacement) of 1 μm. Load is registered with use of a HBM Z12, 200 kN load cell with 0.0016 kN resolution. The test procedure is programmed in DionPro software via which the system is subsequently actuated and the load versus displacement data is recorded. Temperature data is registered via a QuantumX DAQ system through Catman software on a parallel computer.

3.3. Experimental procedure

An initial leak check of the IB sample is performed, by pumping vacuum inside the IB sample followed by the passage of pressurised gaseous He around the sample. The initial leak checks performed have been satisfactory for all measured IBs. Following cryostat mounting and subsequent IB sample including gas dome fixation, the sample is vacuumised. Generally a leak rate level is reached better than $10^{-11} Pa \cdot m^3/s$ prior to thermalisation (better than $10^{-13} Pa \cdot m^3/s$ at 77 K prior to testing). Subsequently liquid nitrogen is introduced into the cryostat and owing to the dome shaped gas shell and thus its unsealed bottom, a direct flow of nitrogen at the IB sample is realised. Once full thermalisation is achieved, resulting from complete submersion of the IB sample, gaseous He is introduced in the dome. This is done by means of introducing pressurised

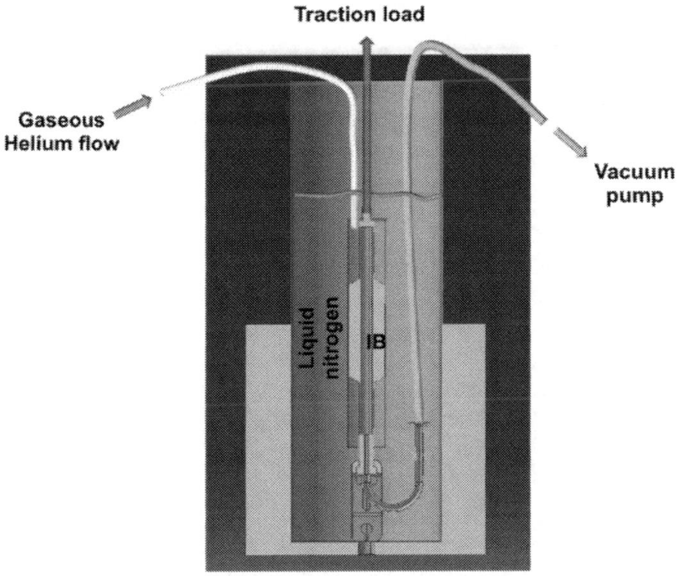

Figure 5: Schematic diagram illustrating the binary test setup design. Note the gaseous helium, directly surrounding the insulation break with internal vacuum, immersed in liquid nitrogen.

gas from the top of the dome, consequently expelling the liquid nitrogen from the bottom. The introduction of the gaseous helium around the IB sample offers a final leak check prior to loading.

A dual test procedure is implemented, consisting of a primary and a secondary test. The primary test holds the load-to-failure measurement, while in the secondary test the nature of the introduced damage is examined. For both tests a low frequency cyclic loading scheme was adopted to discriminate reversible from irreversible flaws. Five consecutive cycles are carried out, where after each cycle load was released down to 2 kN. The loading scheme includes successive cycles to 5 kN, 10 kN, 15 kN, 30 kN and 85 kN, with a loading speed of 50 N/s in both the loading and unloading segment. Charging was interrupted promptly at leak detection to avoid any damage to the mass spectrometer pump.

4. Discussion of selected test results

4.1. Primary test

Data for the primary measurements of two typical 30 kV IBs is presented in figure 6 and 7. The plots represent the leak rate over time, as recorded by the mass spectrometer pump, and the load applied, respectively. The plots show a very consistent behaviour between the two IBs, IB30kV1 and IB30kV2, where early in the load-cycles a rapid increase in leak rate is observed, whereas during the third load cycle the leak rate stabilises.

In both cases, during the fifth load cycle, up to 85 kN, a very sudden rise in leak rate is observed, passing rapidly beyond the imposed failure level. The cycle was immediately interrupted at the recorded sample breakdown to avoid any damage to the mass spectrometer pump. The load at which failure is noted is ~65 kN and ~71 kN, for IB30kV1 and IB30kV2, respectively. These values exceed by far the expected performance of such cryogenic insulation breaks, which lies a factor ten below.

Primarily at the onset of the measurement, some noise is observed in the leak rate for both sample IB30kV1 and IB30kV2. An assumption is made that the observed spikes in the leak rate curve appear as a result of internal pollution inside the IB, present prior to testing. Samples have not been specifically prepared for vacuum application, and can therefore contain some particles which can interfere with the leak rate reading of the mass spectrometer pump. For future application of the test procedure a preparation as such may be beneficial.

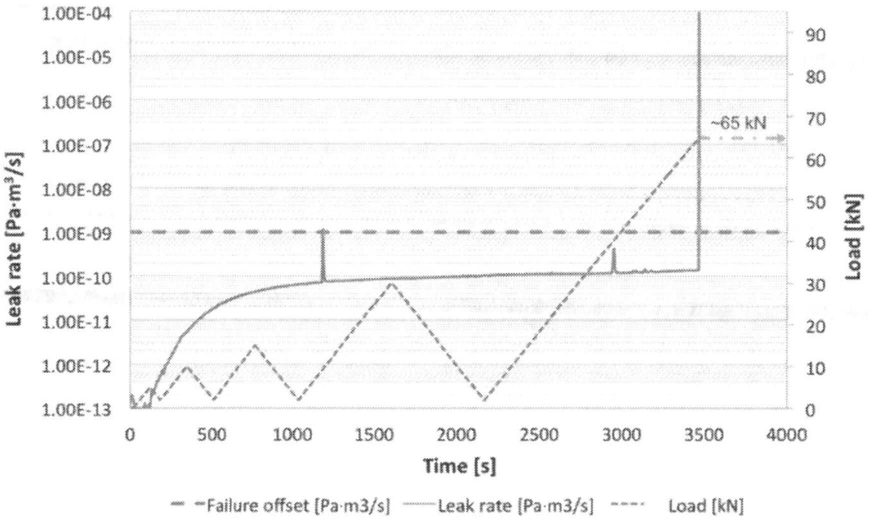

Figure 6: Diagram showing the leak rate and applied load over time for the primary load-to-failure test of IB30kV1 at 77 K, measured following the presented loading procedure. Breakdown was recorded at $\sim$ 65 kN.

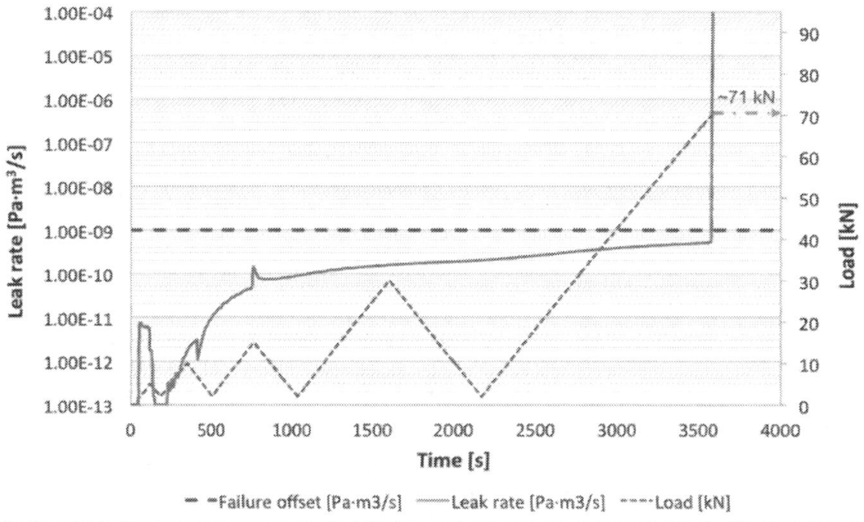

Figure 7: Diagram showing the leak rate and applied load over time for the primary load-to-failure test of IB30kV2 at 77 K, measured following the presented loading procedure. Breakdown was recorded at $\sim$ 71 kN.

4.2. Secondary test

An additional test was performed, following the load-to-failure test, aimed at confirming breakdown and discriminating between reversible and irreversible defects. The leak rate over time, versus the applied load cycles for this secondary test, presented in figure 8 shows a leak rate directly related to the applied load. The leak rate changes abruptly with load, demonstrating a partially reversible feature of the induced defect during the primary test. The imposed failure level is now however exceeded at a significantly lower load, i.e. at $\sim$13 kN, at which the permanent leak becomes apparent.

IOP Conf. Series: Materials Science and Engineering **102** (2015) 012009 doi:10.1088/1757-899X/102/1/012009

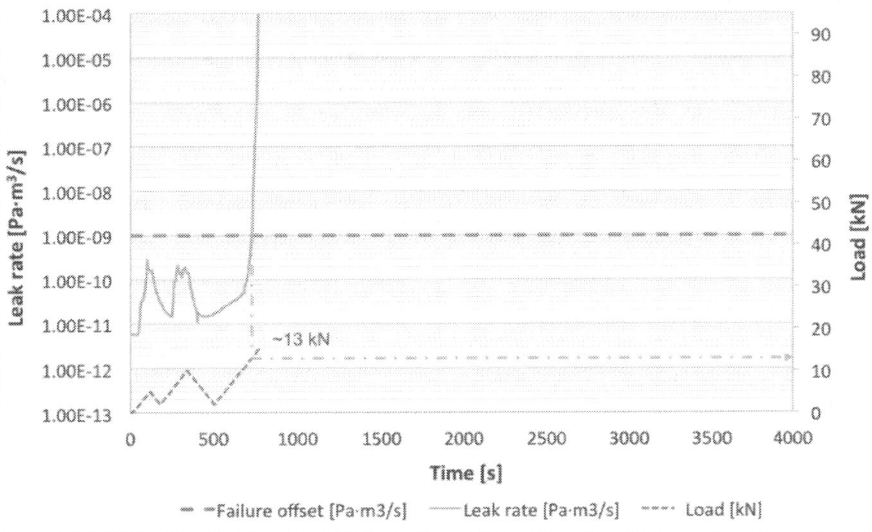

Figure 8: Diagram demonstrating leak rate and applied load over time for the secondary test of IB30kV1 at 77 K. At low applied loads one can note a reversible feature of the leak introduced during the primary test.

4.3. Observed defects

Visual inspection, following the measurements, demonstrated signs of damage on the sample, as shown in figure 9. The damage is located primarily at the interface between the stainless steel end fittings and the first glass-resin layer, or amidst the glass-resin layers, in an interlayer fashion. These defects become more apparent after applying a liquid penetrant followed by a revelator, during penetrant testing (PT), which can be witnessed in figure 10. The image shown is obtained 30 minutes following the application of the revelator, according to ISO standard 3452-1.

Figure 9: Defective regions observed on the as-tested IB30kV2, by visual inspection.

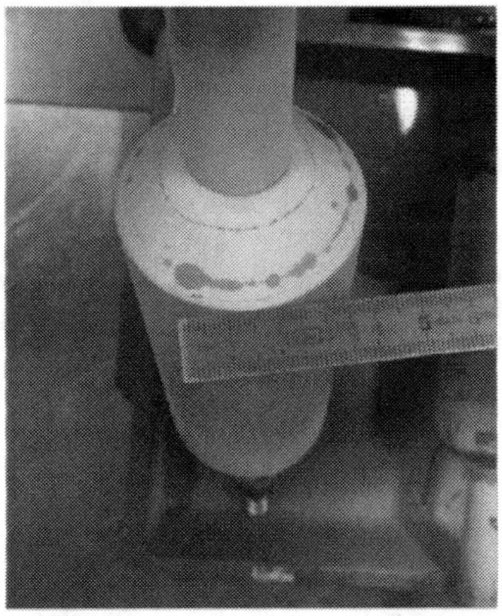

Figure 10: Observation of defective regions on IB30kV2, revealed by the application of liquid penetrant.

IOP Conf. Series: Materials Science and Engineering **102** (2015) 012009 doi:10.1088/1757-899X/102/1/012009

5. Conclusion

A binary test setup was successfully designed, developed and commissioned for load-to-failure tests of axial insulation breaks at cryogenic temperature. Failure was reached, by applying a tensile load until a certain leak rate threshold was exceeded, which was set at $10^{-9} Pa \cdot m^3/s$. Following an initial test campaign, a good test reproducibility was obtained for all the tests performed.

During a primary load-to-failure test, the insulation break samples fail consistently at high tensile loads, i.e. $\sim$ 65 kN and $\sim$ 71 kN for the examples IB30kV1 and IB30kV2 presented, respectively. Failure is accompanied by an extensive damage of the insulation breaks, in an interlayer fashion, and at the interface where the stainless steel end-fitting and the glass-resin composite meet.

Secondary tests, implemented to examine the nature of the introduced damage, have demonstrated that a part of the damage is reversible. It has been noted that with temperature as a result of dilatation differences during cool-down, and with small load increments, damage opens and closes, thereby increasing and decreasing leak rate, respectively. Above a certain load, far below the load necessary for failure during primary testing, the irreversible damage becomes apparent when leak rate increases abrupt beyond the imposed threshold of $10^{-9} Pa \cdot m^3/s$.

5.1. Disclaimer

The above conclusions arise from a limited sample basis, hence they cannot be considered as sufficiently exhaustive to guarantee the performance of the axial insulation breaks at operating conditions, i.e. at supercritical helium temperature and in a repeated loading condition.

The views and opinions expressed herein do not necessarily reflect those of the ITER Organization.

6. References

[1] N. Mitchell, A. Devred, P. Libeyre, B. Lim, and F. Savary, "The ITER Magnets: Design and Construction Status," *IEEE Transactions on Applied Superconductivity*, vol. 22, no. 3, pp. 4200809–4200809, 2012.

[2] B. Turck, D. Bessette, and D. Ciazynski, "Design methods and actual performances of conductors for the superconducting coils of Tokamaks," *Fusion Engineering and Design*, vol. 2, pp. 667–670, 1993.

[3] L. Bottura and C. Luongo, "Superconductors, stability in forced flow," *Wiley Encyclopedia of Electrical and Electronics Engineering*, 1999.

[4] F. Rodriguez-Mateos, D. Evans, and A. Devred, "Essential design, construction and test elements of ITER axial insulation breaks," *Transactions of the International Cryogenic Materials Conference ICMC*, 2014.

[5] J. Kosek, R. Lopez, F. Rodriguez-Mateos, and D. Tommasini, "Non-destructive qualification tests for ITER cryogenic axial insulating breaks," in *AIP Conference Proceedings*, pp. 1026–1033, Jan. 2014.

SOFI/Substrate integrity testing for cryogenic propellant tanks at extreme thermal gradient conditions.

M Haynes[1] and P Fabian[2]
[1] Senior Test Engineer, Composite Technology Development, Inc., Lafayette, CO.
mhaynes@ctd-materials.com
[2]Vice President of Operations, Composite Technology Development, Inc., Lafayette, CO.
pfabian@ctd-materials.com

Abstract. Liquid propellant tank insulation for space flight requires low weight as well as high insulation factors. Use of Spray-On Foam Insulation (SOFI) is an accepted, cost effective technique for insulating a single wall cryogenic propellant tank and has been used extensively throughout the aerospace industry. Determining the bond integrity of the SOFI to the metallic substrate as well as its ability to withstand the in-service strains, both mechanical and thermal, is critical to the longevity of the insulation. This determination has previously been performed using highly volatile, explosive cryogens, which increases the test costs enormously, as well as greatly increasing the risk to both equipment and personnel.

CTD has developed a new test system, based on a previous NASA test that simulates the mechanical and thermal strains associated with filling a large fuel tank with a cryogen. The test enables a relatively small SOFI/substrate sample to be monitored for any deformations, delaminations, or disjunctures during the cooling and mechanical straining process of the substrate, and enables the concurrent application of thermal and physical strains to two specimens at the same time. The thermal strains are applied by cooling the substrate to the desired cryogen temperature (from 4 K to 250 K) while maintaining the outside surface of the SOFI foam at ambient conditions.

Multiple temperature monitoring points are exercised to ensure even cooling across the substrate, while at the same time, surface temperatures of the SOFI can be monitored to determine the heat flow. The system also allows for direct measurement of the strains in the substrate during the test.

The test system as well as test data from testing at 20 K, for liquid Hydrogen simulation, will be discussed.

1. Introduction

Maintaining the Thermal Protection System (TPS), which includes the SOFI, integrity through the thermal, acceleration and frictional variations of launch and pre-launch is challenging. The loss of TPS foam during the shuttle launches, shown in Figure 1, eventually resulted in tragedy, making stress testing of the bond integrity between the tank substrate and the TPS, and between the multiple layers of TPS themselves of greater importance.

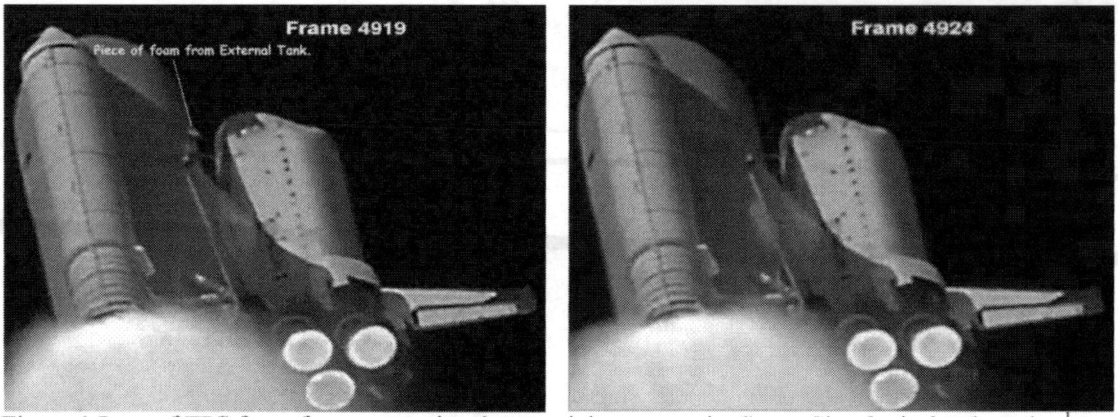

Figure 1 Loss of TPS foam from external tank caused damage to the Space Shuttle during launch. [1]

The test method described here, called the Cryoflex Test,[2] relies on both mechanical strain loading of the tank wall substrate as well as concurrent thermal strain loading of the substrate and the SOFI interface. NASA historically has performed thermo-mechanical testing on these types of systems using liquid Hydrogen [3]. However, the use of liquid Hydrogen is extremely hazardous, which results in a more complex and expensive test that must be conducted and monitored remotely. The lack of direct access to the test specimen during the test is a detriment to evaluating material performance during the test since real time observation of the SOFI is more difficult if done through remote video feeds.

Due to increased interest by various commercial space companies in evaluating the cryogenic performance of SOFI for cryogenic fuel tanks, CTD[4,5] developed a Cryoflex test fixture to address this commercial testing need. Use of flammable or explosive cryogens was not an option for both economic and safety reasons.

2. Test Description

The Cryoflex itself is designed to apply thermal stresses to the substrate, the interface between the insulation and the substrate and to the insulation to simulate the filling of the cryogenic tank with the cryogen. Once the thermal stresses have been applied, additional mechanical stresses are applied that simulate the pressurization of the tank. The initial design called for testing a single part at one time, cooling the substrate material to 20 K, while keeping the TPS foam material exposed to ambient atmospheric conditions.

Based on past NASA Cryoflex history and commercial customer input, specimen dimensions for the substrate insulation and the test fixturing was designed. The specimens were comprised of two proprietary insulating SOFI layers knit together with a proprietary ablative SOFI layer. The substrate was a nominal 46 cm x 5 cm x 3.5 cm thick 2024 aluminum alloy. The substrates had a 36 cm gage section with a reduced width of 3.8 cm with 5 bolt holes on each end to enable gripping for mechanical loading. The SOFI foam layers were nominally 25 mm total thickness with a 12 mm ablative layer on top and a tapered end profile leaving a 30 cm long full thickness foam layer. Figure 2 shows representative Cryoflex test specimens.

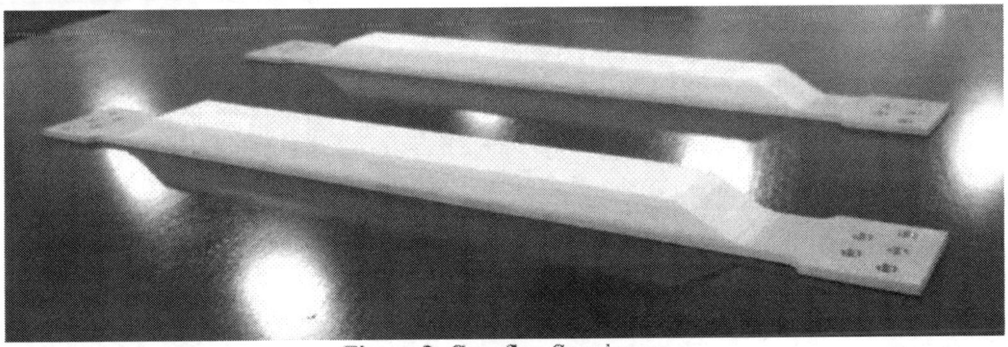

Figure 2. Cryoflex Specimens

3. Cryoflex Test Fixture Design

The primary challenges that needed to be overcome to perform these tests included:

- How to cool the substrate of the specimen below 20 K while keeping the outer surface of the SOFI insulation at ambient conditions;
- How to cool a flat plate substrate to sub-liquid nitrogen (LN_2) temperatures without use of a vacuum jacketed chamber as a thermal break;
- How to monitor the applied strain on the flat substrate while actively cooling the substrate
- How to ensure even cooling of the substrate while monitoring the temperature.

Potential cooling systems that were investigated included cryocoolers, direct LHe contact and use of a flow-through cooling block. Following a review of options, the test parameters were modified to accommodate the testing of two specimens at one time, as shown in Figure 3. Redesign of the concept to this configuration made the definition of the cooling block and grip system more amenable to even cooling and reduced the need for additional insulation and block retention techniques. It also made for a more efficient test since two specimens could be tested at the same time.

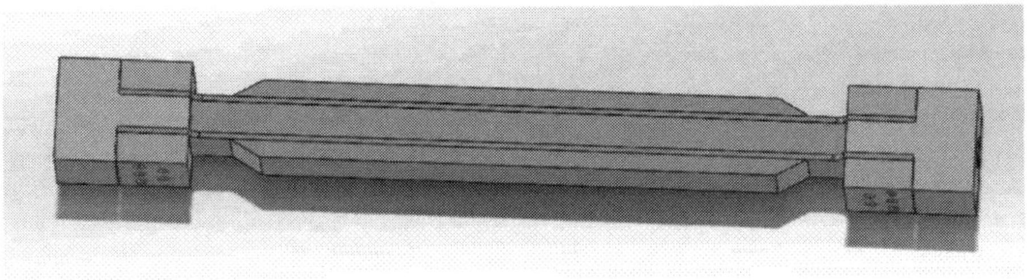

Figure 3: Design model incorporating two specimens. Cooling block is sandwiched between the two test specimens.

The cooling block needed to cool evenly from the top to the bottom of the block, as well as side to side, when the LHe was introduced at one end. Ultimately, the endothermic property of the expanding cryogen was used to achieve an even cooling across the block. By allowing the gas to expand to an ever increasing volume, the cryogen constantly absorbs more heat. In order to reduce the parasitic losses from ambient conditions, the cold exhaust gaseous Helium (GHe) was routed past the outer lateral edges of the

IOP Conf. Series: Materials Science and Engineering **102** (2015) 012010 doi:10.1088/1757-899X/102/1/012010

cooling block, acting as a cold shroud for the cooling portion of the block. These exhaust channels were subsequently insulated using commercially available foam cut to fit tightly at room temperature.

The cooling block was minimally relieved to allow strain gages to be mounted at the center of the gage section to monitor the applied strain within the substrate Pocket relief for two silicon diodes were also created in order to acquire thermal verification of each of the substrate temperatures throughout the testing.

Test specimens were length matched in pairs to achieve more uniform loading within the fixture. Each specimen was gripped independently, ensuring that similar strains could be applied to each specimen. The cooling block was clamped into place by the substrates, cooling the gage sections and utilizing an insulated contact patch to the grip portions of the test fixture, to decrease parasitic losses.

The fixture grips were designed to hold two specimens back to back with enough room between them for the cooling block to remain in intimate contact with both specimens at 20 K. The Cryoflex mechanical test system was designed to fit CTD's 450 kN Servo-Hydraulic load frame, which is capable of acquiring up to 10 channels of strain, 10 channels of thermal input, load and displacement at speeds of up to 100 Hz.

4. Cryoflex Test Procedure

The specimen assembly was installed into the cryostat, the exposed cooling block sides were insulated, and the temperature diode and strain gage functions checked. The specimens were then pre-loaded to 225 N and cooled to 50 K using Liquid Helium. The cooldown from 50 K to 20K (-5 K, +0 K) was performed as quickly as possible, usually within 105 seconds, to simulate the thermal strains encountered with a quick fill scenario for an actual cryogenic fuel tank. When the specimen temperature is within the specified range (20 K, -5, +0 K), the loading portion of the test was started. The specimens saw an average of -0.32% thermal strain during the cool down phase from 295 K to 20 K, which was zeroed out prior to the mechanical stress portion of the test. The test configuration during cool down is shown in Figure 4.

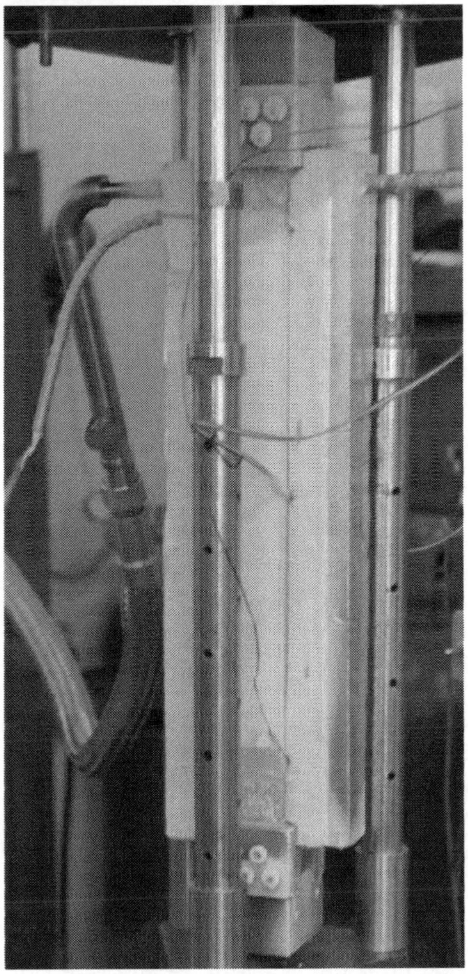

Figure 4: Cryoflex test system during cool down phase

After the initial cool-down, the aluminum substrates of the specimens were kept within the test temperature range (20 K +0, -5 K) for the duration of the mission cycle. Each "mission cycle" typically consists of mechanically loading the two specimens in tension to an initial load of approximately 29 kN with a force hold for 60 seconds prior to being further loaded up to approximately 58 kN and then unloaded to 5.8 kN to end the load cycle. All loading and unloading was performed at a rate of around 4 kN/s. At the 58 kN load level, each specimen substrate saw an average strain of 0.21% strain at a stress of approximately 213 MPa. A typical load cycle might be performed up to 20 times, making up a single "mission cycle" depending on the expected fill/drain cycles of the tank being studied. The example shown in Figure 5 included twelve separate mission cycles that were performed on the same two specimens.

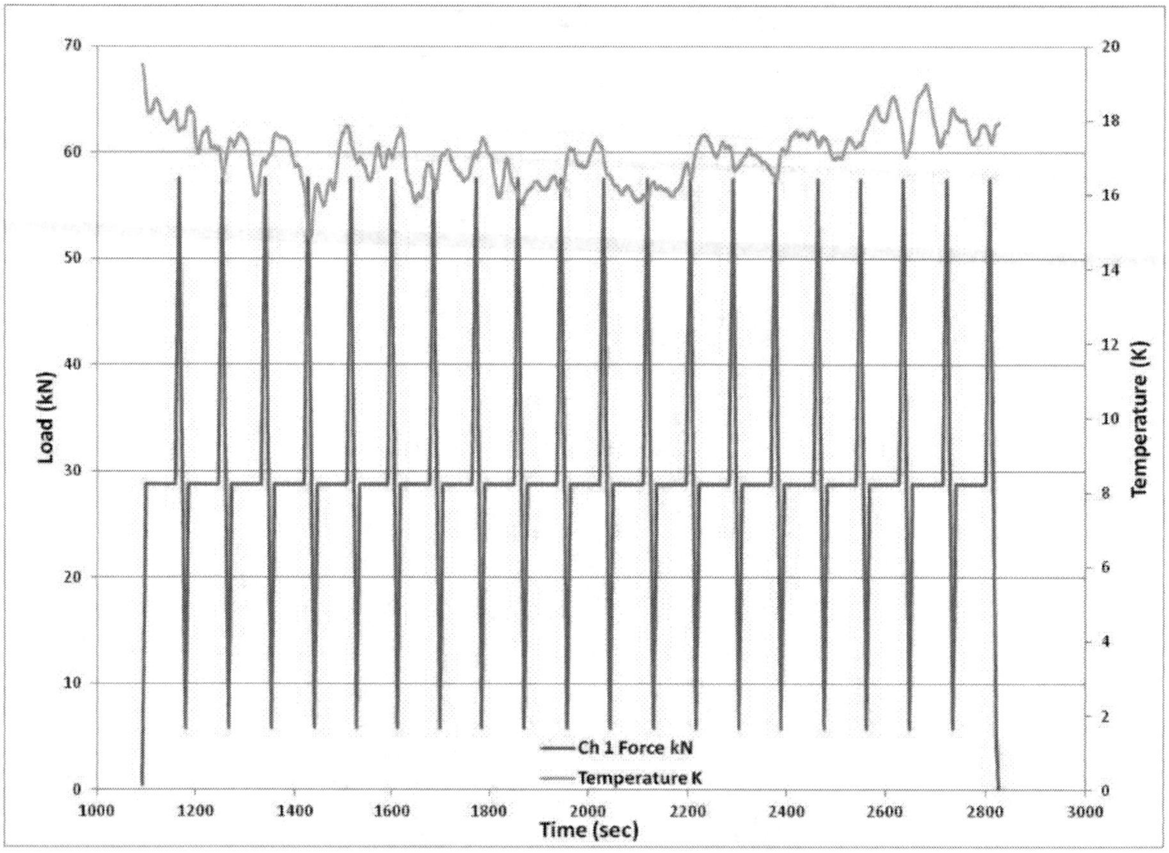

Figure 5: Sample data of Temperature and Load Vs Time

The failure criteria for the SOFI insulation is set by the customer and based on their application and expected failure modes. Failure criteria typically used includes:

1) A crack in the foam perpendicular to the substrate connected to a delamination at the substrate bond line in excess of 7.6 mm.
2) 2 or more cracks separated by less than 25 mm
3) 2 or more cracks with delaminations, where the sum of the delaminations length is greater than 20% of the distance between the two cracks
4) One single delamination at the bondline in excess of 15mm with no cracks

5. Test Results

SOFI surface temperatures were monitored at specified intervals throughout each set of mission cycles through contact measurements within a 12 mm diameter location mid-height on the ablative surface. In the example shown in Figure 6, the lowest temperature recorded on the ablative layer was 190 K, well above the condensation temperature of Oxygen. The average ablative layer temperature seen was 226 K, though the temperature in close proximity to the cracks tended to be lower due to atmospheric condensation in the localized area.

Usually, each crack formation is accompanied by an acoustic emission, and is able to be immediately located and identified. At the test temperature, the cracks are visible to the naked eye both along the width of the insulation and along the edge of the ablative layer. These same cracks are not visible after active cooling had ended, indicating the necessity of having first hand observance during this type of test. Each of the cracks within the foams was additionally identifiable by the accumulation of frost during the warm up period as shown in Figure 6.

Figure 6: Crack in Cryoflex specimen insulation at 20 K

In the test example shown in Figure 6, over the course of one set of twelve mission cycles, only three cracks presented themselves, one on each specimen during mission cycle #2, and a third one during mission cycle #11. Most cracks initially formed through the harder, denser ablative material and during the following cycles they propagated through the SOFI to the substrate. Some cracks which appear during the later cycles went through the foam and on to the substrate immediately, but almost all cracks propagated through both layers of foam (ablative and insulation) to the substrate.

Since the cooling of the specimens is produced through cyclic flow of LHe, observation of the SOFI foam during and immediately after the testing is quite easy. Through visual observation, most cracks are visible while cooling of the specimen is actively occurring. However, once active cooling is ceased and the specimen starts to warm up, thus expanding, the cracks close up and are very hard to distinguish. Therefore, during the test, crack locations are marked through the use of ink marks made at crack onset. Additionally, on warmup, the knit line between the two layers of insulation will usually show signs of frost buildup after cracks have propagated through them, a possible indication of higher thermal transmission at that interface than elsewhere in the SOFI.

6. Conclusion

Thermal insulations are a necessary part of any cryogenic tank application, and the ability to withstand multiple fill and drain cycles without cracking within the insulation is even more critical in the manned space flight arena. CTD's Cryoflex test system has demonstrated the ability to use LHe to achieve an intermediate cryogenic temperature of a metallic substrate and ambient temperature exposure to SOFI foam, while at the same time allowing mechanical loads to be applied to the substrate. This test system can be used to simulate actual conditions in the fill, pressurization, and draining of a cryogenic fuel tank used in aerospace applications.

The ability to directly monitor the condition of the foam and the foam/substrate interface to immediately locate and identify a disjuncture within the foam or the foam to substrate bond enables immediate feedback on the performance of the insulation system to the designers and engineers responsible for the testing and evaluation of these subsystems. In addition, the ability to reduce the safety precautions by using LHe (as opposed to the many precautions necessary with the use of more volatile cryogens) results in additional cost savings to the customer and test program.

7. References

[1] *Columbia Accident Investigation Board (CAIB)/ NASA Accident Investigation Team (NAIT), Working Scenario 2003*
http://history.nasa.gov/columbia/Troxell/Columbia Web Site/CAIB/CAIB Website/news/STS 107 Working Scenario/sts107workingscenario_1.pdf

[2] Gates T, Johnson T, Whitley K 2005.*NASA/TM-2005-213778 Assessment of Technologies for the Space Shuttle External Tank Thermal Protection System and Recommendations for Technology Improvement*
http://ntrs.nasa.gov/archive/nasa/casi.ntrs.nasa.gov/20050241761.pdf

[3] DiMiero B 2008 *NASA Internship PPT, Hydrogen Test Facility, MSFC*
http://www.powershow.com/view/146f6d-YjFjO/The_National_Aeronautics_and_Space_Administration_powerpoint_ppt_presentation

[4] Fabian P, Haynes M *et al* 2005 *Precision Cryogenic Temperature Control for Mechanical Testing of Carbon Composite/Invar Specimens for the James Webb Space Telescope: Space Cryogenics Workshop, Colorado Springs, CO*

[5]Haynes M, Fabian P 2013 *Controlled Temperature Testing Techniques for Composite Materials and Metal Interfaces at Sub-77 K Temperatures for Space Applications: National Space and Missile Materials Symposium, Bellevue, WA*

A cryogenic tensile testing apparatus for micro-samples cooled by miniature pulse tube cryocooler

L B Chen[1,2], S X Liu[1,2], K X Gu[1], Y Zhou[1] and J J Wang[1,3]
[1]Key Laboratory of Cryogenics, Technical Institute of Physics and Chemistry, CAS, Beijing 100190, China
[2]University of Chinese Academy of Sciences, Beijing 100049, China

E-mail: wangjunjie@mail.ipc.ac.cn

Abstract. This paper introduces a cryogenic tensile testing apparatus for micro-samples cooled by a miniature pulse tube cryocooler. At present, tensile tests are widely applied to measure the mechanical properties of materials; most of the cryogenic tensile testing apparatus are designed for samples with standard sizes, while for non-standard size samples, especially for micro-samples, the tensile testing cannot be conducted. The general approach to cool down the specimens for tensile testing is by using of liquid nitrogen or liquid helium, which is not convenient: it is difficult to keep the temperature of the specimens at an arbitrary set point precisely, besides, in some occasions, liquid nitrogen, especially liquid helium, is not easily available. To overcome these limitations, a cryogenic tensile testing apparatus cooled by a high frequency pulse tube cryocooler has been designed, built and tested. The operating temperatures of the developed tensile testing apparatus cover from 20 K to room temperature with a controlling precision of ±10 mK. The apparatus configurations, the methods of operation and some cooling performance will be described in this paper.

1. Introduction

Tensile tests are widely applied to measure the mechanical properties of materials. At present, most of the cryogenic tensile testing apparatus are designed for samples with standard sizes, while for non-standard size samples, especially for micro-samples with an equivalent diameter of hundreds of micrometres to several millimetres or a thickness of hundreds of nanometres to several millimetres, the tensile testing cannot be conducted generally. However, sometimes, micro-samples will be used in the study of materials cryogenics properties inevitably. Two examples: the thickness of the Polyimide films for inertial confinement fusion is only hundreds of nanometres; and the sample for Transmission Electron Microscope (TEM) usually with a diameter of 3 mm and a thickness less than 500 nm. It is practically impossible to conduct tensile test by the present commercial tensile testing apparatus. As to the cryogenic tensile testing apparatus, the general approach to cool down the specimens for tensile tests is by using of liquid nitrogen or liquid helium [1-4], which is not convenient to a certain extent: firstly, it is difficult to keep the temperature of the specimens at an arbitrary set point precisely; secondly, in some occasions, liquid nitrogen, especially liquid helium, is not easily available. Compared with the cryogens, cryocooler has some disadvantages in vibration and long-term stability.

[3] Corresponding author; Tel: + 86 10 82543758; Fax: + 86 10 62564050

However, cryocooler is much easier to operate: just press the On/Off switch then the cooling begins. At present, there are many kinds of cryocooler available in the temperature range of 20 K to 300 K. While most of the commercially available cryocooler working below 20 K is G-M cryocooler or G-M type pulse tube cryocooler, which employ the oil-lubricated compressor to drive the cooler with high electric powers, they have an appreciable vibration and periodic maintenance is generally needed.
To overcome these limitations, a cryogenic tensile testing apparatus for micro-samples cooled by a miniature pulse tube cryocooler will be introduced in this paper. Instead of employing oil-lubricated compressor, a linear dual-opposed compressor, which is characterized by high efficiency, low vibration and long-time stability, has been used to generate the pressure wave to drive the pulse tube cryocooler. Figure 1 is the schematic of the pulse tube cryocooler driven by a linear dual-opposed compressor, which is the same with the cryocooler we published before [5-8].

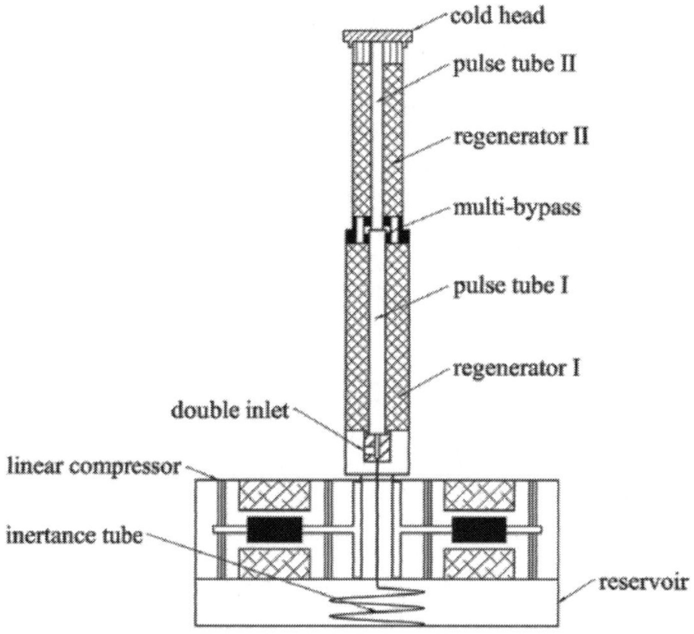

Figure 1. Schematic of the pulse tube cryocooler.

In this paper, a TEM specimen can be tensile tested by the developed apparatus at different set temperature. The original intension of this project is to analyse the deformation mechanisms (dislocation or twins) of samples after tensioned in low temperature, rather than to focus on the tensile test data acquisition. However, it is should be mentioned that disk shaped samples is not an acceptable tension specimen geometry; disks cannot be pulled from their edges to yield suitable test data. Therefore, to yield suitable test data in keeping with best practices established in standards, the sample should be machined out by the standard method for cryogenic tensile testing [9]. But no matter disk shaped samples or the standard shaped samples, the total apparatus configurations, the measures to enhance heat transfer or reduce the thermal loss, the basic design methods of sample holder and so on, are similar, which is the purpose of the paper. In the following sections, the apparatus configurations, the methods of operation and some cooling performance will be described.

2. Designing of a fixture for tensile testing

For the first prototype, the sample with a diameter of 3 mm and a thickness less than 500 nm is designed according to a TEM specimen. As shown in figure 2, the left part and the right part of the fixture are exactly symmetry, and both parts contain one base and one cover plate. The sample should be mounted in the notch of the base and then be compressed by the cover plate through four M2 screws. It should be mentioned that the design of the guide rod between the left and right part is not only to guide the direction in the tensile process, but also to guarantee the centre hole of TEM sample will not be compressed.

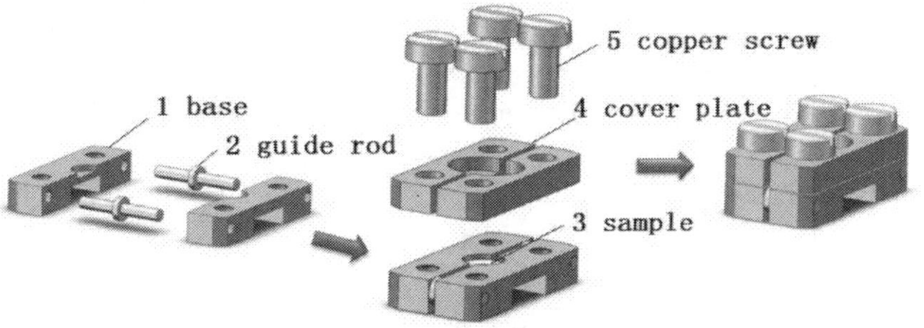

Figure 2. The assembly procedure for the sample.

The sample holder will be moved to the cold head of the cooler after the sample has been fixed. Two sample clamps has been designed to hold the sample holder firmly, and one of them (the right one as shown in figure 3) will be fixed on the cold head tightly by a copper screw to reduce the temperature difference between the cold head and the sample holder, while the other one (the left one as shown in figure 3) will just put on the cold head and then connected to a wire drawing. As shown in figure 3, although the right clamp is fixed on the cold head, it is still need to connect with the vacuum shield by a nylon rope to prevent the cooler from bending in the process of the tensile by the forces of the wire drawing.

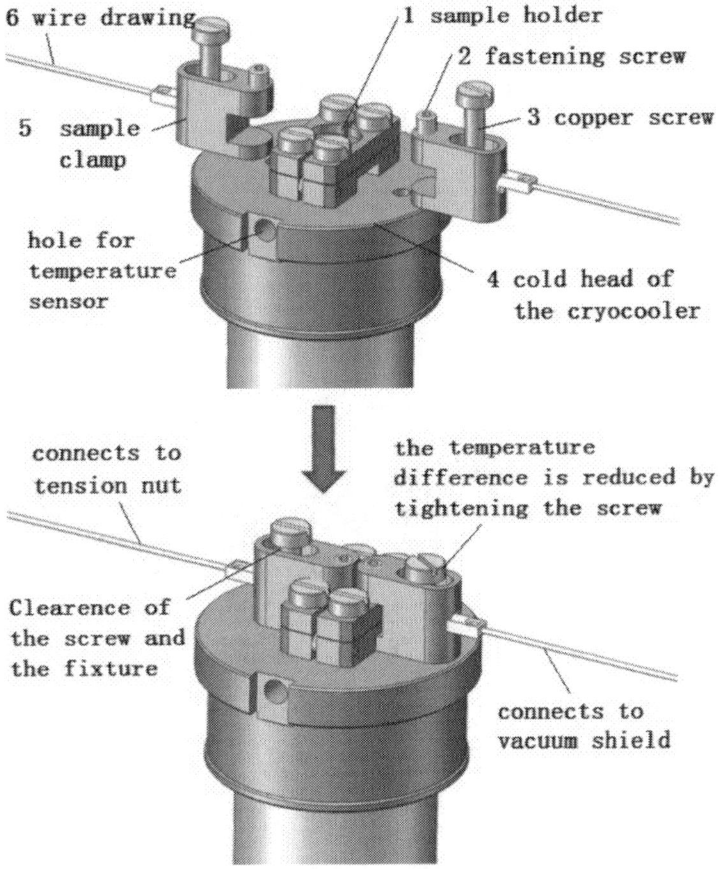

Figure 3. The install process of the sample holder.

As shown in figure 4, there is a thermal shield connected with the cold head outside the sample holder to keep the sample temperature the same with the cold head. A radiation shield has also been installed between the cold head and the vacuum shield to improve the cooler performance [6]. As can be seen from figure 4, the wire drawing connects with a tension spring before fastening with the tension nut; and there is a transparent pipe which is made of Plexiglas outside of the tension spring. The linear deformation of the tension spring is used to estimate the tensile forces.

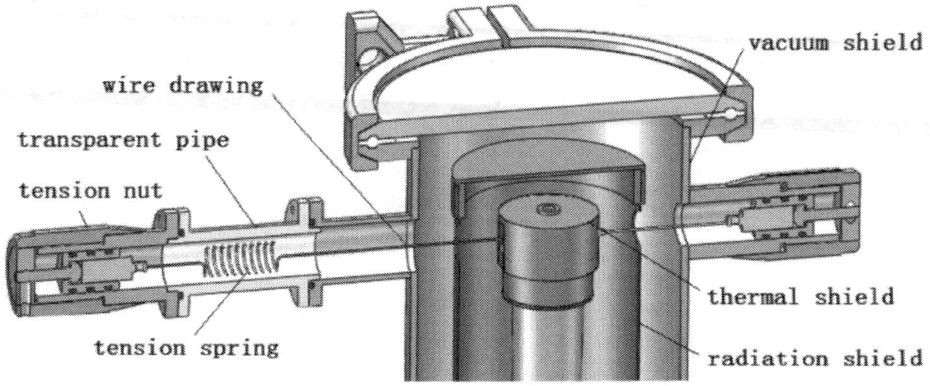

Figure 4. Cutaway views of the cryogenic tensile testing apparatus.

The photo of the developed cryogenic tensile testing apparatus is shown in figure 5. The total weight of the apparatus is about 20 kg and the height is about 0.5 meter, so it is very convenient to move. The cold head of the pulse tube cryocooler is vertical up in figure 5; however, in order to achieve a lower temperature, the cold head should be flipped vertically downward. The temperature of the cold head will be 5-10 K higher when it is tipped with the pulse tube's cold end above its hot end due to natural convection for the cryocooler developed in our team.

Figure 5. Photo of the developed cryogenic tensile testing apparatus.

3. Cooling performance

The developed cryogenic tensile testing apparatus is cooled by a single-stage high frequency pulse tube cryocooler, and a lowest temperature of about 16 K can be achieved with a maximum electric power of 260 W. Figure 6 is the typical cool down curve, it takes about 45 minutes to reach its bottom state.

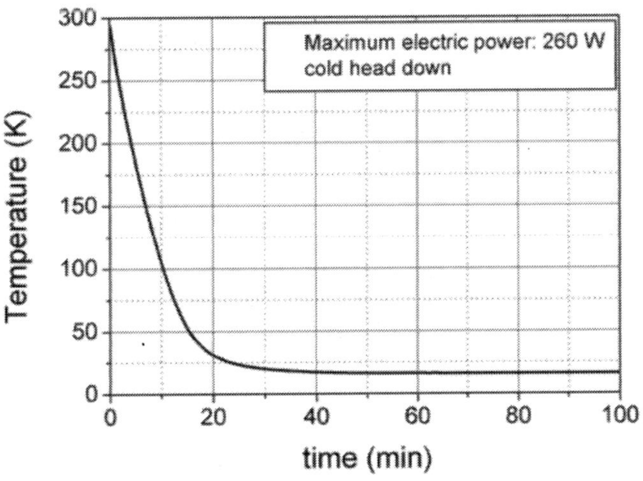

Figure 6. A typical cool-down curve of the developed apparatus.

It is difficult to keep the temperature of the specimens at an arbitrary set point precisely by the traditional LN2/LHe cooling. But to cryocooler, the temperature control can be realized easily. The general temperature control method for traditional cryocooler is to control the heater power through PID adjustor control. As to the developed high frequency pulse tube crocooler in our team, we control the temperature of the cold head by adjusting the input electric power to the compressor, which is an energy-efficient way [8]. Figure 7 are the temperature control curve, a temperature control precision of ±10 mK can be achieved.

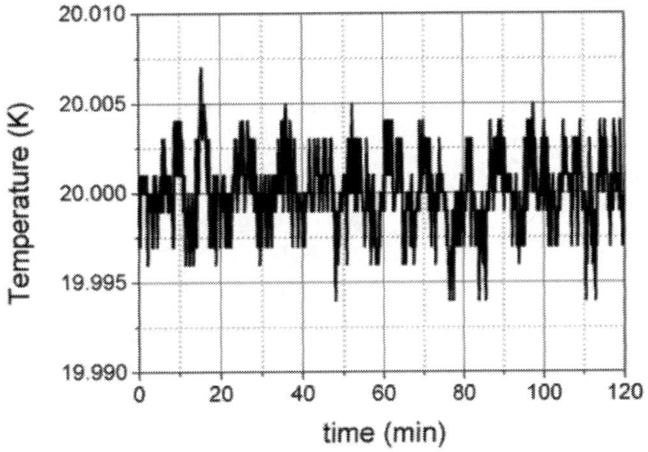

Figure 7. Temperature control curve of the developed apparatus.

4. Considerations for further improvement

As has been described above, the tension force is estimated by the deformation degree of the spring for the present apparatus. However, to some users, the magnitude of the tension force and the tensile displacement are important parameters to study mechanical properties of materials. Therefore, a force sensor will be employed to substitute the tension spring to measure the forces quantitatively, and the displacement of the tension nut will be marked in the next prototype.

The present samples holder is designed according to the size of TEM samples. In fact, kinds of samples with different sizes can be tested by using different sample holder with the same apparatus. To the Transmission Electron Microscope, it is difficult to conduct cryogenic tensile testing in-site, but to Scanning Electron Microscope, it is possible to integrate a cryocooler with a small size to realize the in-site cryogenic tensile test function, which will be our next work

5. Conclusions

A cryogenic tensile testing apparatus for micro-samples cooled by a miniature pulse tube cryocooler has been designed, built and tested. At present, a general TEM sample with a diameter of 3 mm and a thickness of less than 500 nm can be tensile tested. Kinds of samples with different sizes or shapes can also be tested by using different sample holder with the same apparatus by further optimization. The lowest temperature of the present apparatus is about 16 K and the temperature of the sample can be controlled at an arbitrary set point from 20 K to room temperature with a precision of ±10 mK.

Acknowledgements

This research is supported by The National Natural Science Foundation of China (No. 51327806, No. 51427806).

References

[1] Qin J G, Dai C, Liao G J, Wu Y, Zhu X F, Huang C J, Li L F, Wang K, Shen X G, Tu Z P and Ji H 2014 *Cryogenics.* **64** 16-20
[2] Czarkowski P, Krawczynska A T, Brynk T, Nowacki M, Lewandowska M and Kurzydlowski K J 2014 *J. Cryogenics.* **64** 1-4
[3] Tomo Takeda, Weiwei Fan, Qingpin Feng, Shaoyun Fu, Fumio Narita and Yasuhide Shindo 2013 *Cryogenics.* **58** 33-7
[4] Fink M, Fabing Th, Scheerer M, Semerad E and Dunn B 2008 *Cryogenicst.* **48** 497-510
[5] Chen L B, Jin H, Wang J J, Zhou Y, Zhu W X and Zhou Q 2013 *Cryogenics.* **54** 54-8
[6] Chen L B, Zhou Q, Jin H, Zhu W X, Wang J J and Zhou Y 2013 *Cryogenics.* **57** 195-99
[7] Chen L B, Zhou Q, Zhu W X, Xue X D, Wang J J and Zhou Y 2013 *Advances in Cryogenic Engineering Transactions of the Cryogenic Engineering Conference – CEC.* **2013** 982-7
[8] Chen L B, Zhou Q, Zhu X S, Zhou Y and Wang J J 2015 *Physics Procedia.* PHPRO4742
[9] Walsh R P 2000 *Mechanical testing and evaluation* ASM handbook Vol. 8 (Ohio State: ASM International) pp 164-171

Effect of Discontinuities and Penetrations on the Shielding Efficacy of High Temperature Superconducting Magnetic Shields

R Hatwar[1,2], J Kvitkovic[1], C Herman[2], S Pamidi[1]
[1] Center for Advanced Power Systems, Florida State University, Tallahassee, Florida 32310 USA
[2] Heat Transfer Lab, Mechanical Engineering Department, Johns Hopkins University, Baltimore, Maryland 21218 USA

E-mail: rajeev.hatwar@jhu.edu

Abstract. High Temperature Superconducting (HTS) materials have been demonstrated to be suitable for applications in shielding of both DC and AC magnetic fields. Magnetic shielding is required for protecting sensitive instrumentation from external magnetic fields and for preventing the stray magnetic fields produced by high power density equipment from affecting neighbouring devices. HTS shields have high current densities at relatively high operating temperatures (40-77 K) and can be easily fabricated using commercial HTS conductor. High current densities in HTS materials allow design and fabrication of magnetic shields that are lighter and can be incorporated into the body and skin of high power density devices. HTS shields are particularly attractive for HTS devices because a single cryogenic system can be used for cooling the device and the associated shield. Typical power devices need penetrations for power and signal cabling and the penetrations create discontinuities in HTS shields. Hence it is important to assess the effect of the necessary discontinuities on the efficacy of the shields and the design modifications necessary to accommodate the penetrations.

1. Introduction

Sensitive electronic equipment such as SQUIDS in magneto-encephalography [1, 2] and other medical applications require a high level of magnetic shielding to protect them from spurious magnetic fields. Further, devices that generate strong magnetic fields also need to be shielded in order to contain their magnetic fields and ensure proper functioning of other instruments in their vicinity. Conducting materials can be used for AC magnetic shielding applications, but for frequencies lower than 1 kHz ferromagnetic materials, such as mu-metal, are normally used. High temperature superconducting (HTS) materials are usually more efficient in shielding than ferromagnetic materials [3]. HTS materials such as $YBa_2Cu_3O_7$ (YBCO) have been demonstrated to be suitable for shielding both for DC and AC magnetic fields [4, 5]. Second generation (2G) HTS have been successfully developed on a commercial level with excellent superconducting and mechanical properties. Long lengths of 2G HTS are being produced at various widths up to 45 mm [6].

Various studies have been done in the past on magnetic shields made of superconducting bulk material and tapes. Kvitkovic *et. al.* have studied the effect of the geometry, operating temperature, and interlayer separation on magnetic shielding efficacy of shields made of 2G HTS [7-11]. Fabrication of large practical shields require many sections of HTS and magnetic shields needed for large high power density machines will need to allow access to power feedthroughs and sensor wiring. Aging and mechanical stresses will also lead to spots of degraded superconducting properties [12, 13].

The effect of different forms of discontinuities have been studied in the past. Denis *et. al.* studied the shielding properties of a polycrystalline Bi-2223 HTS hollow tubes under axial field [14]. The effect of a cap on the end of the tube and the effect of a hole in the cap was studied in another paper [15]. They further analysed the effect of non-superconducting joints on shielding efficacy[15]. Fagnard *et. al.* analysed the effect of slits in the superconducting bulk materials under DC magnetic field [16].

So far the effects of discontinuities and penetrations, in two parallel rectangular superconducting HTS sheets with hole, on their shielding performance in AC magnetic field have not been studied. Therefore it is important to quantify the effect of discontinuities on shielding efficacy of HTS magnetic shields and examine potential solutions to mitigate the negative effects of the penetrations and discontinuities.

The present work focuses on understanding the effect of a circular hole on the magnetic shielding efficacy of a 2G HTS shield. This paper presents the results of experimental studies of the shielding factor at variable amplitudes and frequencies of an external magnetic field on the shield with and without an opening, and a simple method to mitigate the negative effects of an opening.

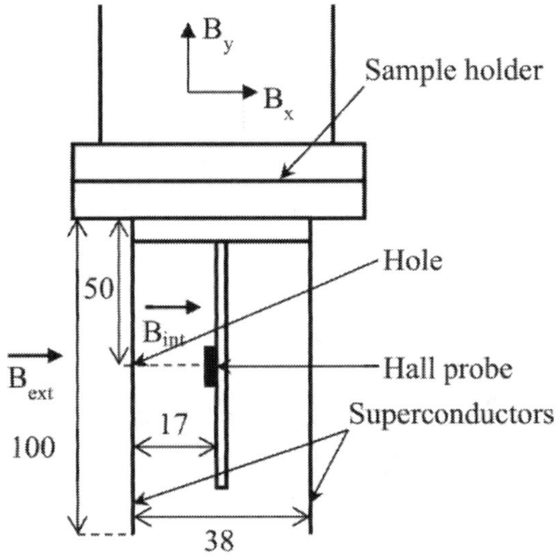

Figure 1. Schematic of the experimental setup and position of the hole. All dimensions are in mm.

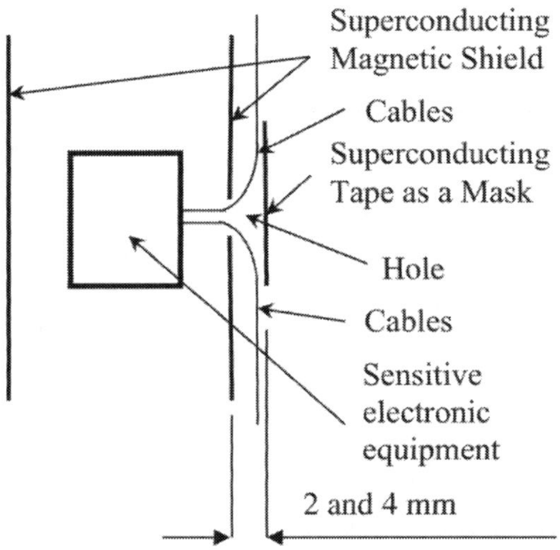

Figure 2. Use of a masking layer to mitigate the effect of a hole in the magnetic shield.

2. Experimental

A helical copper magnet [17] was used to generate a transverse external AC magnetic field. Magnetic shields were kept in a uniform AC magnetic field, which was perpendicular to the surface of the shields. The external magnetic field, B_{ext}, was calculated based on the current supplied to the magnet and the previously measured magnet constant. The internal magnetic field, B_{int}, was measured using a calibrated Hall Probe placed inside the magnetic shield, at mid distance between two parallel rectangular plates and at the centre of the rectangular plates as shown in Figure 1.

The dimensions of the rectangular plates are 60 mm × 100 mm and the distance between the two parallel sheets is 38 mm. The support structure for the magnetic shield is made of G-10 plates. Three layers of YBCO tapes are attached onto the G-10 plates using cryogenic Kapton adhesive tape. Each layer consists of three pieces of 45 mm wide 2G HTS superconducting tape, manufactured by

American Superconductor Corporation [6]. Holes were drilled at the centre of the shield, at the same height as the Hall probe Figure. 1. The size of the hole was varied with diameters of 1 mm, 3 mm, 6 mm, and 12 mm to study the effect of the size of discontinuities and penetrations on the shielding effect. A 45 mm × 45 mm section of the HTS tape is placed on top of the 12 mm hole to mask the opening, with a G-10 fiberglass piece in between them. The distance between the mask and the shield was varied between 2 mm and 4 mm, for the two configurations shown in Figure 2.

The shielding factor, SF, is a measure of the effectiveness of the magnetic shield and it is expressed as:

$$SF\ (\%) = 100 \times \frac{B_{ext} - B_{int}}{B_{ext}}$$

The frequency of the AC magnetic field was varied from 20 Hz to 400 Hz. The experimental setup was maintained at 77 K by immersing in a liquid nitrogen bath. All magnetic field magnitudes are expressed in rms values. Details of the experimental setup are described elsewhere [11].

3. Results and discussion

The variation of shielding factor of the shield without any opening with B_{ext} for different frequencies is shown in Figure 3. It can be seen that for lower B_{ext}, the shielding factor increases with increasing frequency. The shielding factor decreases as the B_{ext} increases. The shielding factor at 20 Hz goes down from 44 % to 1.5 % as the B_{ext} increases from 2.5 mT to 63 mT. A similar trend is observed for other frequencies. This can be explained by the fact that the critical current, I_c, decreases with increasing B_{ext} which leads to a lower shielding factor for higher B_{ext}. So for high B_{ext}, the shielding factor for higher frequency is lower. The shielding factor at 31.5 mT for various frequencies is in the range of 4 to 7%.

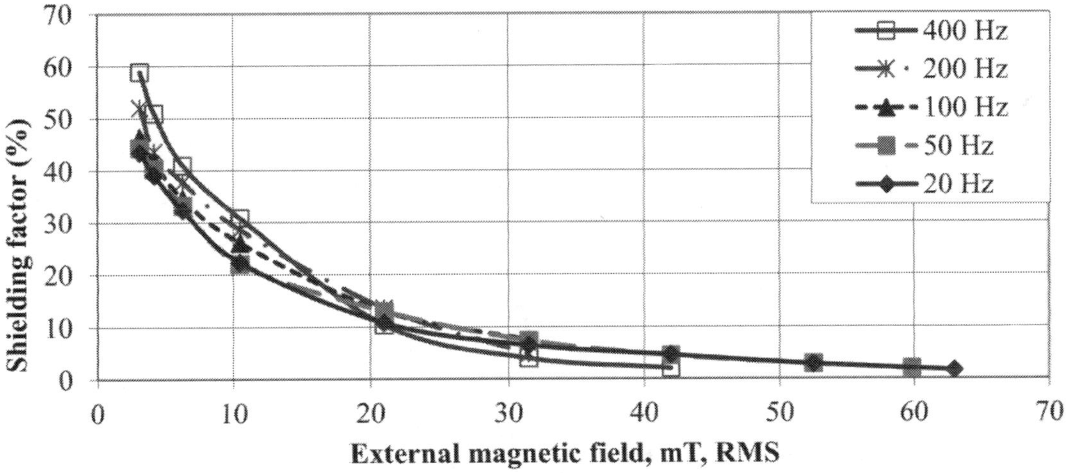

Figure 3. Variation of shielding factor with external magnetic field for 3 layer shield with no holes.

Figure 4 shows the variation in the shielding factor with B_{ext} at 20 Hz for different sizes of holes. Similar to the case of a shield with no opening, the shielding factor for the shield with a hole decreases with increasing B_{ext}. Further, it can be seen that the shielding factor decreases with increasing hole size. For B_{ext} of 2.5 mT, the shielding factor is 44 % with no hole and 37.5 %, 25.5 % and 22 % for 1 mm, 6 mm and 12 mm hole size, respectively. The effect of hole size gradually diminishes as B_{ext} increases. There is a decrease of shielding factor from 10.7 % to 4. 8 % as the hole size increases from no hole to 12 mm hole at 21 mT. At 63 mT this reduction in shielding factor is from 1.5 % to 0.03 %. A similar trend is observed for 400 Hz as shown in Figure 5. The frequency dependence of the shielding factor at different applied magnetic field amplitudes for different hole sizes is depicted in

Figures 6 and 7. It can be seen that at lower B_{ext}, the dependence of shielding factor on frequency is high and as B_{ext} increases the effect of frequency diminishes to a small value.

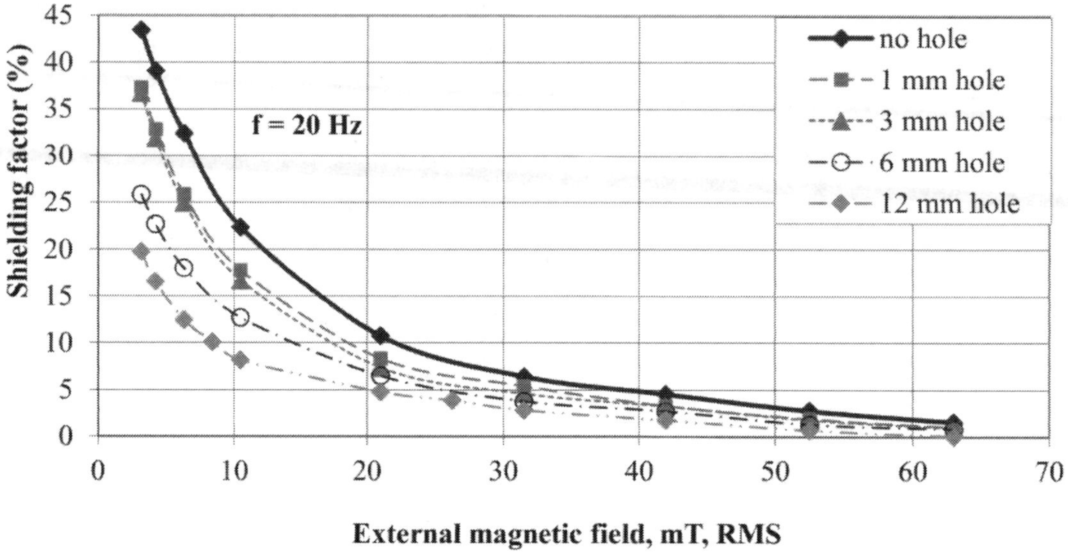

Figure 4. Variation of shielding factor with external magnetic field at 20 Hz for different sizes of holes.

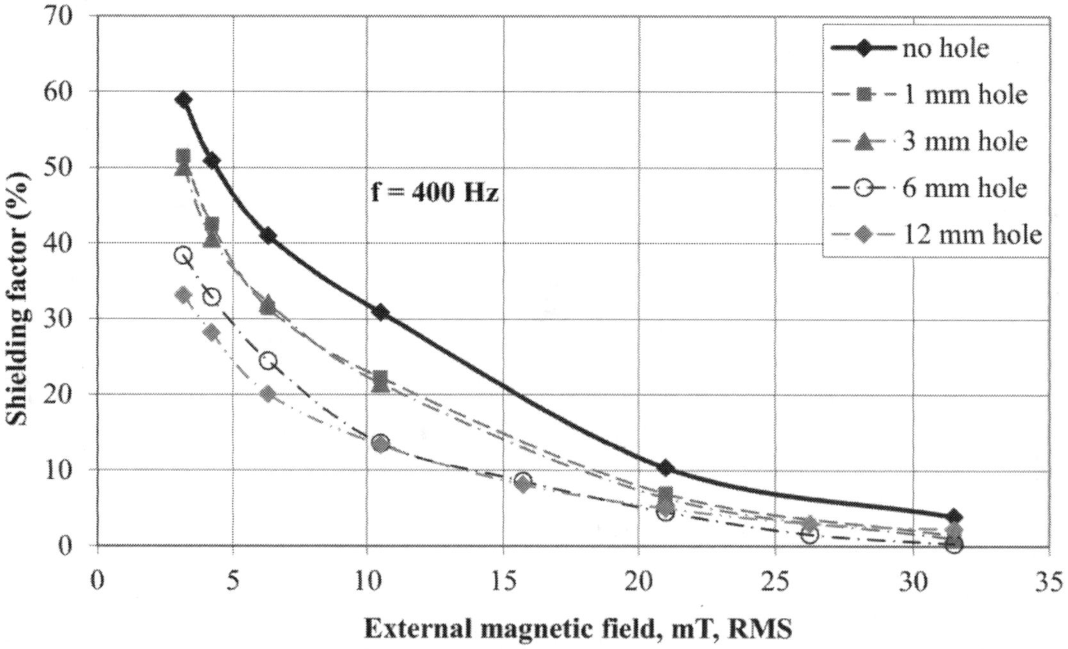

Figure 5. Variation of shielding factor with external magnetic field at 400 Hz for different sizes of holes.

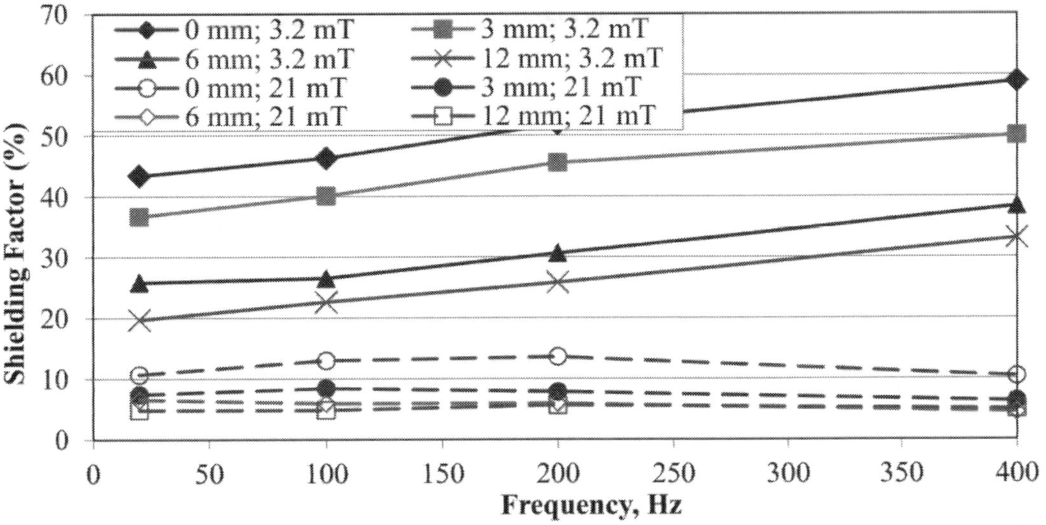

Figure 6. Shielding factor vs frequency for a 3 layer shield for different hole diameters and external magnetic field magnitude of 3.2 mT and 21 mT.

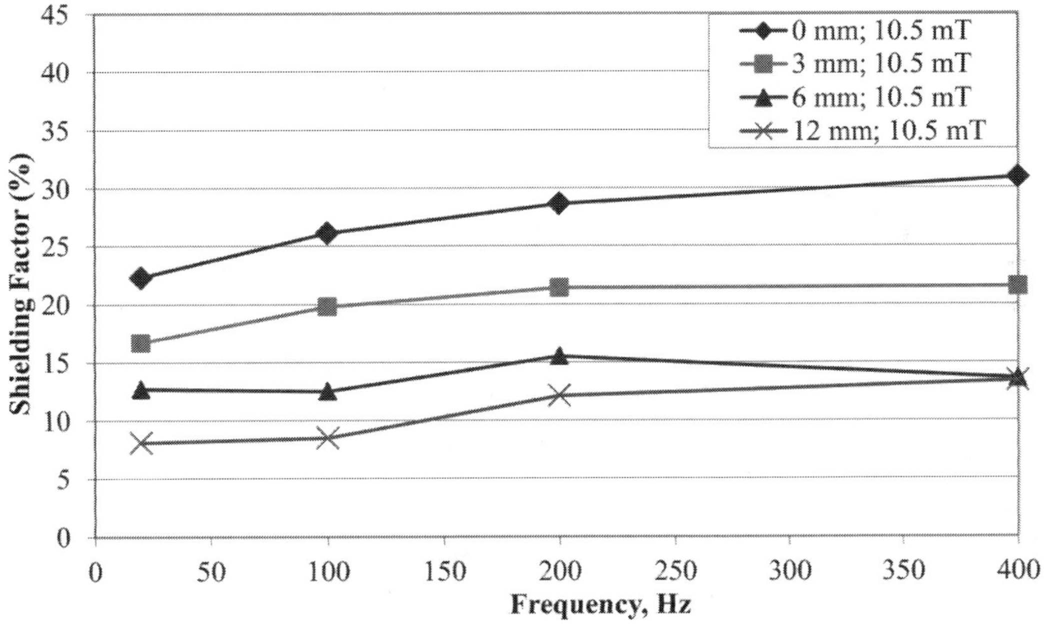

Figure 7. Shielding factor vs frequency for different hole diameters, for an applied magnetic field magnitude of 10.5 mT.

In order to assess ways to mitigate the effect of the hole, a piece of superconducting tape of smaller dimension (45 x 45 mm^2) was used as a mask for the opening. The mask was placed at a distance of 2 mm or 4 mm in front of the 12 mm hole. The effect of the mask layer on the shielding factor at 20 Hz is shown in Figure 8. Both configurations were able to successfully compensate for the 12 mm hole. The shielding factor at 10.5 mT increased from 8.14 % to 27.48 % with the addition of the mask at a distance of 4 mm. This was higher than 22 %, the shielding factor obtained with a sheet without any hole. When the mask was placed at 2 mm the shielding factor further increased to 35%. So the mask not only compensated for the effect of the hole, but also enhanced the shielding. Similar behaviour is observed at 400 Hz as shown in Figure 9. Figures 10 and 11 show the effect of the mask layer, covering the 12 mm hole, on the shielding factor at different B_{ext} for frequencies ranging from 20 to 400 Hz. It can be seen that at lower B_{ext} (3.2 mT), although the shielding factor improves with the mask, it is unable to compensate fully. Whereas for B_{ext} > 7 mT the shielding factors recover completely with the use of the mask and are above the curve representing a sheet without any hole.

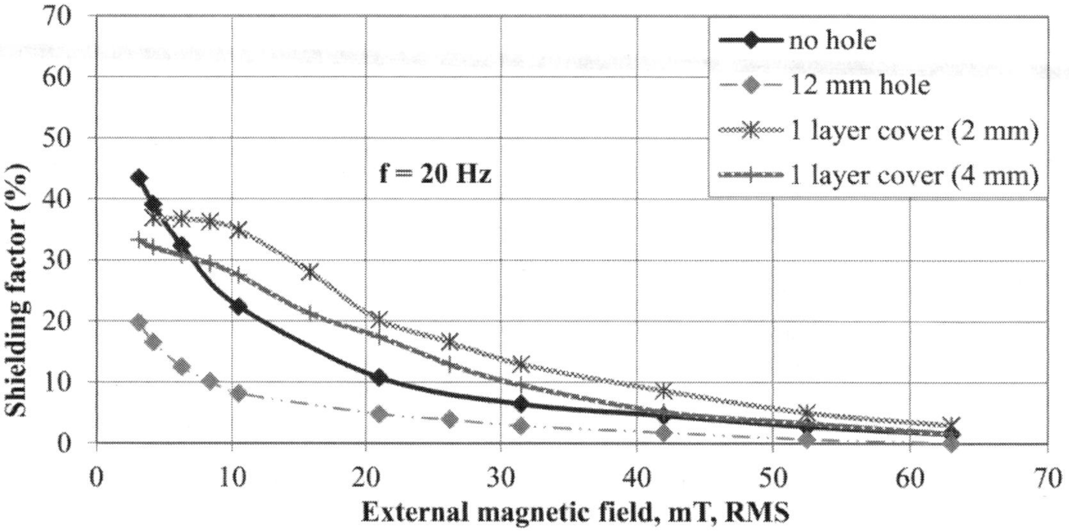

Figure 8. Effect of the mask placed at a distance of 2 mm and 4 mm from the 12 mm hole at 20 Hz.

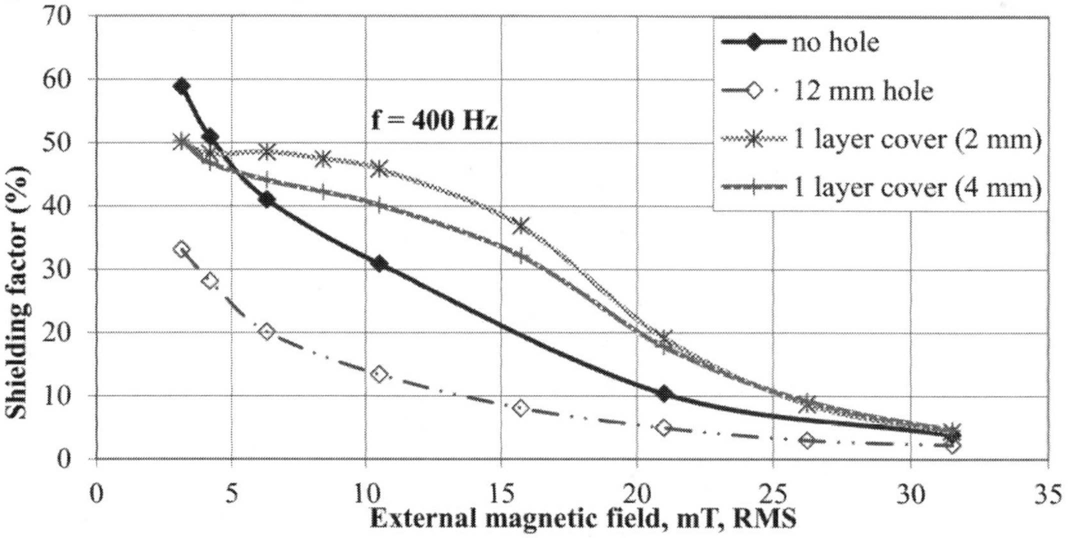

Figure 9. Effect of the mask placed at a distance of 2 mm and 4 mm from the 12 mm hole at 400 Hz.

4. Conclusion

The effect of discontinuities, such as penetrations necessary for pipes and cables, in superconducting magnetic shields fabricated from 2G HTS tapes on the shielding efficacy is studied with the goal finding ways to mitigate the negative effects. It is shown that the decrease in the shielding factor caused by a penetration can be mitigated using a superconducting mask to cover the hole applied at certain distance.

IOP Conf. Series: Materials Science and Engineering **102** (2015) 012012 doi:10.1088/1757-899X/102/1/012012

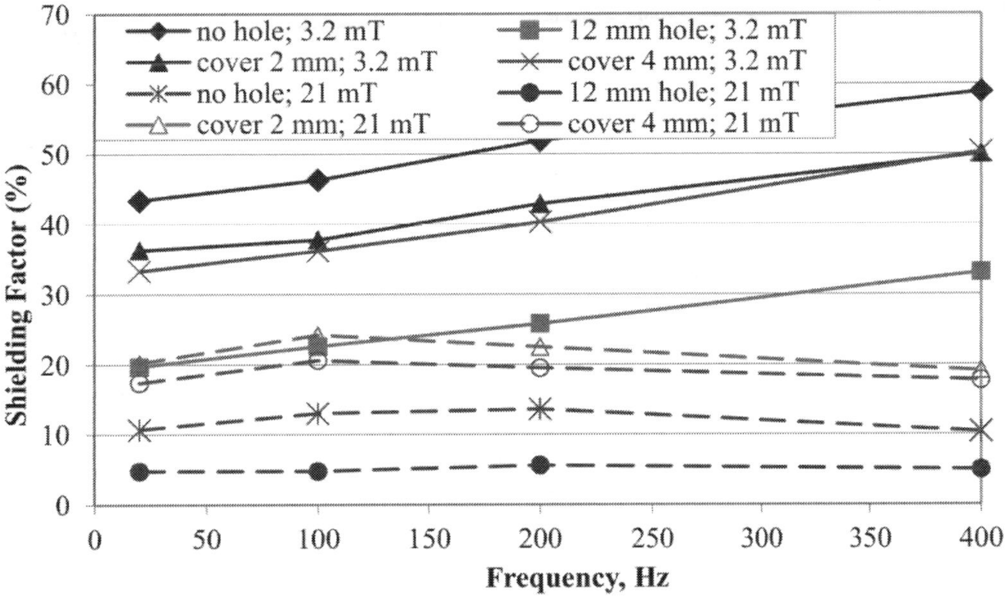

Figure 10. Effect of the mask placed at a distance of 2 mm and 4 mm from the 12 mm hole for 3.2 mT and 21 mT of applied external magnetic fields.

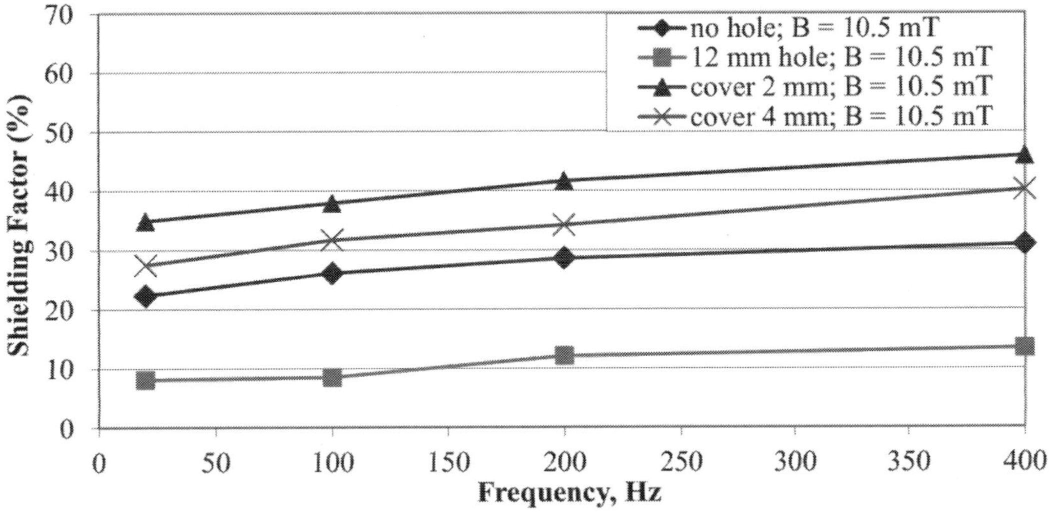

Figure 11. Effect of the mask placed at a distance of 2 mm and 4 mm from the 12 mm hole for 10.5 mT of applied external magnetic field.

The shielding factor showed a strong dependence on the magnitude and frequency of the external magnetic field. The shielding factor decreases with increasing external magnetic field and increases with increasing frequency. The discontinuity in the superconducting shield caused by a hole leads to a drop in shielding factor and the extent of the drop increases with increasing hole size. A superconducting mask is shown to mitigate the negative effect and provides a simple method to manage the need for access to the magnetically shielded space for electrical and signal cables and other piping. Further work is needed to optimize the dimension and location of the mask layer to enable fabrication of practical superconducting shields for large volumes that require penetrations.

Acknowledgements
This work was supported in part by the Office of Naval Research under Grant N00014-14-1-0155 and by the American Superconductor, who provided the superconducting material.

References

[1] Pizzella V, Penna S D, Gratta C D and Romani G L 2001 SQUID systems for biomagnetic imaging. *Supercond. Sci. Technol.* **14** R79
[2] Baudenbacher F, Peters N T, and Wikswo J P 2002 High resolution low-temperature superconductivity superconducting quantum interference device microscope for imaging magnetic fields of samples at room temperatures *Rev. Sci. Instrum.* **73** 1247-54
[3] Pavese F 1998 Handbook of Applied Superconductivity (Bristol: Institute of Physics) p 1461-83
[4] Fagnard J-F, Denis S, Lousberg G, Dirickx M, Ausloos M, Vanderheyden B and Vanderbemden P 2009 DC and AC Shielding Properties of Bulk High-Tc Superconducting Tubes. *IEEE Trans. Appl. Supercond.* **19** 2905-08
[5] Fagnard J-F, Elschner S, Hobl A, Bock J, Vanderheyden B and Vanderbemden P 2012 Magnetic shielding properties of a superconducting hollow cylinder containing slits: modelling and experiment. *Supercond. Sci. Technol.* **25** 104006
[6] www.amsc.com
[7] Kvitkovic J, Voccio J and Pamidi S V 2009 Shielding of AC Magnetic Fields by Coils and Sheets of Superconducting Tapes *IEEE Trans. Appl. Supercond.* **19** 3577-80
[8] Kvitkovic J, Pamidi S and Voccio J 2009 Shielding AC magnetic fields using commercial $YBa_2Cu_3O_7$ -coated conductor tapes *Supercond. Sci. Technol.* **22** 125009
[9] Kvitkovic J, Davis D and Zhang M 2015 Magnetic Shielding Characteristics of Second Generation High Temperature Superconductors at Variable Temperatures obtained by Cryogenic Helium Gas Circulation *IEEE Trans. Appl. Supercond.* **25** 8800304
[10] Kvitkovic J, Patil P, Pamidi S V and Voccio J 2011 Characterization of 2G Superconductor Magnetic Shields at 40–77 K *IEEE Trans. Appl. Supercond.* **21** 1477-80
[11] Kvitkovic J, Davis D, Zhang M and Pamidi S 2013 Influence of Interlayer Separation on Magnetic Shielding Properties of 2G HTS Sheets Made of 46 mm Wide RABiTS Tape *IEEE Trans. Appl. Supercond.* **23** 8200605
[12] Kim J H, Kim C H, Iyyani G, Kvitkovic J and Pamidi S 2011 Transport AC Loss Measurements in Superconducting Coils *IEEE Trans. Appl. Supercond.* **21** 3269
[13] Yuan W, Coombs T A, Kim J H , Kim C H, Kvitkovic J and Pamidi S 2011 Measurements and calculations of transport AC loss in second generation high temperature superconducting pancake coils *Journal of Appl. Phys.* **110** 113906
[14] Denis S, Dusoulier L, Dirickx M, Vanderbemden Ph, Cloots R, Ausloos M and Vanderheyden B 2007 Magnetic shielding properties of high-temperature superconducting tubes subjected to axial fields *Supercond. Sci. Technol.* **20** 192
[15] Denis S, Dirickx M, Vanderbemden Ph, Ausloos M and Vanderheyden B 2007 Field penetration into hard type-II superconducting tubes: Effects of a cap, a non-superconducting joint, and non-uniform superconducting properties *Supercond. Sci. Technol.* **20** 418
[16] Fagnard J-F, Elschner S, Hobl A, Bock J, Vanderheyden B and Vanderbemden P 2012 Magnetic shielding properties of a superconducting hollow cylinder containing slits: Modelling and experiment *Supercond. Sci. Technol.* **25** 104006
[17] Pamidi S, Kvitkovic J, Trociewitz U, Ishmael S, Meinke R and Stelzer G A 2012 Novel Magnet for AC Loss Measurements on 2G Superconductor Rings and Coils in Axial and Radial Magnetic Fields *IEEE Trans. Appl. Supercond.* **22** 9003004

IOP Conf. Series: Materials Science and Engineering **102** (2015) 012013 doi:10.1088/1757-899X/102/1/012013

Material characterisation and preliminary mechanical design for the HL-LHC shielded beam screens operating at cryogenic temperatures.

C. Garion, L. Dufay-Chanat, T. Koettig, W. Machiocha, M. Morrone

CERN, Technology department, 1211, Geneva 23, Switzerland

Cedric.Garion@cern.ch

Abstract. The High Luminosity LHC project (HL-LHC) aims at increasing the luminosity (rate of collisions) in the Large Hadron Collider (LHC) experiments by a factor of 10 beyond the original design value (from 300 to 3000 fb^{-1}). It relies on new superconducting magnets, installed close to the interaction points, equipped with new beam screen. This component has to ensure the vacuum performance together with shielding the cold mass from physics debris and screening the cold bore cryogenic system from beam induced heating. The beam screen operates in the range 40-60 K whereas the magnet cold bore temperature is 1.9 K. A tungsten-based material is used to absorb the energy of particles. In this paper, measurements of the mechanical and physical properties of such tungsten material are shown at room and cryogenic temperature. In addition, the design and the thermal mechanical behaviour of the beam screen assembly are presented also. They include the heat transfer from the tungsten absorbers to the cooling pipes and the supporting system that has to minimise the heat inleak into the cold mass. The behaviour during a magnet quench is also presented.

1. Introduction

In the HL-LHC, the intense flow of debris from beam collisions will be shielded by blocks of Inermet® 180, a sintered tungsten-based composite material with around 3.5 wt% of nickel and 1.5 wt% of copper. The blocks will be assembled on the beam screen [1], which operates in the range 40-60 K, and inserted in the beam pipe (cold bore) of superconducting magnets operating at 1.9 K (figure 1). The overall thermal and mechanical performance depends strongly on the electrical, thermal, magnetic and mechanical properties of the selected W alloy for which no data are available at cryogenic temperatures. Thus, a dedicated study has been conducted to evaluate such properties and finally assess the efficiency and reliability of the system.

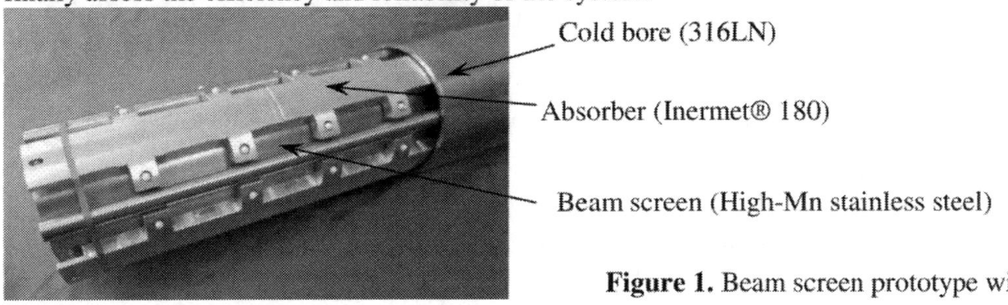

Figure 1. Beam screen prototype with absorbers.

2. Properties of the Inermet® 180 tungsten alloy at cryogenic temperatures

The material used to measure the different properties has been supplied by Plansee, France, according to their specifications. Different batches have been used. The size of the tungsten grains has been measured to around 50-60 μm.

2.1. Mechanical properties

Tensile tests have been carried out on flat machined, dog bone shaped, 2 mm thick samples. They have been performed at room temperature, 77 K and 4.2 K. Samples have been designed according to ISO 6892-1:2009 with a cross section of the active part of 2 mm x 4 mm, fillet radius of 20 mm and a straight length of 62 mm. The displacements are controlled with a strain rate of 10^{-4} s^{-1}. The results (average and standard deviation) of the tensile tests are reported in table 1.

Table 1. Results of the mechanical tensile tests at room and cryogenic temperatures.

Temperature [K]	Number of specimens	Young modulus [GPa]	Yield strength [MPa]	Ultimate strength [MPa]	Deformation to rupture [%]
293	3	330 ± 11	696 ± 18	944 ± 2	9.2 ± 1.1
77	3	320± 40	-	1155 ± 52	0.45 ± 0.03
4.2	4	324 ± 40	-	1048 ± 53	0.34 ± 0.03

Contrary to W-N-Fe alloys that have been deeply studied in the literature, only a few data are available for the W-Ni-Cu materials [2-4]. The tensile strength and the fracture elongation, measured at room temperature, are significantly higher than the values given by the supplier (685 MPa and 3%, respectively) and found in the literature: 660 MPa and 3% in [2], 631 MPa and 0.6% in [4]. The measured values are comparable with reference [3]. The mechanical properties of the Inermet depend on the strain rate at room temperature [4] and the grain size as well. The scattering at cryogenic temperature is rather high. Results should be refined with additional measurements.

The fracture faces of tensile tests done at room temperature and 77 K are shown in figures 2 and 3, respectively. An intergranular facture is observed on specimens tested at room temperature. Some microvoids and quasi-cleavage surfaces are noticed in the copper-nickel matrix. A few cleavages have been observed in the tungsten grains.

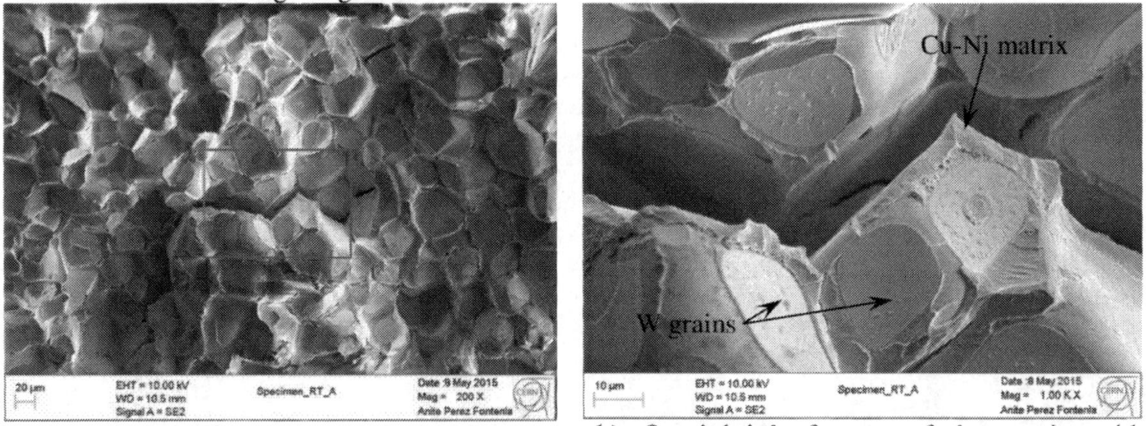

a) Intergranular fracture.

b) Quasi brittle fracture of the matrix, with some local microvoids.

Figure 2. Intergranular fracture surface after tensile test at room temperature.

Examination of the fracture surfaces on specimens tested at 77 K revealed a transgranular brittle fracture dominated by cleavage in the tungsten grains. A debounding between the grains and the matrix is noticed as well.

a) Transgranular brittle fracture.

b) Cleavage in the tungsten grains and debounding of the matrix.

Figure 3. Transgranular fracture surface after tensile test at 77 K.

To determine the thermal contraction of the samples a calibrated dilatometer made out of Pyrex® has been used, which allows the immersion of the mounted sample in cryogenic liquids. The dilatometer is of the compensation type measuring the differential contraction of the sample at the respective liquid temperature of nitrogen and helium at atmospheric pressure saturation condition. Table 2 shows the results of the measurement carried out at room temperature as reference using two specimens precisely machined to 8 mm by 8 mm cross section and a length of 50 mm. To guarantee the mechanical contact to the dilatometer a conical (specimen side) to spherical shape contact is applied. Great care is taken to ensure proper fitting of the sample by measuring the reproducibility on 10 sample mountings.

Table 2. Results of the thermal contraction measurements.

Temperature [K]	Batch reference	Expansion $\Delta l/l$ [%]	Deviation $\Delta(\Delta l/l)$ [%]
293	1	0	± 0.0001
78	1	- 0.086	± 0.003
4.2	1	- 0.096	± 0.007
293	2	0	± 0.0005
78	2	- 0.087	± 0.005
4.2	2	- 0.095	± 0.008

2.2. Electrical resistivity

Depending on the purity, the electrical resistivity of tungsten varies a few orders of magnitude at cryogenic temperature [5]. The electrical resistivity of the Inermet® 180 has been measured between 4.2 and 300 K. Different batches have been tested to assess possible scattering during the manufacturing. The samples have been machined from sintered blocks. Two different geometries have been considered to evaluate the impact of the machining and the local surface modifications. The measurements are given in figure 4. No significant influence of the batch nor of the sample geometry has been observed at cryogenic temperatures.

IOP Conf. Series: Materials Science and Engineering **102** (2015) 012013 doi:10.1088/1757-899X/102/1/012013

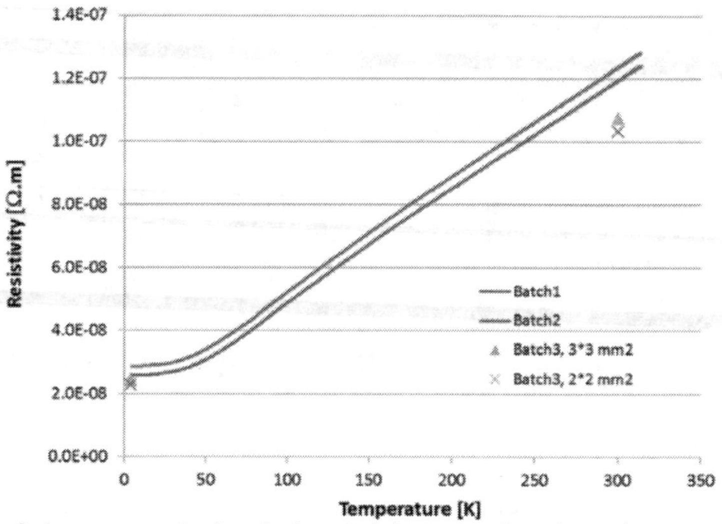

Figure 4. Graph of the measured electrical resistivity as a function of temperature. The sample dimensions of Batch 1 and 2 have been 2 mm x 2 mm x 73 mm.

2.3. Magnetic permeability

The magnetic permeability has been measured at University of Zaragoza using a vibrating sample magnetometer. Three different samples, 8 mm x 4 mm x 2 mm each, have been extracted by electro discharge machining at different position from a machined tungsten piece and have been tested. The measurements have been carried out between 4 K and 300 K with a magnetic field, H, up to 90 kOe (7162 kA.m^{-1}). The magnetic permeability curves are given as a function of the temperature and the magnetic field in figures 5 and 6, respectively. Magnetic permeability is less than 1.0003 at all measured fields and temperatures. No significant deviation has been observed between these three specimens. The magnetic susceptibility is in the order of 10^{-4} in the temperature range of 40-60 K. This is more than one order of magnitude lower than the magnetic susceptibility of the specific high-Mn austenitic stainless steel used for the beam screen wall.

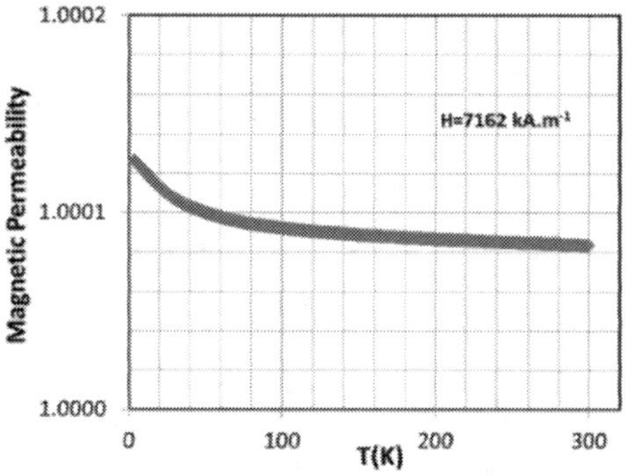

Figure 5. Typical magnetic permeability as a function of temperature. Values obtained from measurement at H=7162 kA.m^{-1}.

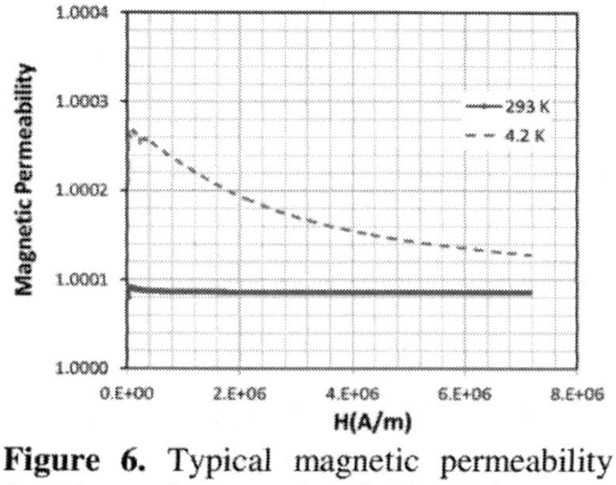

Figure 6. Typical magnetic permeability as a function of magnetic field. Values at room temperature and 4.2 K.

2.4. Thermal conductivity

The absolute measurement method was used to determine thermal conductivity, λ, as

$$\lambda = \frac{\dot{Q} \cdot L}{A \cdot \Delta T} \tag{1}$$

where $\dot{Q}$ denotes the heat flow. L and A stands for the length and the cross section of the sample, respectively. ΔT is the temperature difference along the sample length. The measurement has been carried out between 3.8 K and 280 K on two 8 mm x 8 mm x 50 mm samples. Samples identical to those used for the thermal contraction measurement have been used to determine the thermal conductivity. An active length of 30 mm has been chosen for the measurement of the temperature difference. Figure 7 shows the measured thermal conductivity as a function of temperature for both specimens. No significant deviation has been observed between the two samples. The relative measurement error was estimated as maximum error considering errors in the measurement of the applied heat load Q, the sample length, the sample cross section and the recorded temperature difference along the sample length by equation (2).

$$\Delta\lambda / \lambda = \left|\frac{\partial\lambda}{\partial\dot{Q}}\Delta\dot{Q}\right| + \left|\frac{\partial\lambda}{\partial L}\Delta L\right| + \left|\frac{\partial\lambda}{\partial A}\Delta A\right| + \left|\frac{\partial\lambda}{\partial T}\Delta T\right| \tag{2}$$

The calculated relative measurement error varies from about 2.5 % to 6.5 % depending on the temperature range of measurements. The respective error bars for the thermal conductivity values are exemplarily shown in figure 7. The error for the absolute temperature is related to the uncertainty of the used calibrated temperature sensor of ±8 mK at 4.2 K and ±120 mK at room temperature.

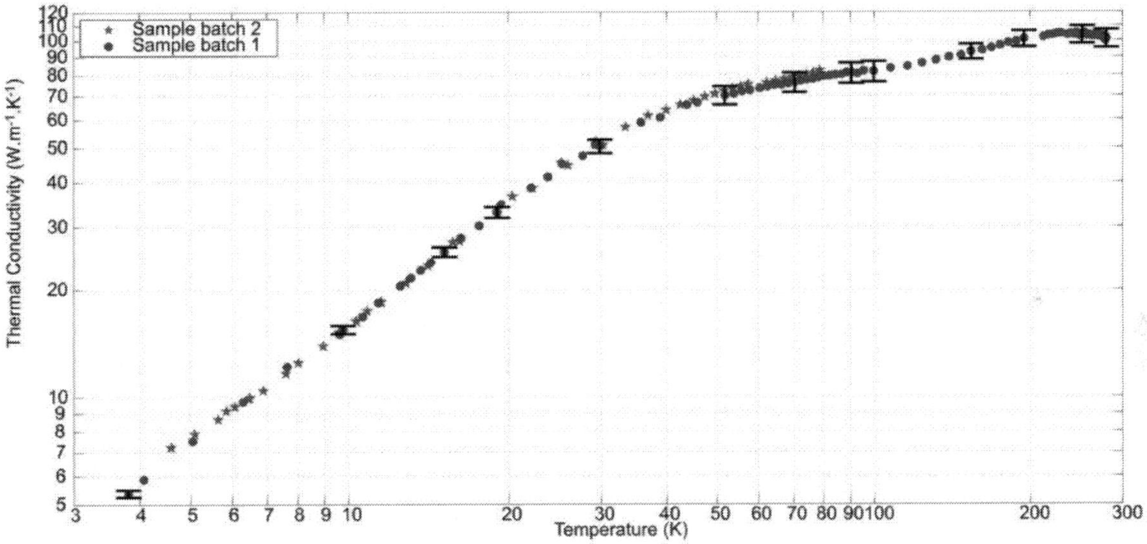

Figure 7. Graph of the thermal conductivity as a function of temperature. The conductivity of one sample, batch 2, has been measured up to 70 K, while for the other (batch 1) the temperature range has been extended up to room temperature.

3. Thermal mechanical behaviour of the High-Luminosity LHC beam screen with shielding

3.1. Mechanical behaviour during a magnet quench

The fast decay of the magnetic field leads to the development of Foucault currents that induce Lorentz forces, especially in high electrical conductivity material such as copper. The phenomenon is governed by the Maxwell-Faraday equation. For a quadrupole magnetic field, characterized by a magnetic gradient G, the specific Lorentz forces 'f' for long and symmetrical geometries are given by the equation (3):

$$f \propto \frac{G \cdot \dot{G} \cdot r^3}{\rho} \tag{3}$$

IOP Conf. Series: Materials Science and Engineering **102** (2015) 012013 doi:10.1088/1757-899X/102/1/012013

ρ denotes the electrical resistivity and r the radial coordinate. The behaviour of the assembly is driven by the 80 μm thick copper layer and the tungsten alloy absorbers; their electrical resistivities have been assumed to be $1.9.10^{-10}$ Ω.m and 3.10^{-8} Ω.m (section 2.2), respectively. The analysis does not take into account the magneto-resistance and is therefore conservative. The specific resultants of the Lorentz forces, per quadrant, are around 230 $N.mm^{-1}$ and 310 $N.mm^{-1}$ for the copper layer and tungsten absorbers, respectively. The beam screen assembly has been designed to be rather elastic and therefore, during a magnet quench, the tungsten absorbers go in contact with the 4 mm thick cold bore, which can withstand the high magnetic forces (figure 8). The contact force between the tungsten and the cold bore is around 370 $N.mm^{-1}$. The maximum Von Mises stress in the cold bore is around 650 MPa (figure 9) which is below the yield strength (860 MPa). The maximum stress in the beam screen wall is around 840 MPa, whereas the yield stress is around 1150 MPa [6]. This value is driven by the initial gap between the tungsten and the cold bore.

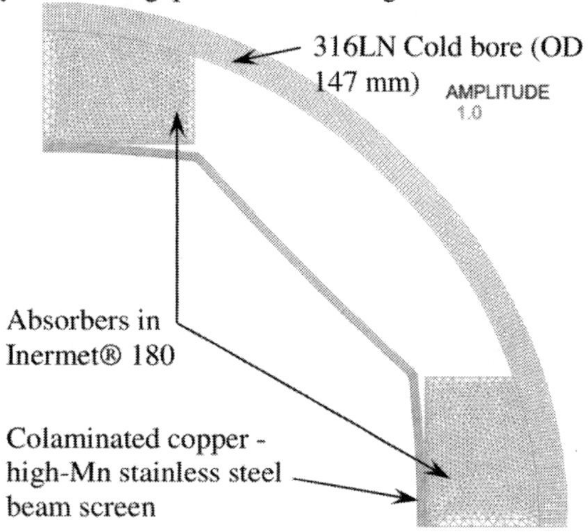

Figure 8. Deformation of the beam screen during a magnet quench.

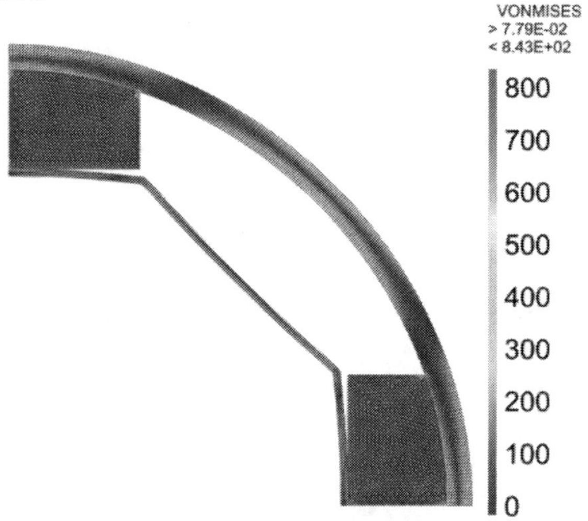

Figure 9. Von Mises stress field in the beam screen and cold bore during a magnet quench.

3.2. Heat transfer from the tungsten absorber to the cooling tube

The heat deposited on the tungsten absorbers is transferred by thermal links to the cooling tubes, in which a helium flow is imposed. The specific thermal load for the whole beam screen is 20 $W.m^{-1}$. A thermal analysis has been carried out based on the model presented figure 10.

The thermal conductivities for the copper, stainless steel and the tungsten alloy have been considered temperature dependent. It is assumed that the copper thermal link is perfectly bounded to the tungsten absorbed and to the stainless steel strip used as the interface with the cooling tube. This interface is welded to the cooling tube on three edges. The heat transfer by convection, between the helium and the cooling tube, has been estimated by several empirical formulae. A conservative value of 150 $W.kg^{-1}.m^{-1}$ has been assumed. A helium flow rate of $1g.s^{-1}$ per cooling tube is considered. The inlet helium temperature is 40 K.

The typical temperature profile is shown in figure 11.The temperature difference between the cooling gas and the tungsten blocs is around 13 K. The helium temperature gradient, along the cooling tube, is 0.5 $K.m^{-1}$. The number of thermal links or their width can be increased to improve the thermal performance (figure 12).

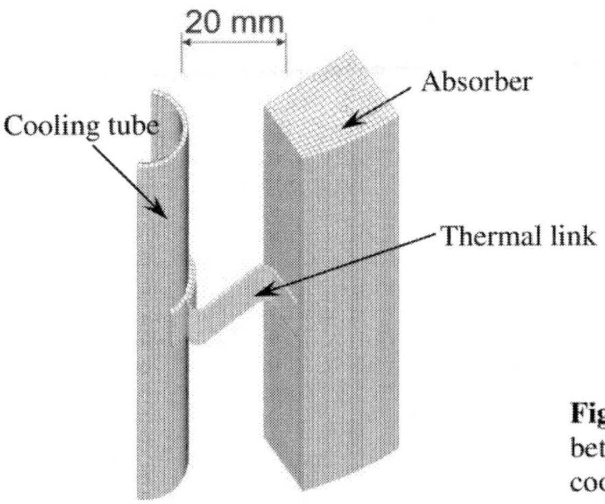

Figure 10. Model of the thermal link between the tungsten absorber and the cooling tube.

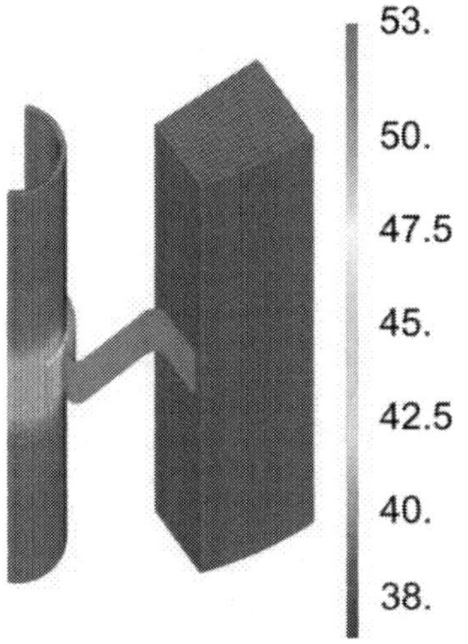

Figure 11. Temperature profile of the heat transfer from the tungsten absorber to the cooling tubes.

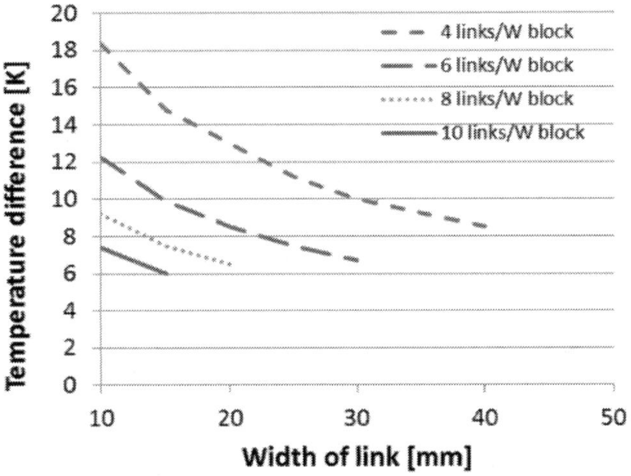

Figure 12. Influence of the number and width of the thermal links on the temperature difference between the cooling He gas and the W absorbers.

3.3. Heat leaks to the cold bore through the supporting system

The beam screen is supported in the cold bore tube. The linear weight is about 50 kg.m^{-1}. The supporting system has to fulfil:

- low thermal conductivity (heat inleak to the cold bore should be lower than 0.5 W.m^{-1}),
- good mechanical reliability,
- good tribological properties under vacuum and at cryogenic temperatures.

Different technical solutions have studied in two steps. The first one consists in a mechanical analysis of the supporting system subjected to the weight of the beam screen. This leads to the estimation of the local stress field and to the determination of the contact area between the components of the supporting system. Then, given the contact area, a thermal analysis is carried out to evaluate the heat transferred from the warm beam screen to the cold bore. The beam screen and the absorbers temperature is conservatively assumed to be 55 K, whereas a temperature of 1.9 K is considered for the cold bore. The thermal conductivity is temperature dependent.

Presently, a solution based on zirconium oxide balls supported by elastic springs made of titanium alloy, grade 5, located in the Inermet absorbers is retained (figure 13). In normal condition, the heat inleak is transferred through the spring and the ceramic ball. Both have a relatively low thermal conductivity coefficient. During a quench, the elastic springs are compressed and the ball comes inside the absorbers. Therefore the high Lorentz forces, which develop in the beam screen wall and in the tungsten blocks, are transferred uniformly to the cold bore.

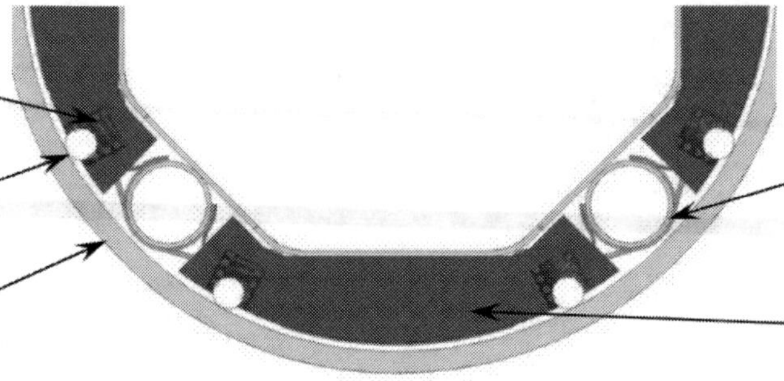

Figure 13. Supporting system of the HL-LHC beam screens with shielding.

High stress field develops between the ball and the cold bore, in which local plasticity occurs. The maximum of the principle stresses in the sphere reaches around 900 MPa and 80 MPa in compression and tension, respectively. The heat leak to the cold mass through the supporting system is estimated to be between 50 $mW.m^{-1}$ in the best case, i.e. contact of the absorber and the spring only at its extremity, to 400 $mW.m^{-1}$ in the worst case, i.e. contact of the absorber and the first spring spire.

4. Conclusion

The tungsten-based composite, Inermet® 180, has been chosen for the absorbers of the final focusing HL-LHC superconducting magnets. Mechanical, thermal, magnetic and electrical properties of this material have been measured at cryogenic temperatures. From mechanical point of view, a brittle behavior is observed at 77 K. Inermet® 180 has also a very low thermal expansion coefficient. The magnetic susceptibility is in the order of 10^{-4} in the temperature range 40-60 K, and the material is therefore suitable in high magnetic field environment. Thermal conductivity and electrical resistivity have been measured at cryogenic temperatures as well.

The thermal behavior of the HL-LHC beam screen has been assessed. The number and geometry of the copper thermal links, used to transfer the heat load from the absorbers to the cooling tube, are being optimized and a validation test of the thermal performance is foreseen on a short prototype.

Acknowledgments

Authors wish to thank the CERN metallurgical laboratory team for the tensile tests.

References

[1] V. Baglin, P. Lebrun, L. Tavian, R. van Weelderen, "Cryogenic Beam Screens for High-Energy Particle Accelerators", ICEC24 – ICMC, Fukoaka, Japan (2012).

[2] P. Lorenzo, M. Miralda, S. Iyengar, S. Melin, Etam Noah, "Fatigue Properties of Tungsten Heavy Alloys IT180 and D176", Int. Jour. of Refractory Metals and Hard Materials, **41**, 250-258 (2013)

[3] M Scapin, "Mechanical characterisation and modeling of the heavy tungsten alloy IT180", Int. Jour. of Refractory Metals and Hard Materials, **50**, 258-268 (2015)

[4] J. Das, G. Appa Rao, S.K. Pabi, "Microstructure and mechanical properties of tungsten heavy alloys", Materials Science and Engineering A, **527**, 7841–7847 (2010)

[5] N.V. Volkenshtein, L.S. Starostina, V.Ye. Startsev, and Ye. P. Romanov, Phys. Met. Metallogr. (USSR) 18, 85 (1964).

[6] S. Sgobba, G. Hochortler, "A new non-magnetic stainless steel for very low temperature applications", European congress; 3rd, Stainless steel: science and market; 391-402 (1999)

IOP Conf. Series: Materials Science and Engineering **102** (2015) 012014 doi:10.1088/1757-899X/102/1/012014

Materials for damping the PTC-induced thermal fluctuations of the cold-head

I. Catarino[1], D. Martins[1], R. Sudiwala[2]
[1] Departamento de Física, Faculdade de Ciências e Tecnologia, Universidade Nova de Lisboa, 2829-516 Monte de Caparica, Portugal
[2] School of Physics and Astronomy, Cardiff University, 5 The Parade, Cardiff, CF24 3AA, UK

E-mail: isabel.catarino@fct.unl.pt

Abstract
The cold head on mechanical Pulse Tube Cryocoolers (PTCs) is subject to substantially less mechanical vibration and electromagnetic interference compared to that typically found in Gifford MacMahon coolers. However, thermal fluctuations at the PTC frequency are still present at the cold-head, typically at a level of 200 mK peak-to-peak at 1.4 Hz for a Cryomech Model PT405 cooler running at 4 K. It is highly desirable to damp out these fluctuations if PTCs are to be used successfully for running systems sensitive to such thermal fluctuations, for example, bolometeric detectors.

We report here the characterization over the temperature range 2.5 K to 6 K of two materials, GOS (Gd_2O_2S) and GAP ($GdAlO_3$), for use as low-pass thermal filters. These materials have antiferromagnetic transitions at around 4 K giving rise to an enhanced heat capacity and have a high thermal conductance. These are two highly desirable properties for thermal dampers in this application. Those materials were fired as ceramic discs to be tested as thermal dumpers. Thermal filter assemblies with discs of diameter 75 mm and thickness 2.5 mm and 1.6 mm (GOS and GAP, respectively) mounted in a PTC show thermal attenuation levels of x0.12 (GOS) and x0.11 (GAP) at 0.01Hz with a clean-side temperature of 4 K; the PTC induced fluctuations at 1.48 Hz are damped completely to within the noise limits (0.2 mK) of the thermometers. Experimentally determined thermal conductance and heat capacity data are reported. For this system, with a PTC cold-head (dirty-side) temperature of 3.3 K, a clean-side power dissipation of up to 30 mW is realized before its temperature rises above 4.2 K.

Introduction

The 4 K cooling method using regenerative cryocoolers is very attractive when compared to the use of a liquid helium bath. Such cryocoolers provide autonomy of the cooling source using a lower cost and greener solution. Intrinsically for their use in thermodynamic cycles, there are oscillations associated to some extent, both thermal and mechanical. Pulse Tube Cryocoolers (PTC) having no moving parts at low temperature significantly reduce the mechanical issue, but not the thermal one.

The addition of a thermal buffer to the cold end is a solution to stabilize the thermal oscillation of the cold source. The trade off between maximizing the heat capacity to stabilize the temperature while having a good conductance to prevent a major loss of performance of the cooler is a challenge. Keeping the system simple is also desirable.

A vibration isolation and thermal damper for a 4 K GM cooler was reported in ref [1], although without details of its composition, showing to be able to reduce the 200 mK at 3.5 K oscillation down to 1 mK at 4.1 K. A report on the addition of 3.6 mm plates of ErNi in a 4 K PTC have shown a reduction from 275 mK to 14.5 mK [2] on the temperature oscillations. The ErNi material exhibits a ferromagnetic transition at about 10.5 K, hence not reflecting its high heat capacity peak on the 4 K temperature range. The cooling power became halved upon addition of such buffer. That work found an effective thermal conductance about a factor of 10 lower than the reported value for the bulk [3]. Other recent work on the same subject explored the heat capacity of a superconductor transition [4]: Lead with its critical temperature of ~8 K was chosen, that metal being combined with others in a stack of thin sheets on the cold head of a GM 4 K cooler and tested up to 50 K. At 4 K the 1.2 Hz oscillation of temperature was reduced from ~30 mK to 1 mK. Adding this buffer led to a double in cooldown time, its lower achieved temperature being 3.4 K instead of 3.0 K.

We investigated the use of bulk ceramic materials developed for use in regenerators, such as GOS = Gd_2O_2S and GAP = $GdAlO_3$ [5-7] as temperature dampers in a 4 K Cryomech PT405 pulse tube cryocooler. Such materials exhibit especially high heat capacity at the interesting 4 – 5 K [8-10] range, due to a para-antiferromagnetic transition, with quite high reported conductivity [5].

This paper presents the experiments performed with GOS and GAP discs as temperature fluctuation dampers. Data on effective thermal conductance for both GOS and GAP samples is reported from 3 K to 6 K. Temperature oscillations at the cryocooler's 1.48 Hz frequency vanishes to the noise level by using either material in. Lower frequency values are tested in an oscillatory heating of one side of each disc, for different temperatures. A gain to frequency Bode-type diagram is reported while the analogy with a standard passive resistor-capacitor circuit *(RC)* type filter fitting is explored. The extrapolation of cutoff frequency values from that diagram yields a preliminary result on the conductance to heat capacity ratio of the materials used.

Experimental setup

The sample assembly is shown in Figures 1 and 2. The Sample (GOS or GAP disc) is sandwiched between two copper plates (Top and Bottom Plates). A G10 ring is used to compress the Sample against the copper plates. The Bottom Plate is bolted directly to the cold finger of a cryocooler (Cryomech PT405). The Sample to Plate interfaces have a smear of vacuum grease. The Bottom Plate to PTC Cold Head is a grease-free interface. Each Plate has a heater (*H*, of 200 Ω nominal value) and a germanium Lake Shore calibrated temperature sensor (*T*).

Two 75 mm diameter discs were used as the Sample, one GAP disc with 1.6 mm thickness and one GOS disc with 2.5 mm thickness. The thicknesses were designed so the samples' conductances could match about 1 W/K conductance [5], regardless of the contact resistance.

The effective conductance (*K*) is measured by applying a small power (0.02 W and 0.03 W) to the Top Plate through heater H^{top} and stepping the power in the Bottom Plate, H^{bottom}. Data is collected at each step for typically 600 s using a AVS47 IEEE 488 interface.

The frequency response of the thermometers is measured using the faster analog output port of the resistance bridge, read by a 24-bit NI DAQ card while modulating the temperature of the bottom plate at a known frequency using H^{bottom}. This is done at four frequencies (0 Hz, 0.01 Hz, 0.02 Hz and 0.05 Hz) and for different DC power levels on H^{bottom}.

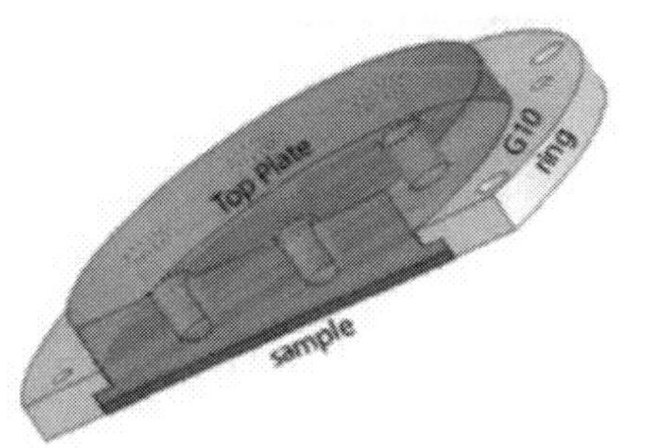

Figure 1. Partial cut of the sample mounting as designed with SolidWorks™.

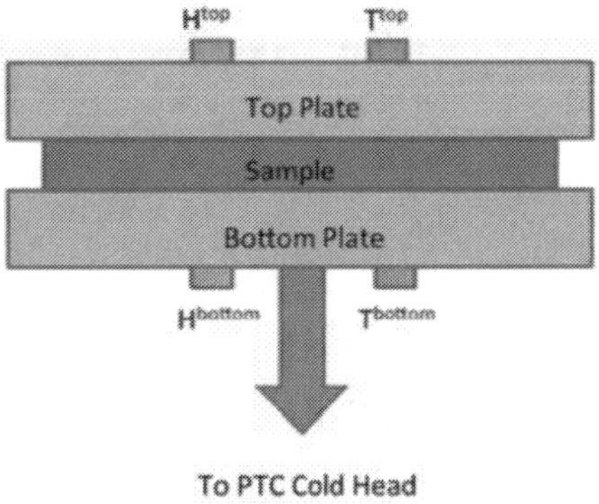

Figure 2. Schema for thermometers and heaters placements.

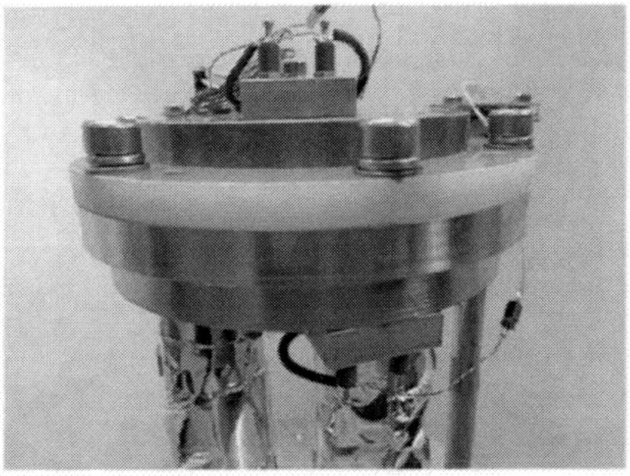

Figure 3. Assembly on the cold head of the PTC.

Experimental results

The ratio of (H^{top} - H^{bottom}) to (T^{top} - T^{bottom}) is assigned as the effective conductance (K) at the average temperature for the arrangement of the measurements performed, and is displayed in Figure 4. Considering that the main thermal resistance is due to the sample disc — i.e., disregarding the copper resistance (about 4 orders of magnitude lower, in a conservative estimate) as well as the contact resistances (that may have a significant role), the experimental conductance along with the sample dimensions allows estimating the conductivity (k) of the sample material, as plotted in Figure 5. Such conductivity comes about two orders of magnitude lower than reported by the manufacturer [5], which is plotted as dashed lines in the same figure.

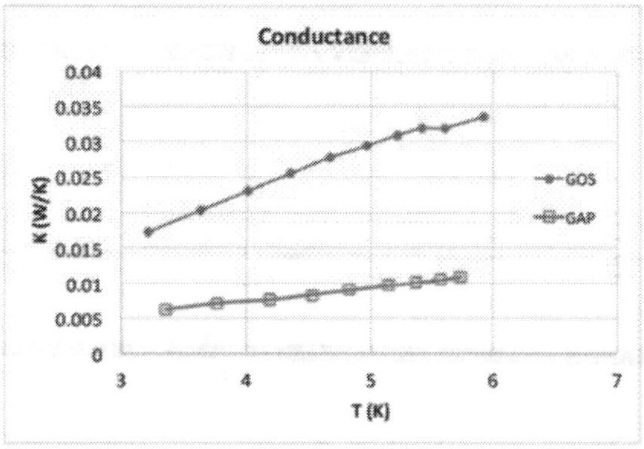

Figure 4. Effective conductance measured.

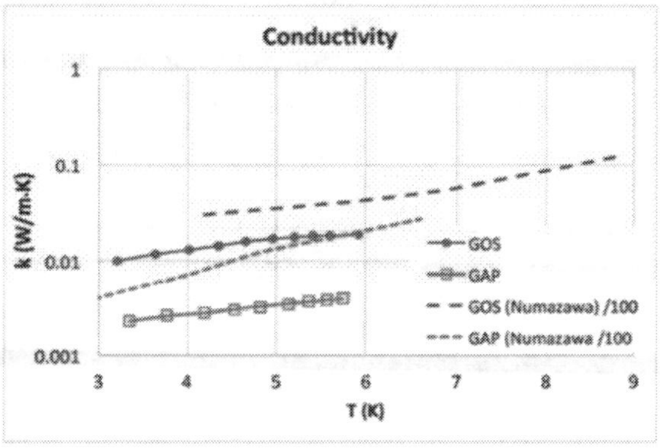

Figure 5. Inferred conductivity for GOS and GAP discs and comparison to the manufacturer data [5].

Figure 6 displays one example of the induced oscillating temperatures logged on both sides of the assembly. The attenuation G takes the ratio of bottom to top temperatures and was measured at different frequencies. The plot of the attenuation as a function of frequency is represented in Figure 7, for GOS and GAP respectively. Such attenuation approximately follows a single pole filter and a corresponding equation 1 fitting was performed. The adjustment of this equation to the attenuation measured at each frequency gives rise to the cutoff frequencies f_0 ($f_0 = K/2\pi C$) plotted in Figures 9 and 10.

$$\frac{T_{top}}{T_{bottom}} = \frac{1}{\sqrt{1+\left(\frac{f}{f_0}\right)^2}} \qquad (1)$$

The phase shift for such first order filter shall follow equation 2, and not surpass -90°:

$$\phi = -\arctan\left(\frac{f}{f_0}\right) \qquad (2)$$

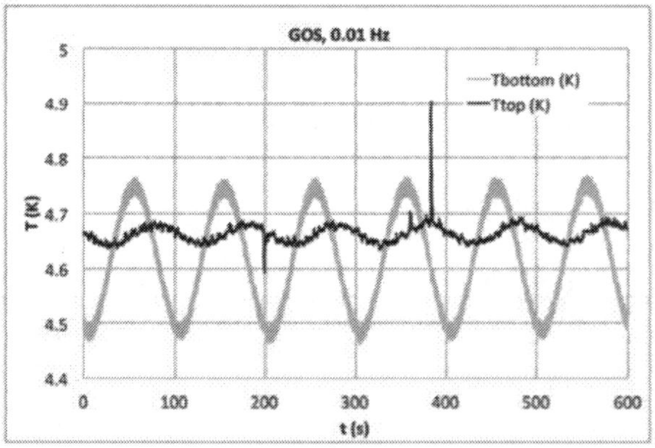

Figure 6. Example of induced oscillations on bottom plate and respective attenuation on top plate.

Attenuation analysis is plotted in Figure 7 as a Bode-type gain plot, both for GOS and GAP. The adjustment of equation 1 to the 0.1 Hz and 0.01 Hz points at each measured temperature determines the cutoff frequency f_0 as plotted in Figure 8. One can see that both sets of values for f_0 are in agreement, while taken from different frequencies. Such adjustment is equivalent to the extrapolation up to $G = 1$ of the fitted slope (corresponding to a first order filter) as in the dashed long line in Figure 7. The best-fitted slope would be slightly steeper than this, which is ascribed to the fact that the current assembly constitutes a distributed system, not exactly a simple RC as in this model. The phase shift analysis yields more than -90° for most points, realizing that the first order model is just an approximation to this system.

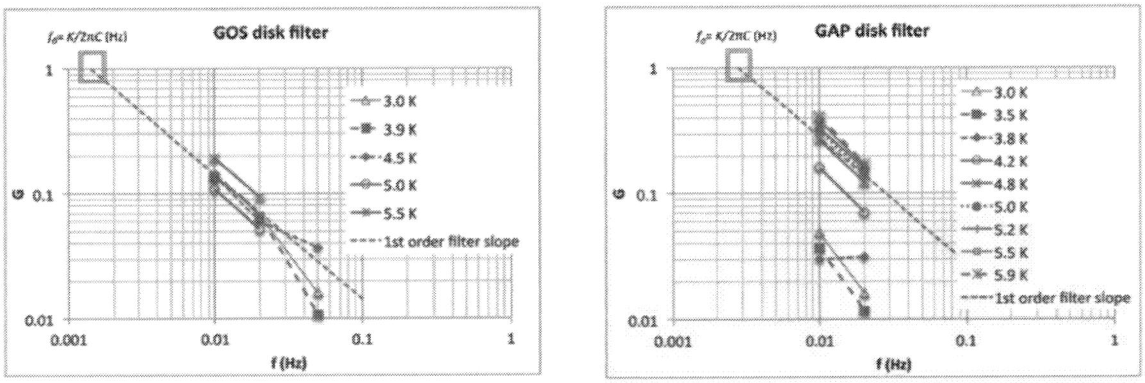

Figure 7. Attenuation of the induced oscillations for GOS (left) and GAP (right) as a function of frequency. The dashed long line emphasises the expected slope for a first order filter — its fitting to each temperature along with an extrapolation back to $G = 1$ confirms the cutoff frequency values determined from equation 1.

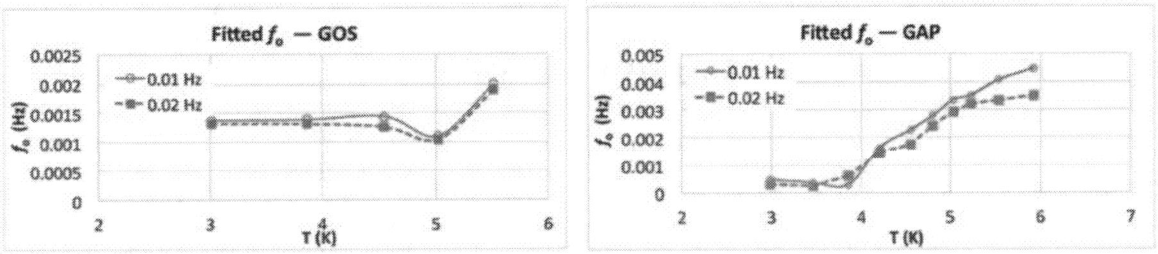

Figure 8. Cutoff frequencies as fitted by equation 1, for GOS (left) and GAP (right).

Using the experimental conductance values (K) along with the cutoff frequencies (f_0), the heat capacity (C) can be calculated and is depicted in Figure 9. For both materials the curve shape reproduces well the reference data [5] but there is about 50% of heat capacity missing. Actually the measured cutoff frequencies provide the information of the heat capacity and the heat conductance of the whole assembly, in a combined circuit.

Each sample disc provides most of the total heat capacity, the copper top plate adding only about 1% at the peak temperatures to the ceramic and the grease layer (1 mm^3 estimated) adds less than 0.1% of the sample heat capacity. This way it is about 98% accurate to consider the peak heat capacity of the whole assembly as the sample property.

The experimental values for thermal resistance add the resistance of the copper blocks to the sample discs as well as all the contact resistances. If the copper block's contribution to the whole resistance is negligible, the same cannot be said about the three contact resistances: copper-grease-sample (twice) and copper-copper. Contact resistances are to be analysed and improved. The observed discrepancy on thermal conductivity may also be due to the fact that the manufacturing process of the discs did not exactly match the same protocol as the materials reported in ref [5]: a heat treatment was missing, which should lead to a grain increase and consequent conductance increase [11]. The related work on ErNi [2] also reported a significant reduction on the experimental effective thermal conductance, suggesting the existence of more than one single phase of the material apart from contact resistance added.

Indeed the thermal system is harder to model, but even so, the performed experiments suggest a promising quite simple new method for measurement of low temperature heat capacity.

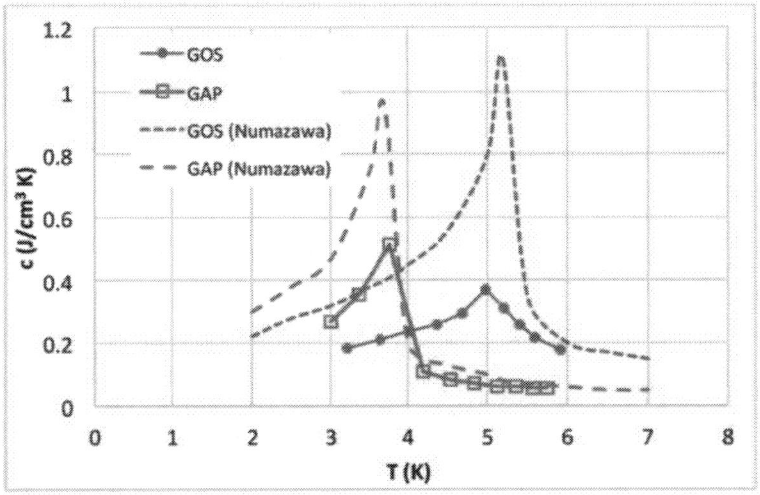

Figure 9. Specific heat capacity c for GOS and GAP materials obtained from the cutoff frequency values and comparison to Numazawa's reference data [5].

Conclusions

Aiming to strongly reduce the temperature oscillations of a Pulse Tube cryocooler, a thin disc (2.5 and 1.6 mm thick) of high heat capacity material, GOS and GAP, was added to the cold finger, acting as a low pass thermal filter. The geometry of the discs was designed to yield about 1W/K conductance at ~4 K but failed by two orders of magnitude. The conductance issue needs improvement, namely on the contact interfaces.

A Bode-type diagram evidences filtering at the cryocoolers' operating frequencies. The approximation to a first order Butherworth filter allows assigning a cutoff frequency as in a single pole thermal RC circuit. The inferred specific heat capacity confirms the abnormal peak on heat capacities but comes reduced to about one half of the expected values.

The coupling of a thin disc of GOS or GAP material to the cold end of such a cryocooler brings several (>2) orders of magnitude reduction in the temperature oscillations at 4 K induced by the normal operating frequencies of such machines.

Acknowledgments
This work was partially supported by *Fundação Luso Americana para o Desenvolvimento* and by *Fundação para a Ciência e Tecnologia* (PTDC/EMS-PRO/1065/2012).

References

1. Mauritsen L, Snow D, Woidtke A, Chase M, Henslee I. Low vibration, low thermal fluctuation system for pulse tube and Gifford -McMahon Cryocoolers. S D Miller and R G Ross J, (null), editors. Cryocoolers 15. Boulder, CO; 2009.
2. Allweins K, Qiu LM, Thummes G, Weisend JG, Barclay J, Breon S, et al. Damping of intrinsic temperature oscillations in a 4 K pulse tube cooler by means of rare earth plates. Adv. Cryog. Engineering. AIP; 2008. pp. 109–16.
3. Ogawa M, Li R, Hashimoto T. Thermal conductivities of magnetic intermetallic compounds for cryogenic regenerator. Cryogenics. 1991 Jun;31(6):405–10.
4. Dubuis G, He X, Božović I. Sub-millikelvin stabilization of a closed cycle cryocooler. Review of Scientific Instruments. AIP Publishing; 2014 Oct 3;85(10):103902.
5. Numazawa T, Yanagitani T, Nozawa H, Ikeya Y, Li R, Satoh T. A New Ceramic Magnetic Regenerator Material for 4 K Cryocoolers. Boston, MA: Cryocoolers 12; 2003. pp. 473–81.
6. Numazawa T, Kamiya K, Satoh T, Nozawa H, Yanagitan T. Status of the Development of Ceramic Regenerator Materials. In: R G Ross J, editor. Cryocoolers 13, Proceedings of the 13th International Cryocooler Conference. p. 373.
7. Masuyama S, Fukuda Y, Imazu T, Numazawa T. Characteristics of a 4 K Gifford–McMahon cryocooler using the Gd_2O_2S regenerator material. Cryogenics. 51(2011):337–40.
8. Numazawa T, Kamiya K, Hirastuka Y, Satoh T, Nozawa H, Yanagitani T. Cooling Performance of Multilayer Ceramic Regenerator Materials. 2007 Jan 31;:1–6.
9. Catarino I, Afonso J, Martins D, Linder M, Duband L, Bonfait G. 6 K solid state Energy Storage Unit. Cryogenics. 2010;50:102–10.
10. NIST Regenerator Materials [Internet]. [cited 2015 May 15]. Available from: http://cryogenics.nist.gov/MPropsMAY/RegeneratorMaterials/RegenPlot.htm
11. T. Numazawa, private communication, 2010.

IOP Conf. Series: Materials Science and Engineering **102** (2015) 012015 doi:10.1088/1757-899X/102/1/012015

Recent NASA/GSFC cryogenic measurements of the total hemispheric emissivity of black surface preparations

J Tuttle and E Canavan

Cryogenics and Fluids Branch
NASA Goddard Space Flight Center, Greenbelt, Maryland, USA

E-mail: james.g.tuttle@nasa.gov

Abstract. High-emissivity (black) surfaces are commonly used on deep-space radiators and thermal radiation absorbers in test chambers. Since 2011 NASA Goddard Space Flight Center has measured the total hemispheric emissivity of such surfaces from 20 to 300 K using a test apparatus inside a small laboratory cryostat. We report the latest data from these measurements, including Aeroglaze Z307 paint, Black Kapton, and a configuration of painted aluminum honeycomb that was not previously tested. We also present the results of batch-to-batch reproducibility studies in Ball Infrared Black™ and painted aluminum honeycomb. Finally, we describe a recently-adopted temperature control method which significantly speeds the data acquisition, and we discuss efforts to reduce the noise in future data.

1. Introduction

High emissivity (black) surface treatments are of great interest in connection with NASA missions. When used on radiators facing toward deep space, they enable relatively large amounts of heat to be removed from cryogenic systems. The power radiated from a surface is described by the Stefan-Boltzmann Equation,

$$\dot{Q} = \sigma A \varepsilon T^4 , \quad (1)$$

where $\dot{Q}$ is the total radiated power in W, T is the surface temperature in K, A is the surface area in m^2, $\sigma = 5.67 \times 10^{-8}$ Wm^{-2}K^{-4} is the Stefan-Boltzmann constant, and ε is the surface's total hemispheric emissivity. An ε value of 1.0 would correspond to the maximum possible power per unit area, referred to as a blackbody. NASA also employs such surfaces on thermal absorbers in large cryogenic test chambers. In general, a surface with a high ε also has a high absorptivity. The total hemispheric absorptivity, α, is the fraction of incoming radiation from all directions which is absorbed by the surface. Shrouds, cooled to low temperature and coated with high-absorptivity surface preparations, are used in ground testing in large vacuum chambers to simulate the deep space environment. The performance of thermal systems destined for space flight, as well as their detailed finite-element thermal models, are verified inside such shrouds.

NASA's Goddard Space Flight Center (GSFC) has a history of measuring surface emissivity. The most common method uses a reflectometer, bouncing monochromatic radiation off a surface and measuring the reflectance as a function of wavelength. These devices, however, are limited to wavelengths that are not long enough to cover the entire relevant blackbody spectrum at the lowest practical space flight radiator temperature of approximately 35 K. The total hemispheric emissivity, which determines the total heat radiated in all directions at all wavelengths, has been measured directly by hanging an object of known heat capacity in a very cold, black-painted shroud in a vacuum chamber and monitoring its cooling rate as a function of its temperature. This rate and heat capacity, along with the object's size, allow one to determine the emissivity. This "transient" technique works well down to temperatures typical of many space flight radiators, but the uncertainty in the indicated emissivity becomes large at the lowest temperatures [1]. A similar, "steady-state" approach was to hang coated plates inside a cold shroud and apply various steady power values to heaters imbedded in the plates. The final steady sample temperatures and the power values were used to back out the emissivity. This method had advantages over the transient approach, but its implementation was expensive, and it also let to significant uncertainty at the lowest temperatures [2].

When the James Webb Space Telescope (JWST) program selected Ball Infrared Black™ (BIRB™) to coat its 35 K deep-space radiators, Ball Aerospace (BASC) re-started its production of this surface treatment. Due to the uncertainty in BIRB's previously-measured emissivity at 35 K and the concern that the new application process would not exactly match that which had been used several years earlier, the project was interested in a new, more precise measurement of BIRB's emissivity. We developed a new, precise and relatively inexpensive method of determining the total hemispheric emissivity down to 20 K. This technique used an innovative geometry in measuring the heat exchange between identically-coated surfaces held at slightly different temperatures, and in 2011 we published the results of measurements taken using samples of the new BIRB [3]. The high emissivity values measured down to 20 K led to a study of the emissivity of thinner BIRB coatings, in an effort to reduce the radiator mass [4]. At the same time we developed a "hybrid" approach to this measurement, using a previously-calibrated BIRB sample plate exchanging heat with previously un-measured samples. We used this approach to characterize three different candidate samples of painted aluminum honeycomb, which had been selected for additional JWST radiators [5]. Since that time we have characterized several additional coatings which will be used on JWST.

In this paper we give a brief description of the emissivity measurement method and a slight correction to earlier-published results. We also present the results of more recent measurements on additional surface preparations. Finally, we describe a recent significant improvement to the technique and discuss new information which may lead to further improvements in the future.

2. Measurement technique

The emissivity measurement approach has been described in detail in the references above, so here we simply provide an overview. The apparatus is shown schematically in figure 1. It used a simple geometry, consisting of a 20-cm-diameter 1100 aluminum "sample" disk suspended inside a "can" made of the same material. The sample's top and bottom surfaces were coated with BIRB, and its outer edge was coated with aluminized Kapton tape. A heater and two thermometers were imbedded inside it, with their leads running out through three equally-spaced holes in its outer edge. The can consisted of top and bottom "cold plates" separated by a spacer ring bolted between the plates' outer edges. In the original work the inner surfaces of the cold plates were coated with the same thickness of BIRB. Thermometers were centered on the outside cold plate surfaces, and a wire heater was wrapped around and epoxied to the outer edge of the spacer ring. The ring's vertical inner surface was polished to a mirror finish. The can was supported on a G10 ring above the cold sink plate of a 7 K cryostat cooled by a Gifford-McMahon cryocooler. The sample was suspended from its thermometer and heater leads, which ran up through slots in the top cold plate to a suspension ring. This ring was mounted above the top spacer ring on a G10 support. During data acquisition it was always controlled at the sample temperature, which almost entirely eliminated heat leaking from the sample via its suspension wires.

A series of steady-state thermal balances was achieved at each average temperature with the sample

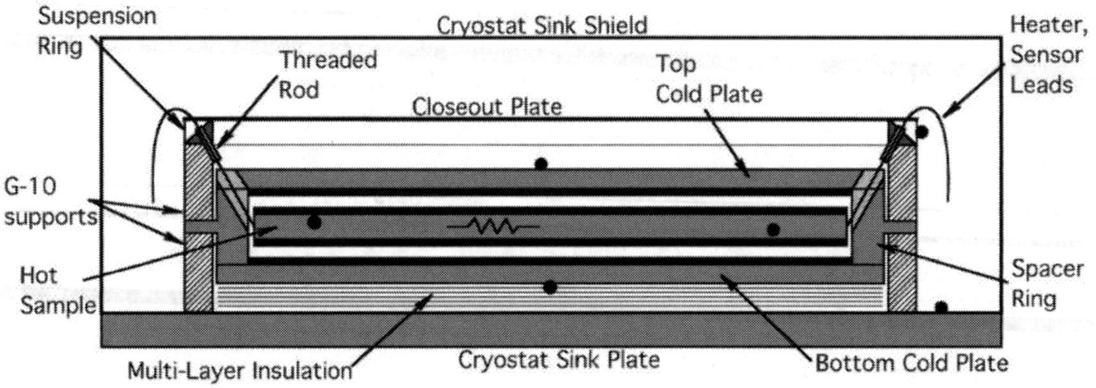

Figure 1. The original test configuration, with four identical BIRB coatings.

controlled at a temperature $T_1 = T_{avg} + \Delta T/2$ and the cold plates controlled at $T_2 = T_{avg} - \Delta T/2$, and $\Delta T = (T_1 + T_2)/2$. For each ΔT value we measured $\dot{Q}$, the sample's control heater power. The spacing between the facing BIRB surfaces was small relative to the diameter, so we approximated a simple textbook case of heat exchange between two infinite parallel surfaces. For small values of ΔT (less than 6% of T_{avg}), we showed that the emissivity could be approximated to better than 0.1% by

$$\varepsilon = \frac{2}{4\sigma A T_{avg}^3\left(\frac{d\Delta T}{d\dot{Q}}\right)+1}. \quad (2)$$

The uncertainty of this indicated emissivity was dominated by the uncertainty of the derivative in the denominator. Since the power was linear in ΔT, the derivative was determined by a linear regression, and we could drive down its uncertainty by balancing at more ΔT values. A small correction to the results of equation (2) was required, because the BIRB area on the sample was slightly smaller than that of the cold plates. We used a Thermal Desktop™ model of the apparatus to determine this correction, which was on the order of 1%. The correction depended on the spacing between the sample and cold plate treated surfaces, so we had to re-calculate it when testing BIRB of a different thickness. In the original campaign we used this method to measure the emissivity of two different BIRB samples from 20 K up to room temperature, as will be discussed later.

One drawback to testing samples in this way is that it takes significant time to produce the sample disk. This process requires tedious wiring of the heater circuit and thermometers to be imbedded inside the sample, and careful epoxying together of two BIRB-coated plates. Another limitation is that the sample plate must have a relatively thin black coating in order to minimize edge effects. To avoid these issues, we adopted a "hybrid" testing approach in which we used a calibrated BIRB sample disk suspended between cold plates with the surface treatment to be tested. In the case of new BIRB samples the configuration was nearly identical to the original setup. For taller samples, such as the painted aluminum honeycomb, custom rings were placed between the cold plates and spacer ring to maintain the desired spacing between the warm and cold sample surfaces. This hybrid approach, configured for painted honeycomb, is shown schematically in figure 2.

Using the hybrid method, we determined the new emissivity ε_1 on the cold plates from

$$\varepsilon_1 = \frac{1}{4\sigma A T_{av}^3\left(\frac{d\Delta T}{d\dot{Q}}\right)+1-\frac{1}{\varepsilon_2}}. \quad (3)$$

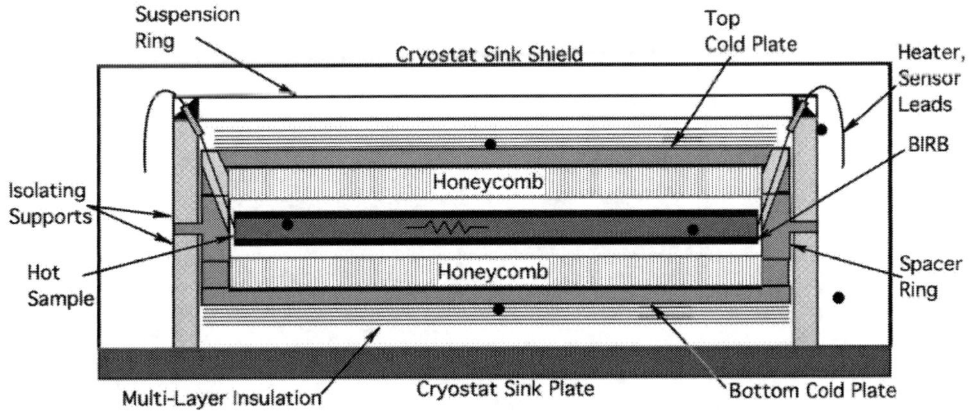

Figure 2. The hybrid test configuration used for characterizing painted aluminum honeycomb.

Here ε_2 is the previously-measured emissivity of the sample disk surfaces, which we curve-fit as a function of temperature. The uncertainty in ε_1 was greater than that of ε_2, since the former includes the uncertainty of the original BIRB measurement and of the hybrid measurement.

In our previous publications on emissivity measurements, the area A in equations (2) and (3) was determined at room temperature. This was an error, however, since the aluminum samples plates are slightly smaller at lower temperatures. We now scale the area values in these equations based on published values of aluminum's thermal contraction [6]. Since we have the same heat exchange for a smaller surface area, the indicated emissivity values are slightly higher when this correction is applied.

3. Results

3.1. BIRB data

In figure 3 we graph the results to-date for BIRB samples. The data from earlier publications are included with their proper corrections for thermal contraction. The 2 mm and 1.5 mm thick BIRB were measured using 4 identical coated surfaces as shown in figure 1. The 2 mm thick BIRB's sample disk was later used in all of the hybrid configuration measurements, and figure 3 includes the curve fit that was used in the hybrid data analysis using equation (3). Clearly there is significant uncertainty in the fit below about 40 K, and this shows up in the hybrid data. For example, the original 0.8 mm thick BIRB data actually rises at lower temperatures below 40 K. We believe this is not accurate, but rather a result of this curve fit. If time allows, we hope to eventually re-calibrate the 2 mm thick BIRB at the lowest temperatures and re-analyze the hybrid data with smaller error bars. The 0.8 mm thick BIRB, which matches the density to be used on the flight radiators, was re-tested after a heat treatment up to 410 K to simulate the affect of launch and ascent. This heating seems to have had a noticeable but barely significant effect on the emissivity. Two additional 0.8 mm thick BIRB samples were tested to determine the repeatability of the application process, and they are shown as upright and inverted triangles in figure 3. A final sample of this thickness, shown as × symbols, was generated when a test target was coated. Except for the 2 mm thick sample, which was the first coating made in the recent JWST campaign, all results match to within about 1 percent. It is possible that the coating process was improved somehow after the 2 mm samples were tested.

3.2. Aluminum honeycomb data

In figure 4 we graph the results to-date for samples of coated aluminum honeycomb. The data from our earlier publications are included with their proper corrections for thermal contraction, and they use the graph's left-hand Y axis. These samples were painted with Chemglaze Z307 paint, and their cell

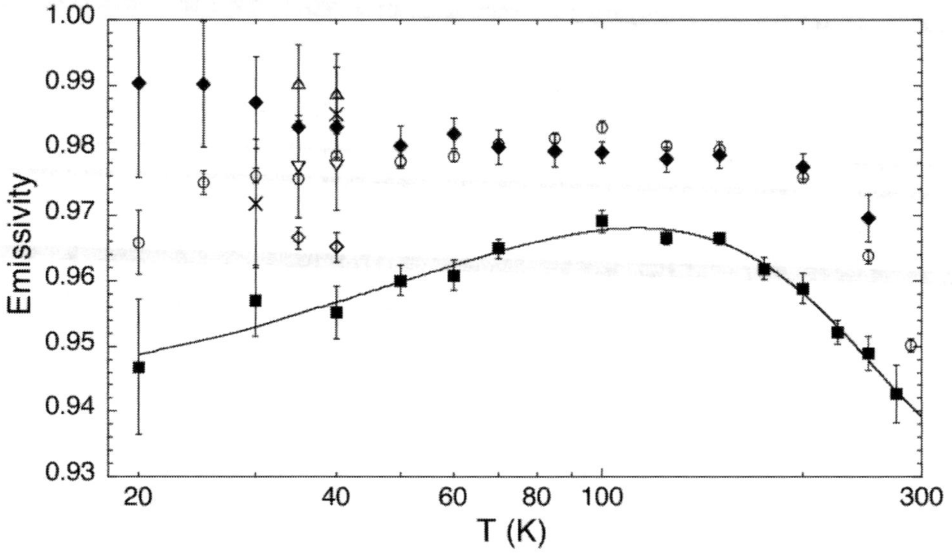

Figure 3. The data for BIRB, as described in the text. The symbols are: squares - 2 mm thick, circles – 1.5 mm thick, filled diamonds – original 0.8 mm thick, open diamonds – original 0.8 mm thick after heat treatment to 410 K, triangles – later 0.8 mm thick, x – test target sample. The solid line is the curve fit to the original 2 mm thick BIRB, which was used in all of the "hybrid" configuration measurements.

diameters were all 3.175 mm. As described elsewhere [5], they were made with progressively shorter cell heights and thinner paint and cell walls in order to reduce the radiator mass. The cell heights ranged from 12.7 to 9.53 mm, the wall thicknesses were between 38.1 and 17.8 μm, and the paint was between 16.3 and 8.4 μm thick. Sample 3, using the thinnest of all these characteristics, was selected as the flight configuration. When the actual flight honeycomb was painted, a new pair of cold plates was made to check for workmanship and repeatability of the process. This flight sample was tested at 35 and 40 K, and the results are included in figure 4. There is no significant variation among any of these honeycomb samples.

More recently, a pair of cold plate samples was prepared to test a cheaper method of blackening aluminum honeycomb for use in cryogenic ground test facilities. The honeycomb was exposed to a liquid alkaline treatment known as Insta-Blak A-380 from Electrochemical Products, Inc. The results for these plates between 35 and 100 K are plotted as open circles in figure 4, and they use the right-hand Y axis. Not only is the emissivity significantly lower than the Z307-painted samples, but it has a much stronger temperature dependence. This is most likely due to the cell walls having a combination of lower local emissivity and specularity for the wavelengths relevant to our temperature range. These results led to the Insta-Blak process being rejected for use in this application.

3.3. Kapton and Z307 paint data

Figure 5 shows data from three additional surface treatments characterized using the hybrid test configuration. Two of these samples were versions of black Kapton considered for use on the JWST observatory. The first was a 69 μm thick sample of Kapton 275XC, bonded to the cold plates with a thin layer of ScotchWeld 2216 epoxy. The second was Kapton 100XC, with a thickness of 25 μm, and bonded to the plates with a high-temperature-cure film adhesive. Of these two samples, the former was known to have a somewhat higher electrical conductivity, and its transmissivity was measured by the GSFC Optics group. They found that between 30 and 300 K it transmitted less than

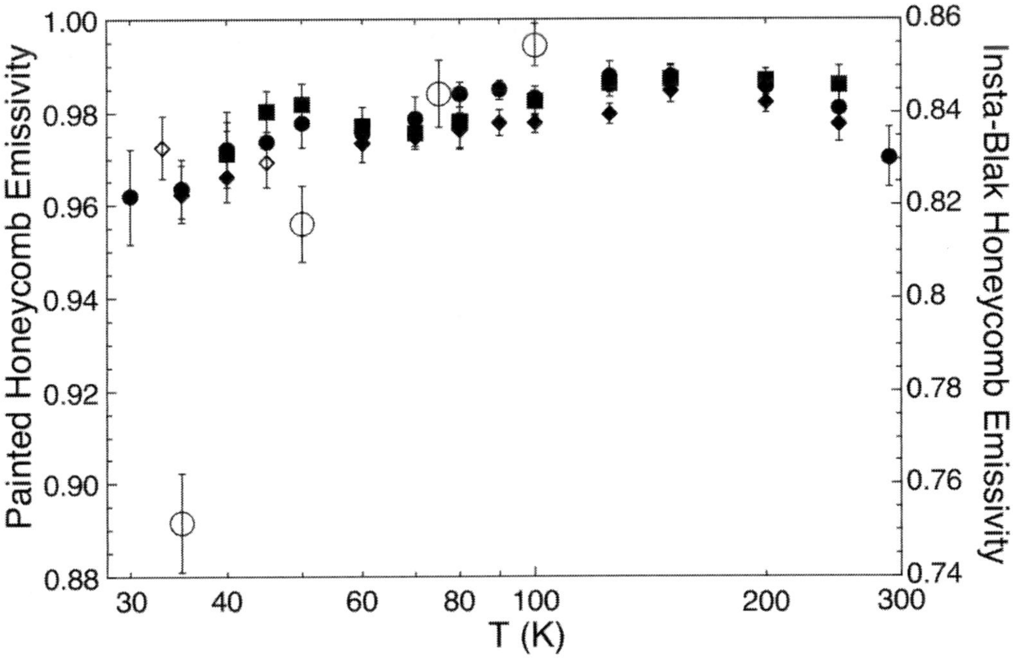

Figure 4. Data for aluminum honeycomb samples, as described in the text. The symbols are: solid circles – sample #1, squares – sample #2, solid diamonds – sample #3, open diamonds – flight sample, open circles – Insta-Blak-treated sample.

1% of wavelengths shorter than 100 μm. Obviously, any radiation transmitted through the Kapton would be absorbed or reflected by the adhesive and aluminum beneath, and our emissivity measurement would not accurately characterize the Kapton. The Kapton 100XC sample had a higher indicated emissivity above 30 K, but its transmissivity was not measured. One significant result if this test was that it demonstrated our hybrid configurations' ability to measure emissivity values below 0.7, which had not been done previously due to our focus on very black surface treatments.

The third data set in figure 5 is for Aeroglaze Z307 paint with measured thickness varying between 61 and 97 μm. These cold plates were painted as witness samples during painting of the walls of the large cryogenic shroud prepared for future JWST hardware testing at the Johnson Space Center. We compared our results with those of 2008 GSFC measurements using the transient cooldown method [7]. These earlier measurements characterized Z307 paint thicknesses ranging from 36 to 117 mm and found no systematic thickness dependence. Our data match these results to within about 1 percent above 50 K, but below this temperature the transient data drop more steeply than ours. At their lowest measurement temperature of 30 K, the transient data ranged from 0.64 to 0.7. We don't understand the difference in results below 50 K, but the transient data had significantly larger reported uncertainty than ours in this temperature range.

4. Process impovements

Our emissivity measurements are significantly more challenging at low temperatures, as the sample heater power for a given ΔT is roughly proportional to T^3. At 20 K the heater power ranges from about 10 to 120 μW. We have seen indications of a source of 10's of μW into the hot sample plate in addition to the applied heater power, and we long suspected that it might be due to vibrations created by the cryocooler. Since we use the slope of power vs. ΔT in our data analysis, we would not be affected by a steady "background" heat of this type. However, our measurements are vulnerable to

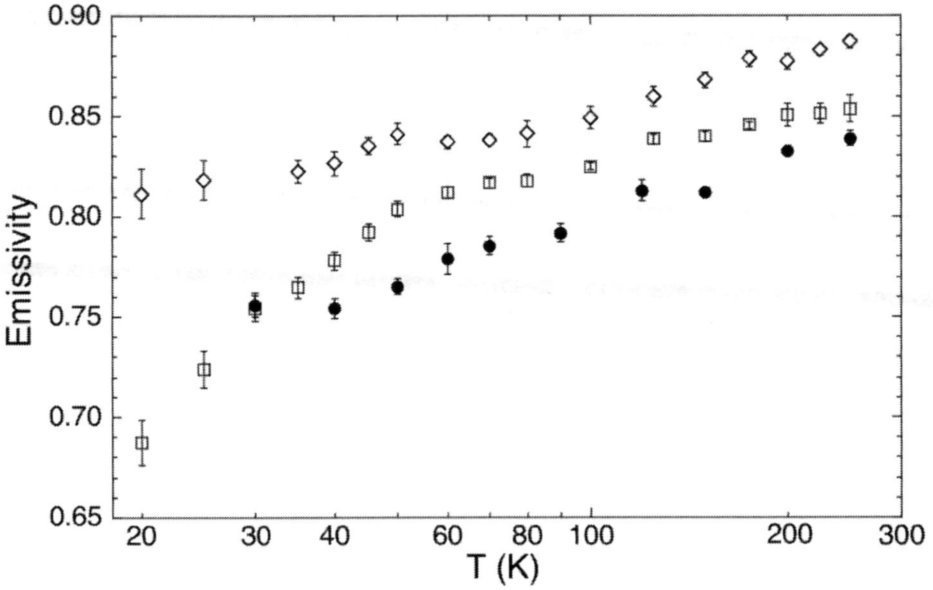

Figure 5. Data for Kapton and Z307 paint samples. The symbols are: filled circles – Kapton 275XC, squares – Kapton 100XC, diamonds – Z307 paint.

variations in this power during the time when we are doing multiple balances at a given average temperature. Often, particularly at 20 K, the power vs. ΔT would be linear for a few balances followed by a shift in the background power. Without understanding the cause of these shifts, we chose to re-run the entire set of balances until we got enough linear points to give a precise slope. Because of the long settling time at 20 K, this approach proved to be very time consuming.

Recently we have taken steps to address this issue. First, we successfully changed our temperature control technique to reduce the time per balance. We had always used temperature controllers to provide independent proportional-integral control of the cold plate can, the suspension ring, and the hot sample. At 20 K the power needed to hold the hot sample at a steady temperature is only 10's of μW, but the sample's heat capacity is relatively large. This results in a system time constant of more than 10,000 seconds, the maximum integral constant allowed by the controllers. To avoid overshooting when approaching a new set-point, we needed a low proportional gain and a very long integral constant, and the settling times were nearly half a day. We developed a faster technique involving switching the proportional gain values, turning off and on the integrator and seeding it during the approach to balance, but it still took several hours per balance.

We now use proportional-only control of the hot sample. With the appropriate gain, its temperature quickly reaches a steady value below the set-point. Then, instead of turning on the integrator, we shift the setpoint upward to bring the sample to within 1 mK of the original setpoint. The size of this set-point shift is proportional to the ΔT value, so we use it to predict the necessary setpoint for the following balance. This has reduced our settling time at 20 K to about one hour, so we can now achieve more balances before a shift in the background heating occurs.

Finally, we attempted to verify that the background heat source was the cryocooler vibrations. The cooler's cold head gives the cryostat's cold plate a horizontal mechanical impulse about twice per second. We installed an accelerometer on the cryostat's vacuum shell and read the signal with a spectrum analyzer. We set the controllers to hold BIRB samples at a steady ΔT value centered at 20 K, and we monitored the vertical and horizontal vibration spectra and the hot sample heater power for 44 hours. The spectra were both "forests" of peaks separated by two Hz, and these peaks' amplitudes

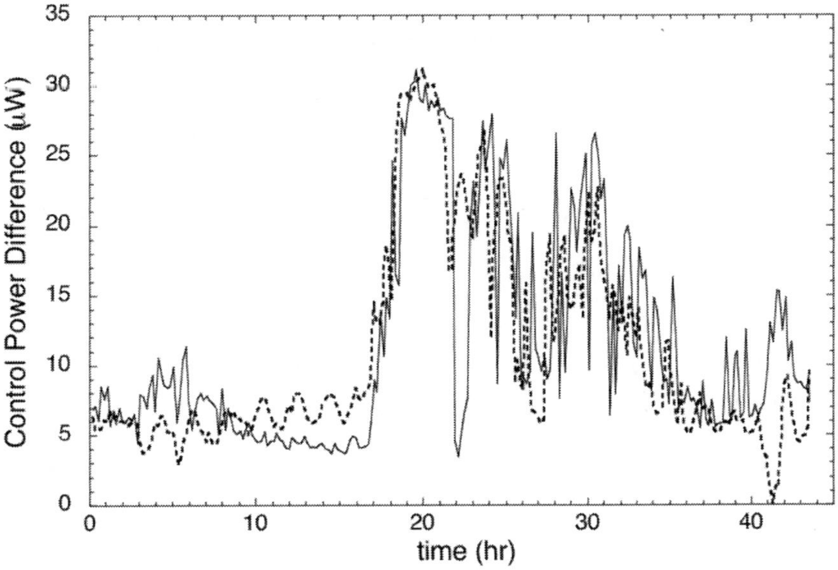

Figure 6. Diagnostic data taken with BIRB samples held at a fixed ΔT, centered at 20 K, for 44 hours. The thick dashed line is the difference between the hot sample heater power and its maximum value during this time. The solid line is the amplitude of a 46 Hz horizontal vibration mode scaled so that its maximum variation matches that of the power data.

changed over time. We were able to identify a horizontal mode at 46 Hz with a strong inverse correlation to the heater power. Figure 6 shows the change in heater power from its maximum value and the 46 hertz horizontal mode amplitude scaled to match the power variations, both plotted against time. The correlation, though not perfect, is convincing evidence that cryocooler vibrations cause the background heating. We plan eventually to modify the cryostat to reduce the vibration amplitude reaching the emissivity apparatus. In the mean time, we may be able to monitor the 46 Hz peak during emissivity measurements and use its variations to select ΔT values for which the background heat was nearly constant.

Acknowledgements
This work was supported by NASA's James Webb Space Telescope program.

References
[1] Kauder L 2005 *Spacecraft Thermal Control Coatings References NASA/TP-2005-212792*
[2] Or C 2011 Edge Space Systems, NASA/GSFC, private communication
[3] Tuttle J *et al* 2012 *Adv. Cryo. Eng.* **57B**, ed J Weisend et al. (NY: AIP) pp 1505-1512
[4] Tuttle J *et al* 2014 *Cryogenics*, **64** pp 240-243
[5] Tuttle J *et al* 2014 *Adv. Cryo. Eng.* **59A** ed J Weisend et al. (NY: AIP) pp 590-596
[6] Touloukian Y 1974 ed *Thermophysical Properties of Matter* **12** (NY: Plenum) p 2
[7] Kauder, L, 2008 internal NASA memo

Cryogenic performances of T700 and T800 carbon fibre–epoxy laminates

Wang Wei[1,2], Huang Rongjin[2], Huang Chuanjun[2], Yuqiang Zhao[2], Shaopeng Li[2], Li Laifeng [1,2]

[1] School of Materials Science and Engineering, Tianjin University, Tianjin 300072, China;
[2] Key Laboratory of Cryogenics, Technical Institute of Physics and Chemistry, Chinese Academy of Sciences, Beijing 100190, China

E-mail: huangrongjin@mail.ipc.ac.cn; laifengli@mail.ipc.ac.cn

Abstract. The temperature dependence of thermal expansion, thermal conductivity and mechanical properties of T700 carbon fibre (T700 CFs) /epoxy composite and T800 CF/epoxy composite were investigated. The mechanical and thermal properties of the unidirectional composite material laminates (0°/90°) at low temperature were studied. The results show that comparing the composite material T700 CFs with T800 CFs, the thermal expansion and thermal conductivity performances of T800 CFs (0°/90°) are all smaller than those of T700 CFs. Typically, the coefficient of thermal expansion (CTE) of T800 CFs in 0° is very low in the temperature range of 120-300K, which reaches as low as -0.4×10^{-6} K^{-1}. The value of thermal conductivity of this material at 0° is about 3.2 $W.(m.K)^{-1}$ at room temperature. Tensile and compression tests indicate that the tensile strength of T800 CFs in 0° direction at 77K reaches 2310 MPa, while the compressive strength is about 852 MPa. This composite material may possibly be exploited to design the critical components for practical applications such as hydrogen storage tanks.

1. Introduction

Composite materials have been extensively applied to the areas of aerospace, aircraft, sports and military industries because of their advantages over traditional materials, for example, a possibility to obtain a required combination of properties, which cannot be achieved for unfilled polymers [1]. Advanced composite materials have progressed from a laboratory curiosity to a production reality. It is known several carbon fibres, such as T300, T700, T800, have several advantages over the traditional metal structures in a wide temperature range. Meanwhile, the high specific strength, the high specific stiffness, and the very strong design ability of the composite material play an essential role in the weight reducing of aerospace structures. Commercial aircraft applications have ranged from small flight-control surfaces to more and more primary structures [2-5].

It is noteworthy, however, that owing to the extreme inertness of surface caused by the alignment of graphitic crystallites, high performance carbon fibres such as T700/T800 always failed to reach the expected properties in their reinforced composites [6]. Therefore, the ability to control the properties of composites over a wide range, by adjusting the reinforcing elements and the degree of filling, is the key for increased interest in composite materials. Modification of the interfacial behaviors between carbon fibre and matrix is a feasible way to improve the mechanical properties of the high performance carbon fibre reinforced polymers [7-8].

In addition, optimization of the composite materials is very important when used in some severely loaded structures such as hydrogen storage tanks[9]. The behavior and performance of composite materials cannot be explained in terms of the specific properties of the constituents alone. The interface that exists between the fibres and the matrix is also a major component of the composite [10]. A good adhesion between the fibre and the matrix is a precondition for stress transfer and influences mechanical properties such as the interlaminar shear strength, fatigue resistance and impact toughness [11]. Therefore, many scientific efforts have been devoted to modify carbon fibres by a variety of methods such as gas-phase, liquid-phase and continuous anodic oxidation [12–15], and then to apply a very thin (usually on nanometer scale) coating of a prepolymer or resin to the modified carbon fibre surface for the purpose of improving the interfacial properties between the carbon fibres and matrix [16], and meanwhile to prevent the fibres from damage through the process of manufacture. Many investigations on the nature of the fibre/sizing and sizing/matrix interphases have proved that the nano-scale interface plays a dominant role in the interfacial adhesion between the carbon fibre and the resin matrix [17]. The functional groups of the sizing (interface) can react and/or interact with the matrix, which can give rise to strong adhesion between the fibre and the matrix.

Also, the thermal expansion due to the anharmonic behavior of atomic vibrations is a usual property exhibited by most materials but a most critical design parameter in advanced industrial applications. The mismatch of the thermal expansion of different components undergoing temperature change will cause thermal stress and then deformations that will lead to reliability problems. For hydrogen storage tanks, for example, some structures undergo thermal shock when temperature varies between high and cryogenic temperatures, resulting in residual stress and deformation [2, 18]. So the thermal expansion property of composite materials should be taken into consideration carefully.

Many investigations pay the attention on the properties of composites at room temperature. Seldom work, however, has been reported that the cryogenic performances of carbon fibre–epoxy composites. The purpose of this study is, thus, to fabricate carbon fibre composite materials and to investigate the mechanical and thermal properties of the unidirectional T700 CF/epoxy composites and T800 CF/epoxy composites (0°/90°) at low temperature.

2. Experiment

The material used was T700, T800 carbon fibre and the diglycidyl ester of aliphatic cyclo (DGEAC) type epoxy resin. This epoxy is one of the most widely used matrices for carbon fibre reinforced composite materials by virtue of its good impregnation and adhesion to carbon fibre. The DGEAC type epoxy resin was provided by Tianjin Jindong chemical factory (epoxy value, 0.85) and both T700 and T800 CFs were obtained from Toray Co., Japan. The resin casts and unidirectional composite laminates with 60% volume content of carbon fibres were cured at 80 °C/2 h + 120°C/2 h + 150 °C/4 h[19].

The linear thermal expansion data ($\Delta L/L(300K)$) were characterized using a dilatometer (L75 PT Vertical, LINSEIS, Germany) in the temperature range of 120 - 300 K in helium gas with a heating rate of 2 K/min. The thermal conductivity was measured by means of a heat and sink steady-state method, which is to establish a stationary temperature gradient, and supplying heat at one end of the sample while the other end is maintained at a constant temperature by measuring the temperature difference between a given distances. The temperature difference of the two ends of the sample was stabilized with an accuracy of better than 0.01 K. The thermal conductivity can be calculated by means of

$$K = \dot{Q}\frac{L}{S \cdot \Delta T} \quad (1)$$

Where K is the thermal conductivity, Q, the power supplied, L, the distance of sample, S, the section of the sample and ΔT, the measured temperature difference. Tensile and compression tests were conducted with a MTS SANS CMT 5000 model test machine (load capacity, 100 kN). The tensile and compression strength under monotonic loading were measured by stretching and compressing the specimen at a constant rate of 0.5 mm/min until failure.

It should be noted that the tensile strength (σ) can be determined as

$$\sigma = \frac{P}{tw} \tag{2}$$

Where P is the ultimate burst force recorded in Newton (N) and t and w are the thickness and width of the sample in millimeter (mm), respectively. More than six comparative specimens with dimensions of 20 mm×6 mm×2 mm were tested for each experiment to get an average value [20]. All the mechanical properties of the composites were determined according to ASTM 3039 test standard specifications [21].

3. Results and Discussion

3.1. Thermal expansion properties

The property of the composites may be influenced by the thermal expansion mismatch between fibre and matrix during a thermal excursion. So we studied the thermal expansion property of the carbon fibre composite material in a wide temperature range. Figure 1 shows temperature dependence of $\Delta L/L$ data of the T700/DGEAC and T800/ DGEAC. It is observed that all samples show a linear expansion as function of temperature at room temperature and the slope decreases gradually with decreasing temperature, which is similar to that observed in most materials. The $\Delta L/L$ curves are near-linear, and we calculated the CTE as shown in table 1. The results show the CTE of T800 CFs(0°/90°) are all smaller than those of T700 CFs. Typically, the CTE of T800 CFs (0°) is very low in the temperature range of 120-300K, which reaches as low as -0.4×10^{-6} K^{-1}. The reason for the wide differences of CTE between the composites in 0° and 90° direction is that in 0°direction, the CTE is dominated by the carbon fibres, while in 90° direction, the CTE is dominated by the resin.

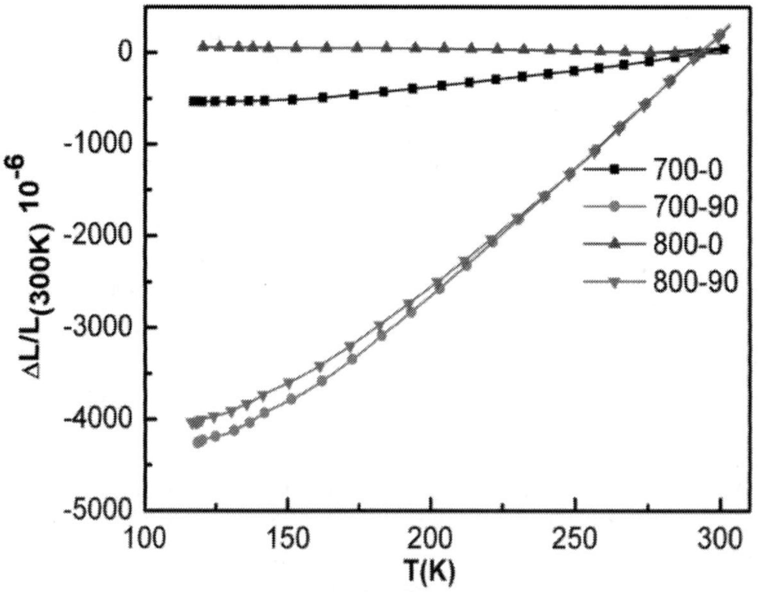

Figure 1. The temperature dependence of the linear thermal expansions $\Delta L/L(300K)$ for the T700/ DGEAC and T800/ DGEAC composites.

Table 3. The coefficient of thermal expansion (CTE) for all samples

Samples	Direction(°)	CTE(10^{-6} ×K^{-1})
T700/ DGEAC	0	3.1
T700/ DGEAC	90	24.7
T800/ DGEAC	0	-0.4
T800/ DGEAC	90	23.4

3.2. Thermal conductivity

Thermal conductivity is basically the property of a material that indicates its ability to conduct heat. Low thermal conductivity is desirable for some applications such as hydrogen storage tanks in order to avoid heat loss. The temperature dependence of the T700 and T800 composites thermal conductivity are shown in Figure 2. The thermal conductivity increases monotonically with increasing temperature over the test temperature range. From the thermal conductivity curves, we can see that there is no very obvious difference between the T700-90° and T800-90° while the thermal conductivities of T700-0° are all larger than that of T800-0° more or less from 25 to 300K. The results show that comparing the composite materials in 0°and in 90°, the 0° are all larger than 90° just because of the higher thermal conductivity of the carbon fibres compared with the resins. And the gap between 0° and 90° become smaller from 300K to 20K, which is beneficial to the practical applications such as hydrogen storage tanks[22].

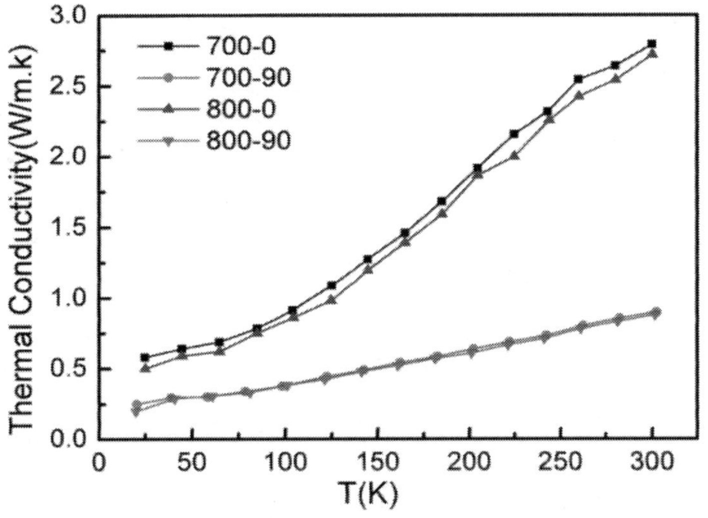

Figure 2. The temperature dependence of the thermal conductivity for the T700/ DGEAC and T800/ DGEAC composites.

3.3. Mechanical properties

In order to evaluate the mechanical properties of carbon fibre composite materials, tensile and compression tests were carried out at 77K and room temperature. The loading rate of the testing machine was 0.5mm / min. Load-displacement curves data were obtained from the tests and were used to determine the tensile and compressive strength. The load - displacement curves of the T700/ DGEAC and T800/ DGEAC laminates by the tensile test at 77K are shown as Fig. (3). Take the T800/ DGEAC in 0° for example, it can be recognized in the curve that the whole test curve of the laminate is changed as a line. As observed by the test, when the experimental load reached 25000N, the specimen started to make a cracking sound. As the load continued, the whole laminate was seriously deformed, and then, completely broken. The load is decreased rapidly after reaching a maximum of 33090 N, and by then the laminate is completely damaged[23-24].

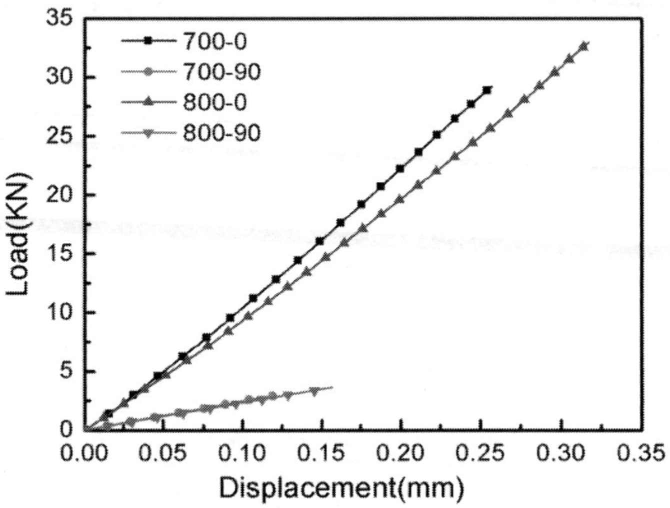

Figure 3. Tensile load - displacement curves of the T700/ DGEAC and T800/ DGEAC laminate.

From the stress-displacement curves we obtained the tensile and compressive strengths for the T700/ DGEAC and the T800/ DGEAC laminates at 77K and room temperature. The results of tensile and compression tests of the composite materials are shown in Fig. 4 and Fig. 5 respectively. We can see both the tensile strength and the compressive strength of the composites all increases monotonically with decreasing temperature. Typically, the value of tensile strength of the T800/ DGEAC in 0° direction at 77K is about 2310 MPa, while at the same condition, the tensile strength of the T700/ DGEAC is about 2293 MPa. The value of compressive strength of the T800/ DGEAC in 0° direction at 77K is about 852 MPa, while at the same condition, the compressive strength of the T700/ DGEAC is about 813 MPa. As is known from the tensile and compressive tests of the T700/ DGEAC and T800/ DGEAC laminates, under the condition of same physical dimension, the tensile and compressive properties of the T800/ DGEAC are slightly higher than those of the T700/ DGEAC. Considering the thermal properties of the materials, the whole properties of the former are apparently better than those of the latter. Also, just as the thermal properties, the differences of strengths between the composites in 0° and 90° direction are still very apparent.

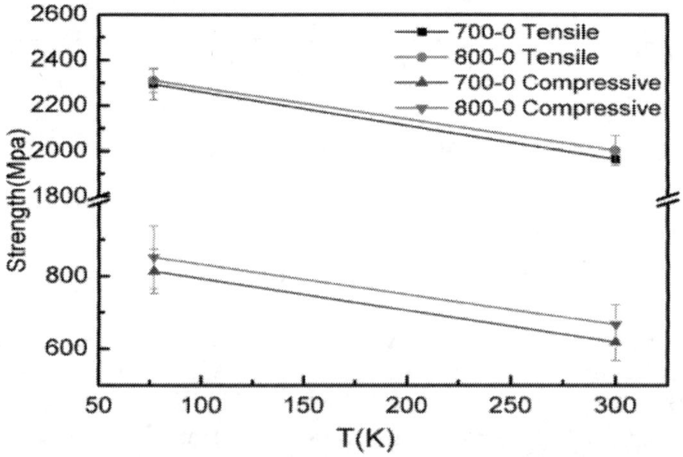

Figure 4. The tensile and compressive strengths for the T700/ DGEAC and the T800/ DGEAC laminates in the 0°direction.

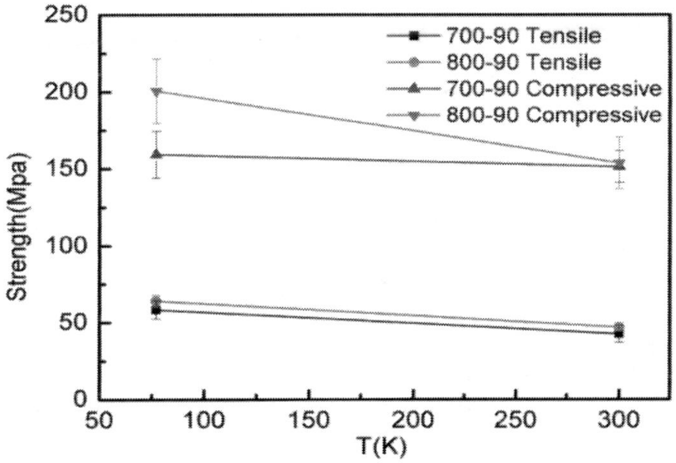

Figure 5. The tensile and compressive strengths for the T700/ DGEAC and the T800/ DGEAC laminates in the 90° direction.

4. Conclusions

1) The T700 and T800 carbon fibre/epoxy systems were fabricated and the cryogenic thermal expansion and tensile/ compressive strengths of T700 and T800 carbon fibre–epoxy laminates were compared.
2) The CTE of the T700/ DGEAC and T800/ DGEAC carbon fibre/epoxy systems in 0° is smaller than that in 90° and in addition. The CTE of the T800/ DGEAC is very low in the temperature range of 120-300 K, which is -0.4×10^{-6} K^{-1}.
3) Under the same condition, the thermal and mechanical properties of the T800/ DGEAC are apparently better than those of the T700/ DGEAC. Typically, the value of tensile strength of the T800/ DGEAC at 77K is about 2310 MPa, while the thermal conductivity is 0.5 W/m.k.

Acknowledgements
This work was supported by the National Natural Science Foundation of China (Grant Nos. 51401224, 51232004, 51377156, 51322605, 51271192), the National Magnetic Confinement Fusion Science Program (Grant No. 2011GB112003), and the fund of the State Key Laboratory of Technologies in Space Cryogenic Propellants, SKLTSCP1204.

References

[1] Neradka VF, Chang Y, Grady JE and Trowbridge DE 1992 *Johns Hopkins APL Tech. Digest.* **113** 415
[7] Szytula A and Leciejewicz J 1989 *Handbook on the Physics and Chemistry of Rare Earths* vol 12, ed K A Gschneidner Jr and L Erwin (Amsterdam: Elsevier) p 133
[6] Cheng G X 2001 *Raman and Brillouin Scattering—Principles and Applications* (Beijing: Scientific)
[2] Schwartz M.M 1997 *Composite Materials: Properties, Nondestructive Testing and Repair*, vol 1, (New Jersey: Prentice- Hall Inc.)
[3] Baker A.A, Callus P.J, Georgiadis S, Falzon P.J, Dutton S.E and Leong K.H 2002 *Composites: Part A* **33** 687
[4] Varna J, Joffe R, Berglund LA 1996 Compos *Sci Technol.* **56** 1105
[5] King TR, Blacketter DM, Walrath DE and Adams DF 1992 *Compos. Mater.* **26** 558

[6] Shalin R.E 1988 *Military handbook 17B. Polymer matrix composites*, vol 1, (London: Chapman and Hall)
[7] Kelly A and Tyson WR 1965 *J. Mech. Phys. Solids.* **13** 329
[8] Fitzer E and Weiss R. 1987 **Carbon 25** 455
[9] Severini F, Formaro L, Pegoraro M and Posca L 2002 *Carbon* **40** 735
[10] Bogoeva-Gaceva G, Ma¨der E, Haüssler L and Dekanski A 1997 *Composites A* **28** 445.
[11] Hughes JDH 1991 *ComposSci Technol* **41**13
[12] Gao SL, Ma¨der E and Zhandarov SF 2004 *Carbon* **42** 515
[13] Ping C, Xigao J, Hui C, Julong G, Bing H and Xingsong Z 2002 *Acta Mater Comp Sinica*;**19** 24
[14] Drzal LT, Micheal RJ, Koenig MF and Lloyd PF 1983 J Adhes **16** 133
[15] Banakar P, Shivananda H.K and Niranjan H.B 2012 *Int. J. Pure Appl. Sci. Technol* **9** 61
[16] Deogonda P. and Chalwa N 2013 *IJSER* **1** 2347
[17] Verma C.S, Chariar V.M and Purohit R 2012 *Inter. J. Eng. Res. Appli.* **2** 1253
[18] Kaleemulla KM and Siddeswarappa B 2009 J. Rein. *Plas. and Com.* **29**12
[19] Chen W, Yu Y, Li P, Wang C, Zhou T and Yang X 2007 *Composites Science and Technology* 2261
[20] Liu L, Huang Y, Zhang Z, Jiang B and Nie J 2001 *J. Appl. Polym. Sci.* **81** 2764
[21] *Annual Book of ASTM Standards 2003 ASTM D3039* **15** (PA: American Society for Testing and Materials)
[22] Krueger R, Cvitkovich M, O'Brien T and Minguet P 2000 *J Comp Mater* **34** 1263
[23] Hartwig G, hubner R and Knaak S 1998 *Crygenics* **38** 75
[24] McNally T, Boyd P, McClory C, Bien D, Moore I, Millar B, Davidson J and Carroll T 2008 *J. Appl. Polym.* Sci. **107** 2015

Qualification of electron-beam welded joints between copper and stainless steel for cryogenic application

C Lusch[1,2], M Börsch[3], C Heidt[1,2], N Magginetti[3], J Sas[1], K-P Weiss[1] and S Grohmann[1,2]

[1] Karlsruhe Institute of Technology, Institute for Technical Physics, Hermann-von-Helmholtz-Platz 1, 76344 Eggenstein-Leopoldshafen, Germany
[2] Karlsruhe Institute of Technology, Institute for Technical Thermodynamics and Refrigeration, Kaiserstrasse 12, 76131 Karlsruhe, Germany
[3] WEKA AG, Schürlistrasse 8, 8344 Bäretswil, Switzerland

E-mail: `klaus.weiss@kit.edu`, `steffen.grohmann@kit.edu`

Abstract. Joints between copper and stainless steel are commonly applied in cryogenic systems. A relatively new and increasingly important method to combine these materials is electron-beam (EB) welding. Typically, welds in cryogenic applications need to withstand a temperature range from 300 K down to 4 K, and pressures of several MPa. However, few data are available for classifying EB welds between OFHC copper and 316L stainless steel. A broad test program was conducted in order to qualify this kind of weld. The experiments started with the measurement of the hardness in the weld area. To verify the leak-tightness of the joints, integral helium leak tests at operating pressures of 16 MPa were carried out at room- and at liquid nitrogen temperature. The tests were followed by destructive tensile tests at room temperature, at liquid nitrogen and at liquid helium temperatures, yielding information on the yield strength and the ultimate tensile strength of the welds at these temperatures. Moreover, nondestructive tensile tests up to the yield strength, i.e. the range in which the weld can be stressed during operation, were performed. Also, the behavior of the weld upon temperature fluctuations between room- and liquid nitrogen temperature was tested. The results of the qualification indicate that EB welded joints between OFHC copper and 316L stainless steel are reliable and present an interesting alternative to other technologies such as vacuum brazing or friction welding.

Keywords: Electron-beam welding, copper, stainless steel, cryogenic application, 316L, mechanical test, hardness, liquid nitrogen, liquid helium

1. Introduction

The joints between oxygen-free high conductivity copper (OFHC) and 316L play an important role for cryogenic application. A possibility to join these materials is electron-beam (EB) welding. Upon the impact of the electron beam on the materials, the kinetic energy of the electrons is transformed into heat and causes the materials to melt. To prevent dispersion of the electron beam and to avoid oxidation of the metal, the welding is mostly performed in a vacuum. One of the many advantages of this method are the deep and narrow welds as a result of the high energy density. This method is often used in aerospace industries since EB welding is well suitable for titanium and titanium alloys, which are the most widely used materials in aerospace industries [1].

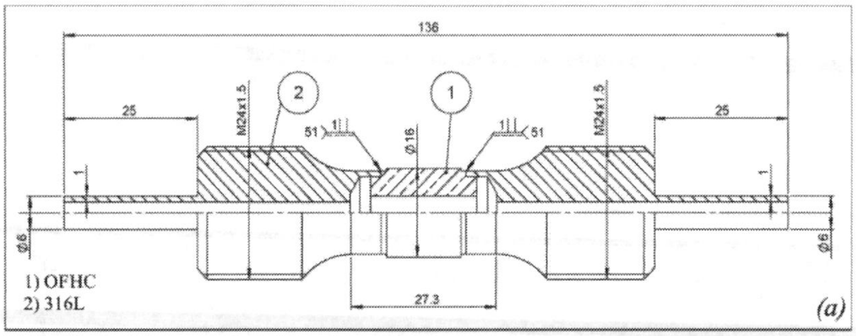

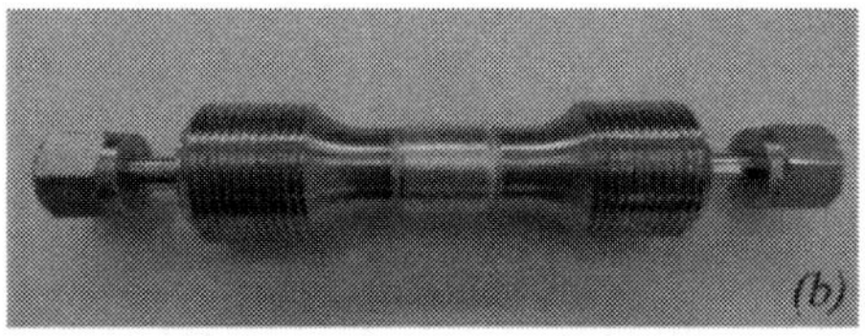

Figure 1. Used samples to qualify the weld. *(a)* shows a technical drawing of the sample, *(b)* shows a picture of the produced sample and *(c)* shows a modified sample, which was used for the hardness test. The OFHC block is numbered with 1 and the 316L block is numbered with 2 in the technical drawing.

EB welding between 316L and OFHC will be used for a new cryogenic mass flow meter for the determination of mass flow rates in gases and liquids. The method of this new mass flow meter is explained in [2]. A technology transfer project together with WEKA AG was started in order to develop this new mass flow meter for a working range of 300 K down to 4 K and with pressures up to 50 bar.

As few data are available for classifying EB welds between 316L and OFHC, specially designed samples were produced and tested between 300 K and 4 K, and with high pressures. The geometries were close to the original design of the mass flow meter. Figure 1 shows a technical drawing and pictures of the samples used for the tests.

In the first chapter, this paper discusses the test program. Hence, the results of the test program are presented and discussed. Finally, conclusions on the qualification of the EB weld for cryogenic application are made.

2. Test program and methods

The test program shown in figure 2 was created and conducted under consideration of the special conditions for cryogenic application. First, a hardness test was conducted to check the influence of the welding on the microstructure of the materials in the area of the weld. The Vickers HV0.2 hardness test[1] was conducted according to ISO 6507-1 [3]. Second, helium pressure and leak tests at room temperature (RT, 300 K) were conducted with the aim to prove the tightness and pressure resistance of the weld. Helium pressure and leak tests at RT and at cryogenic temperature were conducted according to EN 1779 [4]. The integral leak rate for the whole sample (shown in figure 1*(b)*) was measured and helium was used as test gas[2]. Third, destructive tensile tests at RT, at liquid nitrogen (LN_2) temperature (77 K), and at liquid helium (LHe) temperature (4 K) were performed to reveal information on the EB weld such as yield strength σ_y and ultimate tensile strength σ_u.

[1] The hardness test was performed with the NEXUS 4000 hardness tester from the INNOVATEST Europe BV.
[2] The leaking helium was detected by a leak indicator of the type Leybold UL 200 dry with scroll pump SC 5 D from the Oerlikon Leybold Vacuum GmbH.

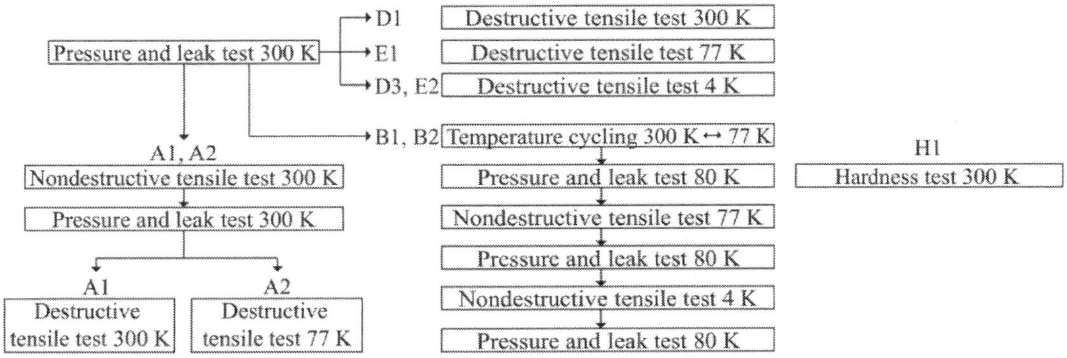

Figure 2. Test program per sample in the order of the performed tests. For a better overwiew, the samples were numbered (e.g. A1, A2, ...).

Since plastic deformation of the sample is avoided during operation, nondestructive tensile tests up to yield strength were conducted, also at RT, at 77 K, and at 4 K in order to see if the weld will be influenced. Both destructive and nondestructive tensile tests, were performed in the CryoMak laboratory at the Institute for Technical Physics (ITEP) whose facilities are explained in [5]. The tests were conducted according to ISO 6892-1 [6] at RT, ISO 15579 [7] at 77 K, and ISO 19819 [8] at 4 K. The deformation during the tests was measured by two extensometers and it was intended to have an initial gauge length of $L_0 = 22$ mm (see figure 3). Therefore, the measured deformation includes the deformation of the different materials in the various areas (316L, OFHC, weld). Since the EB weld is to be qualified for cryogenic application, it is important to know how the weld reacts to temperature fluctuations. Therefore, a temperature cycling test was conducted. Two samples (sample design is shown in figure 1*(b)*) were put into a box filled with LN_2, where they were cooled down to 77 K. Afterwards, the samples were reheated to RT by a dryer. This cycle was repeated ten times. Subsequent to the nondestructive tensile tests and the temperature cycling test, helium pressure and leak tests were performed to prove that the weld was not damaged by the previous tests. These pressure and leak tests took place at cryogenic temperature.

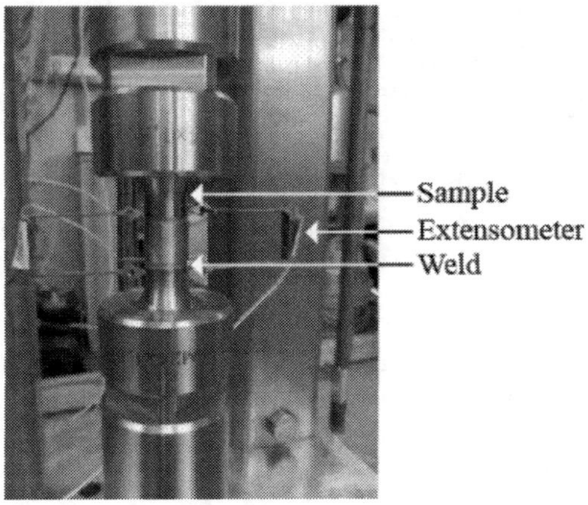

Figure 3. For the tensile tests, two extensometers are mounted on the sample with an initial gauge length of $L_0 = 22$ mm.

IOP Conf. Series: Materials Science and Engineering **102** (2015) 012017 doi:10.1088/1757-899X/102/1/012017

3. Results and discussion

3.1. Hardness test

Figure 4 shows the results of the measured hardness for the indentations of line 4. As can be seen, the hardness was not constant in the weld region. This is the result of an inconsistent microstructure in this area, which was caused by the welding and is a usual behavior. The transparent gray zone in figure 4 was found to have a width between 100 and 150 µm in 316L and between 75 and 100 µm in OFHC. In this area, the welding also changed the microstructure of the materials. This zone is wider in 316L because of its lower thermal conductivity compared to OFHC. Next to this gray zone, the hardness was found to be constant on both sides and corresponds to literature values of the used materials. Therefore, it could be shown the welding had no influence in this part. The same results were found for the other lines.

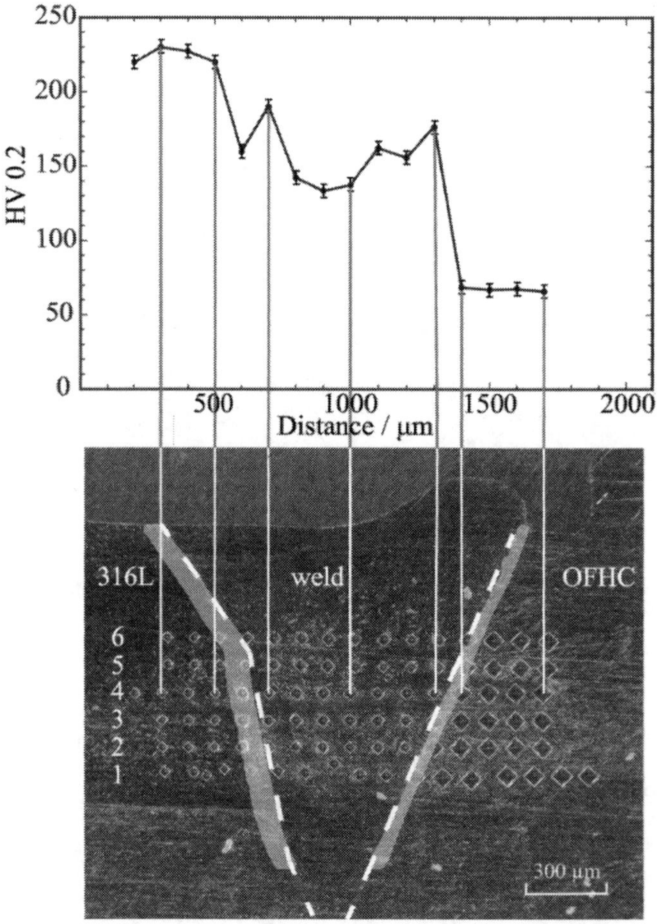

Figure 4. The measured hardness of each indentation of line 4. The dashed white line shows the border between the weld and the pure materials on both sides and the transparent gray zone indicates the area, where the welding had an influence on the microstructure of the pure materials.

3.2. Helium pressure and leak tests

The samples were always pressurized between 130 bar and 160 bar, which was the maximum available pressure depending on the available pressure of the helium bottle and the weld withstood this high pressure.

Several pressure and leak tests at RT were conducted and the measured leak rate was found to be in the range of $\leq 3.2 \cdot 10^{-9}\text{mbar l s}^{-1}$ and $\leq 1 \cdot 10^{-10}\text{mbar l s}^{-1}$ for all tests. Moreover, several pressure and leak tests at about 80 K were performed and the leak rate was measured in the range of $\leq 7 \cdot 10^{-10}\text{mbar l s}^{-1}$ and $\leq 1 \cdot 10^{-10}\text{mbar l s}^{-1}$ for all tests. These results were in the minimum detectable range of the used test setup. Also, external influences had an impact on the results and consequently the background determines the result and therefore the results were like expected. The results of the pressure and leak tests demonstrated that EB welding can join materials without leakages and with sufficient tightness even at high pressures and cryogenic temperatures.

3.3. Temperature cycling test

An external change was not apparent after conducting the temperature cycling test and also the subsequent pressure and leak test showed no difference to previous pressure and leak tests. Consequently, the EB weld seams to be resistant against temperature fluctuations in the temperature range between RT and 77 K.

3.4. Destructive tensile tests

Figure 5 shows the resulting stress-strain curves for the destructive tensile tests at different temperatures. Since the fracture caused by the destructive tensile tests was in the weld for all tests, the stress was measured with an assumed cross-sectional area A_{CS} in the part of the weld ($A_{CS,weld} = 44\,\text{mm}^2$). The breaking strain decreases with lower temperature since the materials are more sensitive to fail at lower temperature. Moreover, the 0.2 % proof stress $\sigma_{0.2}$ is higher with a lower temperature as well as the yield strength σ_y. σ_y is in the range of 40-50 MPa and consequently the weld should not be stressed higher during the real operation.

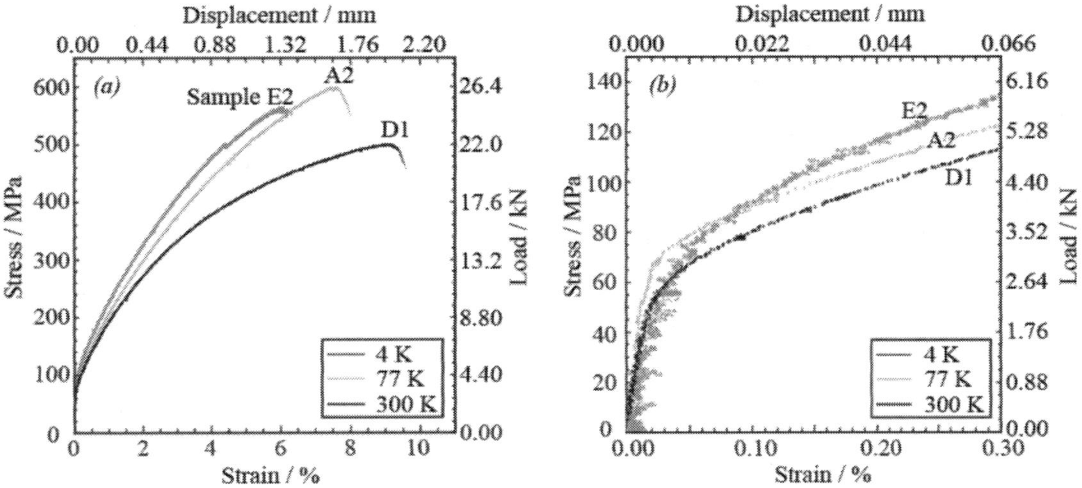

Figure 5. Stress-strain curves for the destructive tensile tests at RT, at 77 K, and at 4 K. *(a)* shows the curves in total and *(b)* shows a zoom into the elastic zone. The curve for the test at RT was measured with sample D1, for the test at 77 K with sample A2, and for the test at 4 K with sample E2.

IOP Conf. Series: Materials Science and Engineering **102** (2015) 012017 doi:10.1088/1757-899X/102/1/012017

$\sigma_{0.2}$ of the used OFHC is in the range of the measured $\sigma_{0.2}$ of the weld. Therefore, the OFHC is the weak point of the joint. Since the slope of the three curves in the elastic zone is in the same range, the modulus of elasticity E hardly changes with temperature. In total, the shape of the curves show the expected results.

The measured values for the mechanical parameters and their uncertainties, calculated according to the *Guide to the expression of uncertainty in measurement* [9], are shown in table 1.

Table 1. Results of the destructive tensile tests at RT, at 77 K, and at 4 K

Parameter	unit	RT	77 K	4 K
σ_u	MPa	500 ± 1.5	600 ± 1	565 ± 1.25
A	%	9.35	7.75	5.8
σ_y	MPa	40 ± 1.5	45 ± 1	50 ± 1.25
$\sigma_{0.2}$	MPa	105 ± 1	114 ± 1	126 ± 3
E	GPa	260 ± 35	270 ± 22	260 ± 144

3.5. Nondestructive tensile tests

The destructive tensile tests indicated a yield strength of about $\sigma_y = 45\,\mathrm{MPa}$, giving the operational limit of the weld. In order to see if stress below σ_y already influences the weld, nondestructive tensile tests at RT, at 77 K, and at 4 K up to σ_y have been conducted. The resulting stress-strain curves can be seen in figure 6.

Since it was tested in the lower force range of the machine, the maximum stress that occured on the weld was difficult to control. Consequently, the weld was stressed into the plastic region for the test at RT and 77 K and a permanent deformation occured. No plastic deformation occured for the test at 4 K (see figure 6*(e)* and *(f)*). The scattering of the measuring values is wider for the test at 4 K because of the thermal convection at lower temperature, but the resulting uncertainty is included in the results. As the curves show, the slope in the elastic zone varies for tests at same temperatures, especially at 77 K (see figure 6*(c)* and *(d)*). The reason is that the positioning across the EB weld varies slightly leading to different fractions of 316L and OFHC across the gauge length. Furthermore, slight differences in the shape of the weld - differences in A_{weld} - influence on the results and effect a different slope in the elastic zone.

Subsequent to the nondestructive tensile tests at different temperatures, helium pressure and leak tests were perfomed. The results of these tests showed that the nondestructive tensile tests did not damage the weld since the measured leak rate was found to be in the range of the previous pressure and leak tests (between $10^{-10}\mathrm{mbar\,l\,s^{-1}}$ and $10^{-9}\mathrm{mbar\,l\,s^{-1}}$). In consequence, the EB weld is resistant against stresses in the elastic region.

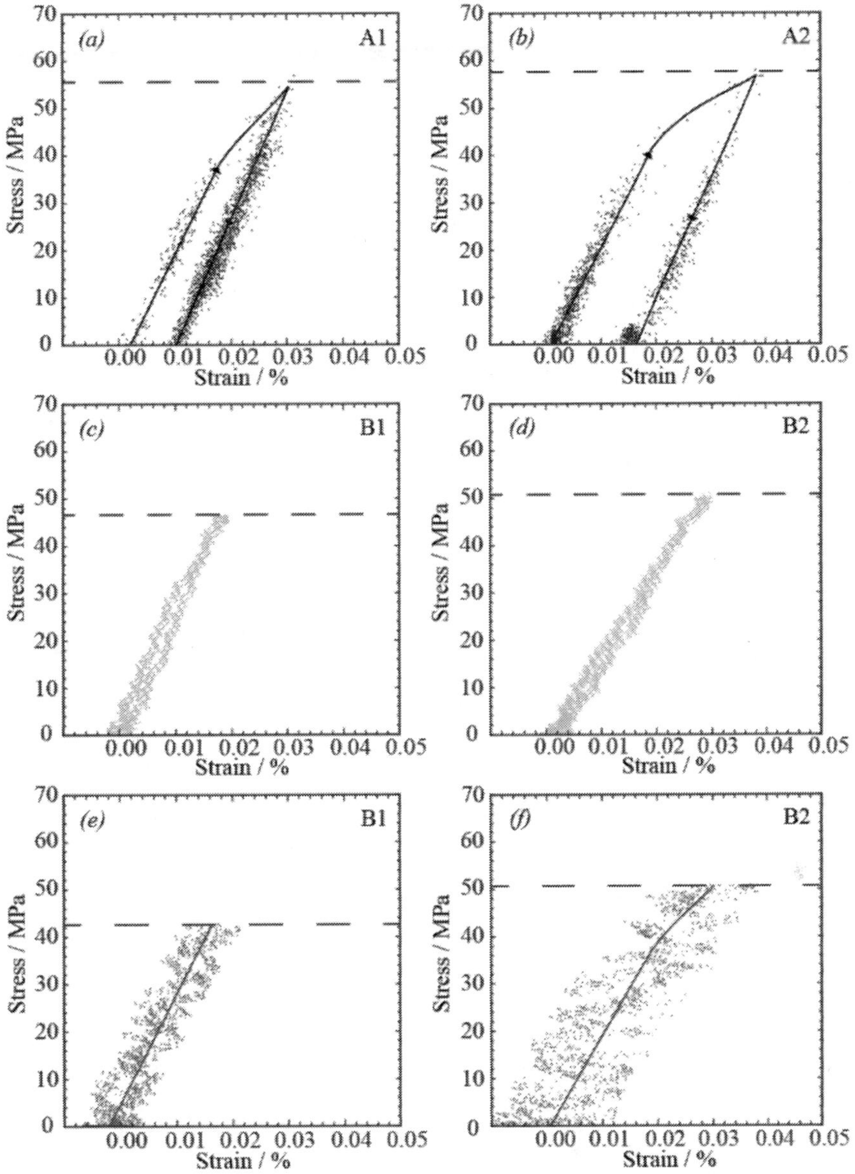

Figure 6. Resulting stress-strain curves of the nondestructive tensile tests at RT, at 77 K, and at 4 K. *(a)* and *(b)* show the curves for the tests at RT with sample A1 and A2, *(c)* and *(d)* show the curves for the tests at 77 K with sample B1 and B2 and *(e)* and *(f)* show the curves for the tests at 4 K with sample B1 and B2. For the tests at RT and 77 K a plastic deformation occured.

4. Conclusion

The performed test program with the hardness test, the helium pressure and leak tests, the destructive- and nondestructive tensile tests and the temperature cycling test yields the desired results. The EB weld was resistant against pressures of about 160 bar and also the weld was found to be leak-tight with measured leak rates smaller than 10^{-9}mbar l s^{-1}. The elastic region

IOP Conf. Series: Materials Science and Engineering **102** (2015) 012017 doi:10.1088/1757-899X/102/1/012017

was measured between 40 MPa and 50 MPa for all performed tensile tests at RT, 77 K, and 4 K and the temperature cycling from RT to 77 K did not damage the weld. Therefore, EB welding is a suitable method for cryogenic application.

References

[1] Kaur A, Ribton C and Balachandaran 2015 Electron beam characterisation methods and devices for welding equipment *J. Mater. Process. Tech.* **221** 225-232
[2] Grohmann S 2014 A new method for flow measurement in cryogenic systems *Cryogenics* **60** 9-18
[3] ISO 6507-1 2005 Metallic materials - Vickers hardness test - Part 1: Test method
[4] EN 1779 1999 Non-destructive testing - Leak testing - Criteria for method and technique selection
[5] Bagrets N, Weiss E, Westenfelder S and Weiss K-P 2012 Cryogenic Test Facility CryoMak *IEEE T. Appl. Supercon.* **22** 9501204-9501204
[6] ISO 6892-1 2014 Metallic materials - Tensile testing - Part 1: Method of test at room temperature
[7] ISO 15579 2000 Metallic materials - Tensile testing at low temperature
[8] ISO 19819 2004 Metallic materials - Tensile testing in liquid helium
[9] ISO/IEC Guide 98-3 1995 Uncertainty of measurement. Part 3: Guide to the expression of uncertainty in measurement

IOP Conf. Series: Materials Science and Engineering **102** (2015) 012018 doi:10.1088/1757-899X/102/1/012018

Dielectric Properties of Cryogenic Gas Mixtures Containing Helium, Neon, and Hydrogen

L Graber[1], W J Kim[1], P Cheetham[1], C H Kim[1], H Rodrigo[2], S V Pamidi[1]
[1] Center for Advanced Power Systems, Florida State University, Tallahassee, FL
[2] Dielectric Sciences, Inc, Chelmsford, MA

E-mail: graber@caps.fsu.edu

Abstract. Past efforts of cooling high temperature superconducting (HTS) power cables by gaseous cryogens focused exclusively on helium. The limited dielectric strength of helium gas necessitated alternatives that could be used in the temperature range suitable for HTS power applications. This paper presents the benefits of gas mixtures containing helium with small concentrations of hydrogen or neon to mitigate the limited dielectric strength of pure helium gas. The expectation was that such gas mixtures could improve dielectric characteristics while maintaining the thermal, non-flammable and non-corrosive properties of pure helium gas. The AC breakdown voltage of helium gas mixtures containing 4 mol% neon or 4 mol% hydrogen respectively have been measured and compared to those of pure helium and pure neon. All measurements were performed at 77 K at gas pressure levels between 0.5 and 2.0 MPa. While the 4 mol% neon mixture did not result in any improvement over pure helium, the 4 mol% hydrogen mixture resulted in 80% higher breakdown strength. This is expected to enable higher operating voltages for gas cooled HTS power devices.

1. Introduction

High temperature superconducting (HTS) power cables are expected to enable electric power distribution of unprecedented power density as is required on future all-electric ships. However, shipboard applications prohibit the use of liquid nitrogen (LN2) since there exists a risk for asphyxiation hazards in case of a LN2 leak. The US Navy considers helium gas (gHe) as a viable option to cool HTS power cables without the risk for asphyxiation hazards since the gas inventory is smaller by several orders of magnitude and helium does not accumulate in the lower sections of the ship.

Cooling the HTS cable with gHe has a number of additional advantages. Most importantly, it allows a wide temperature range including temperatures at which LN2 solidifies. The temperature range below the LN2 range increases the critical current in the individual HTS tapes and therefore the power density of the HTS cable. Another benefit is the tunability, allowing certain interconnects to operate temporarily at lower than nominal temperature during times of high demand.

Past efforts of cooling HTS power cables by gaseous cryogens focused exclusively on pure helium gas [1]. An investigation was undertaken to determine the benefits of mixing small amounts of hydrogen or neon gas to helium to mitigate the limited dielectric strength of pure helium gas. The authors expected that this could to potentially improve dielectric characteristics while maintaining the thermal, non-flammable and non-corrosive properties of pure helium gas.

From the dielectric point of view, hydrogen gas is far superior to helium and neon. Hydrogen gas has approximately 50% of the dielectric strength of nitrogen gas at room temperature [2]. The noble gases helium and neon exhibit a dielectric strength 15% and 25% respectively that of nitrogen gas at room temperature. It is known from other gas mixtures studies at room temperature that even a small amount of a superior gas can considerably improve the dielectric strength of the mixture more than the ratio would suggest. Typical examples for such investigations is the addition of small fractions of SF_6 into N_2 [3] or CO_2 to a SF_6-N_2 mixture [4]. Theoretical models to explain and predict the dielectric strength of gas mixtures are currently being developed [5].

The dielectric properties of pure helium gas at temperatures in the range of 50 K to 80 K have been studied in past HTS cable projects at Florida State University. The challenges with pure helium's low breakdown voltage have led to the investigation of potential alternative cooling gases. The dielectric strength of helium gas mixtures containing 4 mol% of neon and 4 mol% of hydrogen have been measured and compared to pure helium and pure neon. The 4 mol% mixture ratio of hydrogen was selected to ensure that the gas mixture was inflammable in air. All experiments were performed at 77 K at pressures between 0.5 and 2.0 MPa.

2. Methodology

The dielectric properties of the gases were examined by completing alternating current (AC) breakdown tests at various pressure levels while at cryogenic temperatures. The AC breakdown tests were performed by determining the voltage required to arc across a pair of electrodes installed inside a pressure vessel, filled with the various gas mixtures and immersed in a cryostat filled with LN2. The electrodes have a profile similar to a Bruce profile, are highly polished, and consist of 316 stainless steel. The gap distance was adjustable, however was kept at 2 mm as this appeared to be a good trade-off between voltage resolution and upper voltage limitations of the setup. The experiment installed within the pressure vessel is schematized in Figure 1.

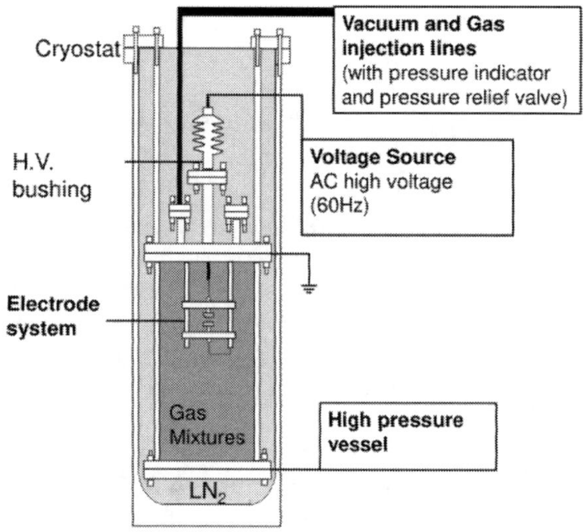

Figure 1. Schematic of the experimental setup to characterize cryogenic gas mixtures (not drawn to scale).

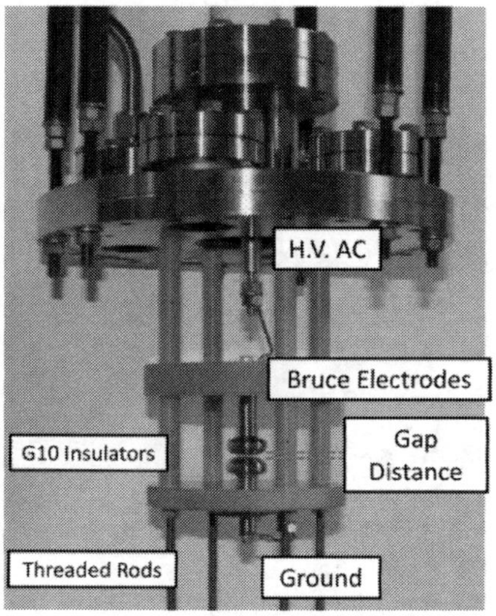

Figure 2. The electrode system with uniform field in the gap between the electrodes.

In Figure 2 it can be seen that the top electrode was connected to the voltage source and the bottom electrode was grounded. The gap between the electrodes was set to 2 mm using a precision gauge block. The electrodes were not changed throughout the duration of the experiment to ensure that the results obtained from the gas mixtures were comparable. On completion of the experiment, the gap distance between the electrodes was re-measured and confirmed to be 2 mm. This shows that the electrodes did not become loose while performing the experiments at cryogenic temperatures. However, the distance between the electrodes is expected to be slightly different from 2 mm due to the thermal contraction of the materials at 77 K. The electrodes as well as all the other parts inside the pressure vessel were thoroughly cleaned by isopropyl alcohol before closing the vessel.

Before the beginning of the experiments, the pressure vessel was evacuated to 10^{-4} mbar or lower for at least 12 hours using a turbomolecular pump. The vessel was subsequently flushed twice with industrial grade helium gas to ensure all contaminants were removed before being filled with research grade helium gas of 99.999% purity. The pressure vessel was then immersed in a cryostat filled with LN2 and the pressure inside the pressure vessel was adjusted to 2.0 MPa. The AC breakdown tests were not undertaken until equilibrium had been reached for the gas pressure, which ensured that the temperature of the helium gas and all components reached 77 K. Throughout the experiment LN2 was added to the cryostat to ensure that the whole pressure vessel including the bushing was immersed in liquid nitrogen.

Initially, ten AC breakdown tests were performed to "season" the surfaces of the electrodes. It is a well-known effect that new electrodes tend to show lower breakdown voltages for the first few breakdown events. After seasoning the electrodes, fifteen AC breakdown tests were performed at 2.0 MPa with these results being recorded. The pressure within the vessel was reduced to 1.5 MPa, then 1.0 MPa and finally 0.5 MPa with fifteen AC breakdown tests being performed at each pressure level. Once the AC breakdown tests at 0.5 MPa had been completed, the pressure within the pressure vessel was released and a dry scroll vacuum pump was connected to remove the remaining helium gas from the vessel.

This process was then repeated for the 4 mol% neon, pure neon and 4 mol% hydrogen gases. A lack of sufficient quantities of pure neon gas restricted the maximum operating pressure to 1.8 MPa.

It should be noted that there are some indications that helium and neon are mixed very slowly at 77 K due to slow diffusion at cryogenic temperatures. Therefore, the gas mixtures were ordered pre-mixed in high pressure steel tanks from a gas supplier. While the exact mixing procedure is unknown to the authors, there is no reason to assume the mixture would be inhomogeneous. Once the gases are mixed at room temperature, there were no indications that they would separate and stratify later during the experiment at cryogenic temperature.

On completion of the experiment, the electrodes were inspected by scanning electron microscopy (SEM) to check if the discharge activity created craters that might have locally enhanced the electric field, therefore resulting in a decrease of the breakdown voltage. While the craters were visible in the SEM images, they are of similar size as the machining marks (surface roughness) and thus expected not to play an essential role. The energy in the discharge pulses is very limited due to a high source impedance.

3. Experimental Results

The result of the average AC breakdown voltage for the pure helium, pure neon, helium with 4 mol% neon and helium with 4 mol% hydrogen gas mixtures can be seen below in Figure 3. The error bars in Figure 3 show the range of voltages recorded (maximum and minimum) at each pressure level for each gas mixture.

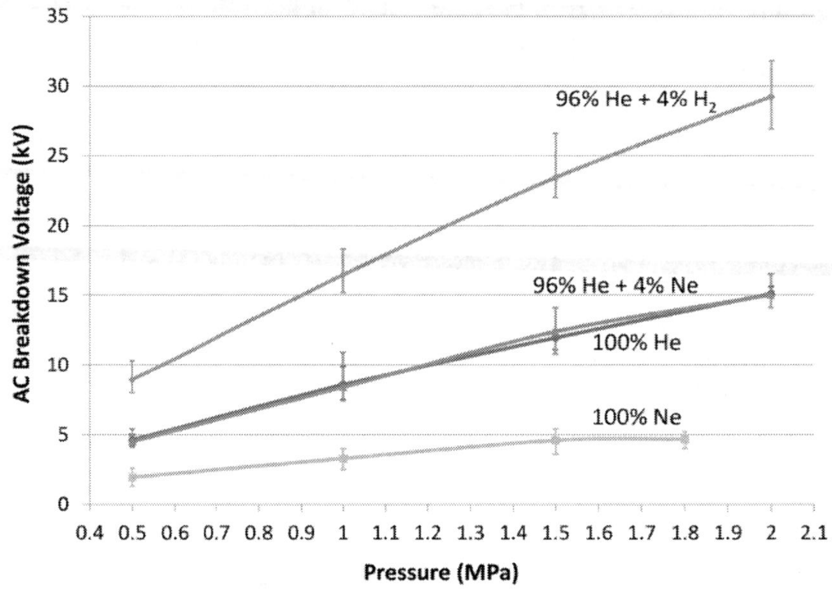

Figure 3. AC breakdown voltage as a function of pressure at a 2-mm gap distance in homogeneous field for helium, neon, 4 mol% hydrogen in helium, and 4 mol% neon in helium mixture.

Analysis

From Figure 3, two distinct trends were observed when helium gas was mixed with neon and hydrogen. Firstly, when a small amount of a stronger dielectric gas is mixed with a weaker dielectric gas there is an increase in AC breakdown strength. Adding 4 mol% hydrogen to helium, the AC breakdown voltage increased by approximately 80% for all pressure levels tested. Hydrogen is a diatomic gas with a dielectric strength far superior than those of noble gases. The second trend is the addition of a small amount of weaker dielectric gas being mixed with a stronger dielectric gas does not have any significant effect on AC breakdown voltage. The dielectric properties of neon are significantly lower than those of helium; however, the AC breakdown voltage for the pure helium and 4 mol% neon gas mixture is approximately identical with the pure helium gas for all pressure levels. While we do not see a reason to add neon gas to the mixture, it is good to know that a small amount of (unintended) impurities does not have a negative effect on the dielectric strength.

4. Conclusion

Breakdown experiments in homogeneous electric fields at a temperature of 77 K and pressure levels between 0.5 MPa and 2.0 MPa suggest that adding 4 mol% hydrogen by volume to helium increases the dielectric strength by approximately 80% while the mixture is still considered non-flammable at any ratio with air. This is promising for the application of high temperature superconducting power apparatus such as cables when a gaseous cryogen is preferred. A mixture of 96 mol% helium with 4 mol% neon did not indicate any change of the dielectric properties compared to those of pure helium gas. More experiments are planned to investigate different mixture ratios of helium with hydrogen. Additional experiments are currently being performed to investigate the thermal properties of such a mixture. A patent for cryogenic gas mixtures is pending [6].

Acknowledgments

This project was funded by the United States Office of Naval Research through the grants N00014-14-1-0346: "Basic research on cryogenic dielectric and thermal aspects pertinent to high voltage high temperature superconducting devices for Navy applications," and N00014-14-1-0377: "Exploring helium gas mixtures for enhanced cryogenic dielectric characteristics".

References:

[1] Pamidi S V, Kim C H, Graber L, Crook D, Ranner S L, Trociewitz B, Dale S, Knoll D, Willen D, Thidemann C, and Lentge H 2013 *Installation and Testing of Helium Gas Cooled Superconducting DC Cable at FSU-CAPS: 11th EPRI Superconductivity Conference (Houston, USA 28-30 October 2013)*
[2] Vijh A K 1977 *IEEE Trans. Electr. Insul.* **12** 313-315
[3] Malik N H and Qureshi A H 1979 *IEEE Trans. Electr. Insul.* **14** 1-13
[4] Ohtsuka S, Nagara S, Miura K, Nakamura M and Hikita M 2000 *IEEE Int'l. Sympos. on Electr. Insul.(Anaheim, USA 2-5 April 2000)* p 288-291
[5] Zhao H, Li X, Jia S, Murphy A B 2014 *J. Phys. D: Appl. Phys.* **47** 325203
[6] Graber L, Kim C H and Pamidi S V 2015 *Dielectric Properties of Cryogenic Gas Mixtures Containing Helium, Neon and Hydrogen (invention disclosure filed on 23 June 2015)*

Insulation system for high temperature superconductor cables

P C Michael[1], A E Haight[2], L Bromberg[1], and K Kano[2]

[1] MIT Plasma Science & Fusion Center
77 Massachusetts Avenue, Cambridge, MA 02139, USA

[2] Composite Technology Development, Inc.
2600 Campus Drive, Suite D, Lafayette, CO 80026, USA

E-mail: michael@psfc.mit.edu

Abstract. Large-scale superconductor applications, like fusion magnets, require high-current capacity conductors to limit system inductance and peak operating voltage. Several cabling methods using high temperature superconductor (HTS) tapes are presently under development so that the unique high-field, high-current-density, high operating temperature characteristics of 2nd generation REBCO coated conductors can be utilized in next generation fusion devices. Large-scale magnets are generally epoxy impregnated to support and distribute electromagnetic stresses through the magnet volume. However, the present generation of REBCO coated conductors are prone to delamination when tensile stresses are applied to the broad surface of REBCO tapes; this can occur during epoxy cure, cooldown, or magnet energization. We present the development of an insulation system which effectively insulates HTS cabled conductors at high withstand voltage while simultaneously preventing the intrusion of the epoxy impregnant into the cable, eliminating degradation due to conductor delamination. We also describe a small-scale coil test program to demonstrate the cable insulation scheme and present preliminary test results.

1. Introduction

The high energy physics and the thermonuclear fusion communities are actively developing high-current capacity cables using 2nd generation REBCO coated conductor tapes for large-scale, high-field magnet applications. REBCO coated conductors are one of the few high-temperature superconductors (HTS) capable of providing the approximately 20 T peak magnetic flux that may be required by these devices and the only one presently available in large scale production. Potential applications include the LHC Energy Upgrade [1], with peak magnetic flux density in the range from 16 T to 20 T, and compact fusion devices like ARC [2], with peak magnetic flux density in the range from 20 T to 22 T. The REBCO cable types under development include: conductor on round core (CORC) [3], twisted, stacked-tape cable (TSTC) [4] and Roebel bar [5]. In addition to operating in a high electromagnetic stress environment, the conductors will also be exposed to ionizing radiation. The anticipated radiation dose on the ITER magnets is 1×10^{22} neutrons/m^2 [6], while the dose rate on subsequent fusion magnets will likely be up to an order of magnitude higher.

A principal risk factor for HTS coils is the low through-thickness tensile strength of present generation REBCO tapes. Therefore, it is necessary to control the thermo-mechanical stress transmitted to the conductor when a large, high-field REBCO-based coil is cooled to cryogenic temperatures and energized [7]. The thermo-mechanical stresses develop in part because of the different thermal contractions and mechanical stiffnesses of typical epoxy potting compounds and the REBCO tapes. The through-thickness thermal contraction of typical fiberglass/epoxy insulation is about 1%, whereas the thermal contractions of the REBCO tapes and steel former are on the order of 0.25% [8, 9]. Large differential stresses develop as the coil is cooled to low temperatures, and these stresses can potentially delaminate the REBCO tapes. Through thickness delamination of the REBCO tape has been reported at tensile stresses as low as 10 MPa, and has been cited as the cause of degraded performance for several epoxy impregnated small coils [10]. One approach to address delamination due to differential thermal contraction during cooldown has focused on the use of filled epoxies [11, 12].

To support the development of next-generation, high-field magnets, we sought to develop a reliable means to electrically insulate and epoxy-impregnate high-current capacity cables using presently available REBCO coated conductor tapes. Our approach was to develop a combination of epoxy potting and barrier insulation materials that, when combined in a magnet, can absorb most of the through-thickness tensile strains applied to the REBCO tapes when a magnet is cooled and energized. The high dielectric strength barrier insulation we selected can be tightly sealed to the surface of the REBCO cable to prevent the intrusion of epoxy in direct contact with the tapes. The candidate epoxy impregnant was selected based on its anticipated high radiation resistance. The epoxy impregnant also provides suitable stiffness, structural strength, low thermal expansion, and relatively high thermal conductivity to react the mechanical and thermal loads applied to a magnet, while preventing delamination due to differential thermal contractions during resin curing and magnet cooldown. Our criterion for success was to demonstrate negligible degradation to the cable's superconducting performance following insulation, epoxy impregnation, cool-down, and operation.

2. Test configuration

We utilized an inductive charging method to evaluate the performance of the cable samples to eliminate the need for high-capacity current leads. Figure 1 shows a solid-model image of the arrangement. The cable sample was in the form of a small, partially slit-tape coil [13]. The upper-most, water-jet-cut aluminum plates used to shape and support the coil were suppressed in Figure 1 to emphasize the conductor shape. The coil was also equipped with a small AISI 1018 iron pole piece to increase the magnetic flux density captured by the coil.

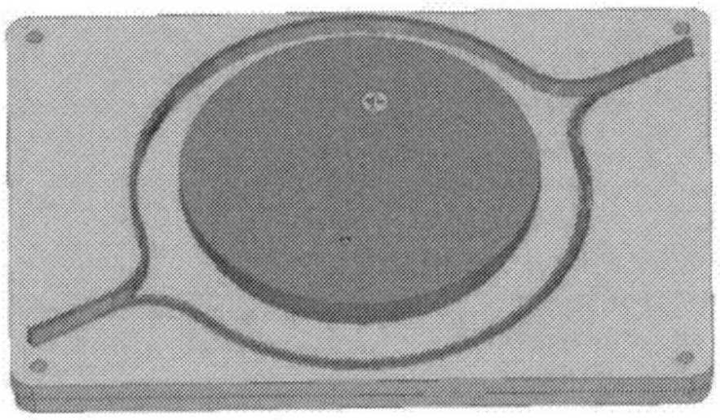

Figure 1. Solid model image of test sample, consisting of an assembled stack of partially slit REBCO tapes, embedded in aluminum support plates, and surrounding an AISI 1018 pole piece. The top aluminum plate was removed to show details of the sample assembly.

To form each test coil, a set of between twenty and thirty, 305 mm long by 12 mm wide REBCO tapes from SuperPower, Inc. was laser slit along the tape centerlines over most of their length, leaving a short section of full width tapes at each end. The tapes were then stacked together to form an approximately 2 kA current capacity cable and installed in the aluminum support plates, with the top halves of the tapes deformed to a roughly semi-circular shape in one direction, and the bottom half of the tapes deformed in the other direction to form a continuous closed loop. Two types of SuperPower HTS tapes were used during the project. The first sample used SuperPower type SCS-12050-AP tapes with 10 μm thick electroplated copper applied to each side, while the second sample used SuperPower type SF-12050-AP tapes with 2 μm thick electroplated silver applied to each side. The minimum 77 K self-field critical current for copper-clad tape was 328 A, while that for silver-clad tape was 298 A.

A slit-tape coil geometry was used to permit direct comparison of cable performance both before and after application of the barrier insulation and epoxy impregnation. Energization of the coil was performed by flux trapping, during which the coil was placed at room temperature in an energized background field coil, cooled to 77 K by liquid nitrogen transfer, and subsequently energized by removal of the background magnetic field. The coil configuration and test procedure were selected to minimize confounding effects such as stress concentrations at current lead breakouts, and variation in the distribution of current among cable tapes from one test sequence to the next. The conductor's critical current is easily determined by applying external field before cooldown greater than the sample's anticipated, maximum self-magnetic field. After the sample is cooled and the external field is removed all conductor tapes are at their critical condition, with zero applied external magnetic field.

The expected critical current for each sample was determined using a COMSOL Multi-Physics [14] model of the test setup. The model investigated the distribution of magnetic flux density over the cable cross-section near the ends of the slits, where the upper half of the cable loop passes directly over the lower, and near the center of each half of the cable. The cable critical current at each location was iteratively determined by: imposing a cable current at zero background field in the COMSOL model, determining the distribution of magnetic flux density over the cable cross-section, evaluating the effect of local magnetic flux density (intensity and direction) on expected critical current density, integrating over the cross-section to determine the expected cable critical current, and adjusting the applied current in the model until the applied and expected current matched. In our model, we relied on Takayasu's recent low-field, 77 K measurements on SuperPower advanced pinning site tapes [15] to model the effect of magnetic flux density on the local critical current density in the cable. The COMSOL model also calculated the expected magnetic field across the surface of the sample's iron pole piece. The surface of the pole piece was equipped with F.W. Bell model BHT-921 Hall probes mounted at the center and near the edge as a means to indirectly determine the sample current.

Several samples were fabricated during the development of the insulation scheme. Because of the high cost of superconductor, development began using a surrogate material, namely 22 mm wide by 50 μm thick Inconel tape. The first conductor sample was made using the copper-plated REBCO tapes, while the second sample was made using silver-plated tapes. Copper-plated tapes are more relevant for large-scale magnet applications, where the copper provides an alternative current path in the event of superconductor quench. However, the copper-plated tapes were more prone to the development of a significant, built-up edge of rapidly resolidified material during laser slitting, which interfered with the regular stacking of the tapes during sample fabrication. The switch to silver-plated tapes, which do not demonstrate this effect to an appreciable extent, was made, because the primary focus of the study was on the development of the insulation scheme. All tape samples were partially laser slit at Laser Edge Technology in Merrimac, MA.

During processing of the REBCO conductor samples, the critical current of the uninsulated sample was first determined at the Plasma Science and Fusion Center at MIT. The sample was transferred to Composite Technology Development, Inc. for insulation application. The sample was returned to MIT for re-evaluation of its critical current and finally returned to CTD for sectioning and visual examination of the insulation quality.

3. Insulation scheme

Various approaches have been developed to electrically insulate individual REBCO coated conductors, including the application of electrodeposited polyimide [16] or encapsulation within heat shrink tubing [17]. The insulation would be applied to the conductor before magnet winding and epoxy impregnation. The objective of both approaches is to surround the conductor with a continuous, tight-fitting but poorly adhering insulating layer that would electrically insulate but mechanically decouple the conductor from tensile stresses which might be applied to the REBCO layer. The approach presented here is conceptually similar, but intended for use on high-current capacity cables fabricated from multiple REBCO tapes. Our proposed approach for insulating long-length REBCO cables is to continuously wrap them with thin polyimide shrink tape and then pot with a highly-filled epoxy for enhanced mechanical strength and radiation resistance. A significant advantage of the approach described here is its compatibility with continuous manufacturing processes that might be used during cable production.

3.1. Material Selection

We selected polyimide tape because of its known high dielectric strength and radiation resistance. Tape wrapping was chosen because the processing method can be performed on long lengths using an automated system. Shrink tape was chosen rather than adhesive-backed tape because of the known radiation sensitivity of the acrylic adhesives and the tendency of adhesive tapes to wrinkle during wrapping, forming gaps through which resin could flow.

The polyimide shrink tape used in this study is capable of up to 13% shrinkage when processed at 350 °C. However, the tape was processed at lower temperature in this study to accommodate the maximum recommended processing temperature for the REBCO conductors. Our results show that when the tape is applied with high tension initially to form a snug wrap, the 8% shrinkage that can be achieved at 150 °C is acceptable for forming a tight barrier.

CTD-101G resin was chosen as the potting material in the test samples based on its relatively low cure shrinkage, high cryogenic strength and stiffness, and anticipated high radiation resistance. CTD-101G is an anhydride-cured, bisphenol-A based resin with a high concentration of filler added to reduce shrinkage and improve thermal conductivity. For a 5 hr cure at 110 °C, followed by 1.5 hr at 135 °C, the linear cure shrinkage is 0.55±0.04% when evaluated per ASTM D 2566. The resin viscosity during processing at 60~80 °C ranges from 200 cP to 1000 cP when measured using a Brookfield viscometer, which is low enough to be used for potting coils. Its pot life of 1.5~3 hr is long enough for processing small coils, but may need to be optimized for use in large-scale magnets. When tested per ASTM D 638 at 77 K, the neat resin (without fiberglass) has a tensile modulus slightly above 19 GPa and a tensile strength exceeding 95 MPa. All resin properties were determined at CTD.

3.2. Preliminary Irradiation Testing

For fusion magnet applications, radiation stability of the insulation scheme is critical. To evaluate the radiation resistance of the selected materials, polyimide shrink tape tension specimens and CTD-101K and CTD-101G short-beam-shear specimens were irradiated at the Massachusetts Institute of Technology Nuclear Reactor Laboratory in their 2PH1 facility. The particular tests conducted were chosen as they most dramatically demonstrate any material property changes that arise from radiation exposure. The short-beam-shear specimens were fabricated with S2-glass. Radiation exposures of two and four times the ITER design fluence were used in this work, which equates to 2 x 10^{22} and 4 x 10^{22} neutrons/m^2, respectively. The specimens were returned to CTD for film tension testing at 295 K (per ASTM D882) and short-beam-shear testing at 77 K (per ASTM D2344). The test results are shown in Figure 2, where it can be seen that the polyimide film retains 79% of its tensile strength

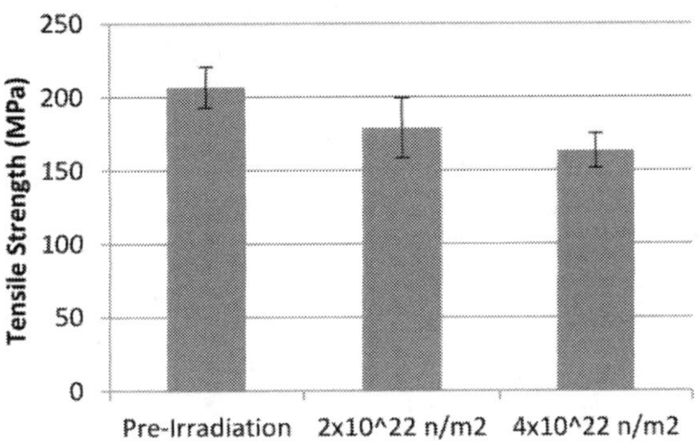

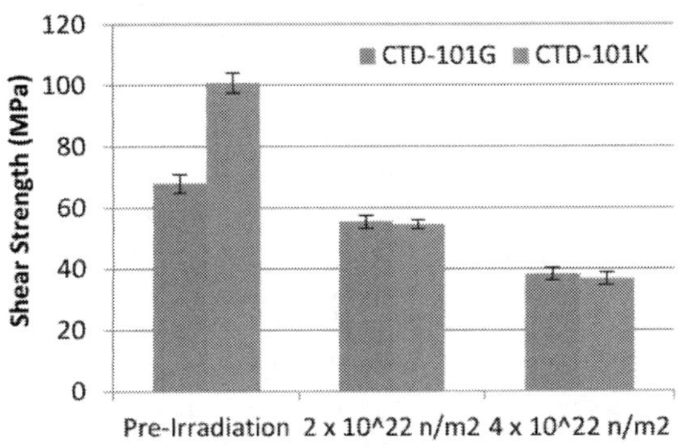

Figure 2. Pre- and post-irradiation test results. a) Polyimide shrink tape tensile test at 295 K and b) resin/S2-glass short-beam-shear test at 77 K.

after exposure to 4x the ITER design fluence. The CTD-101K and CTD-101G specimens had a higher degree of strength reduction after exposure to the high irradiation levels, but future work will focus on improving the potting compound radiation stability.

3.3. Insulation of test coils

Figure 3 shows initial efforts to apply the polyimide shrink tape to a stack of 32 partially slit Inconel strips. This effort was performed to develop the process that would subsequently be used to insulate the significantly more expensive superconductor coils that were subjected to critical current evaluation. The Inconel strips were stacked and held together with clamps while the polyimide shrink tape was tightly wrapped with a ½ lap around the strips (Figure 3a). The ends of the shrink tape were held in place with Kapton tape. Once the conductor loop was completely wrapped with polyimide, it was reinserted into the aluminium support plates and heated in an oven to 150 °C for 30 minutes to shrink the polyimide and form a tight wrap around the Inconel strips. The polyimide-insulated sample is shown in Figure 3b. Unlike the processing for a continuous cable length, the complex geometry near the ends of the slits prevented straightforward wrapping of the polyimide tape. Our ability to successfully seal these regions in the final test sample (to be discussed below) confirms the robustness of the proposed approach. The insulation process was similar for each succeeding sample, however, slight improvements were made as we gained experience.

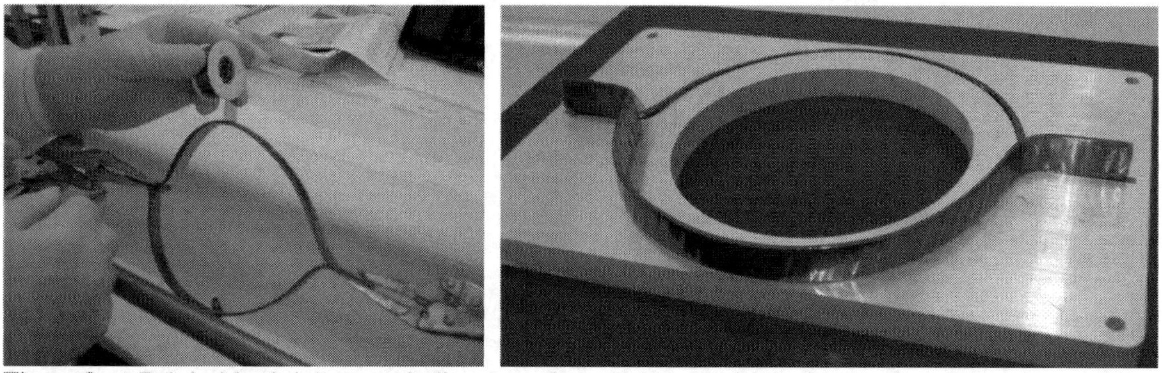

Figure 3. a) Polyimide shrink tape winding procedure, and b) polyimide-insulated sample reinserted in its aluminium support plate (right).

To complete the test sample fabrication, the polyimide-wrapped conductor was removed from the aluminum plates, which were then used as a potting mold to apply and cure the resin. The machined slot in the aluminum plate was sealed at the bottom using a 2.5 mm thick silicone sheet as a gasket. The slot was half filled with resin, warmed to 80 °C, and degassed in a vacuum oven. The lower half of the conductor was reinstalled in the plate and additional resin was applied to completely fill the slot. The resin was cured using the following schedule: 3 hr ramp to 110 °C, soak at 110 °C for 5 hrs, 1.5 hr ramp to 135 °C, soak at 135 °C for 1.5 hrs, 2 hr cool to 22 °C. The upper half of the coil was then potted by adding a second sheet of silicone between the upper and lower plates. For the second half of the coil, pre-warmed, degassed resin was injected into the gap between conductor and plate using a syringe to completely fill the slot. The resin cure cycle was then repeated for the resin applied to the second half of the coil. The thermal process cycle was verified by Super Power to have no detectable impact on expected critical current of their REBCO conductors [18]. The silicone sheet gasket between the aluminum support plates was not part of the original sample fabrication plan and was not present during critical current testing of the uninsulated samples.

Figure 4 shows results from the sectioning analysis of the sample made using a stack of 20 partially laser-slit copper-plated REBCO conductors. The locations of the cuts are shown in Figures 4a and 4b, while representative microscopic images of conductor cross-sections are shown in Figures 4c and 4d. Figure 4c shows a conductor cross-section taken at sample location 1 in Figure 4a. The free surface of the sample is to the right, while the silicone gasket between support plates is to the left. The conductor stack is not well aligned and has a slightly trapezoidal shape due to the presence of the built-up edge along the edge of the laser cut slit. The conductor stack shows considerable resin infiltration, most likely due to the irregular packing on the stack which introduces gaps between the polyimide tape wraps. The approximately 20% reduction in measured critical current for this sample from the uninsulated to insulated conductor tests (discussed below) is attributed to the presence of resin within the conductor stack, which may have damaged some of the tapes. Figure 4d shows the conductor cross-section taken at sample location 3 in Figure 4b. The silicone gasket between the plates is towards the right in this figure. It shows similar trapezoidal shape, but is well compacted, with resin intrusion limited towards the periphery of the stack.

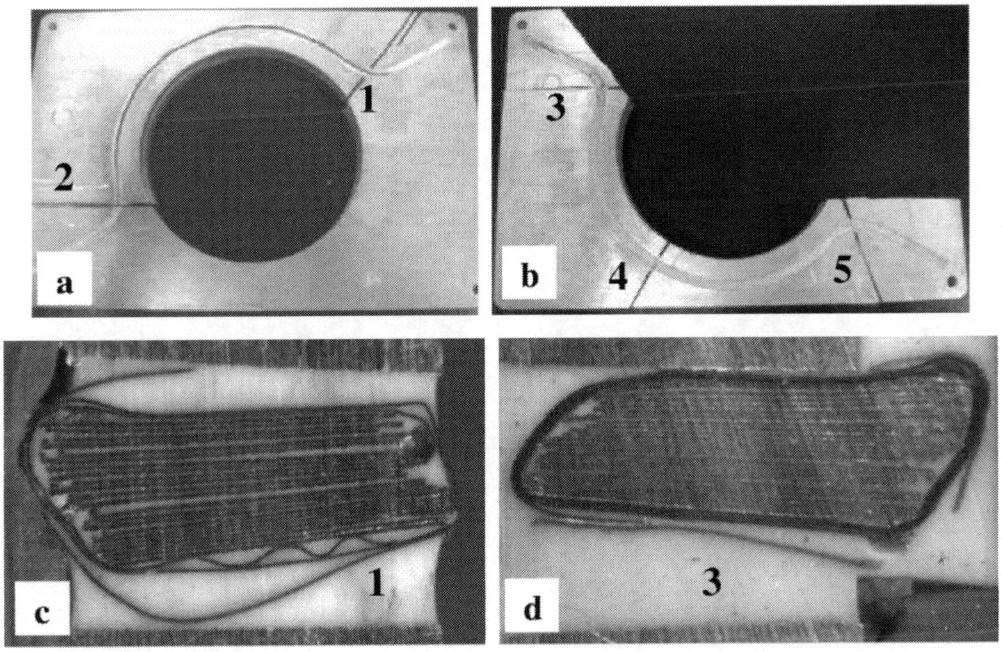

Figure 4. Section views of the sample made using the copper plated REBCO conductors. a) and b) location of sample section lines. c) and d) representative cross-section views.

Figure 5 shows results from the sectioning analysis of the sample made using a stack of 29 partially laser-slit silver-plated REBCO conductors. The locations of the cross-sections are the same as those shown in Figures 4a and 4b. Figure 5a shows the conductor cross-section taken at sample location 3, while Figure 5b shows the conductor cross-section taken as sample location 5. The microscopic images at these locations show well compacted, neatly arranged conductor stacks, with near rectangular shape. No resin flow can be seen between tapes within the conductor stack and only negligible resin can be detected within the polyimide wrap towards the left hand side of Figure 5a.

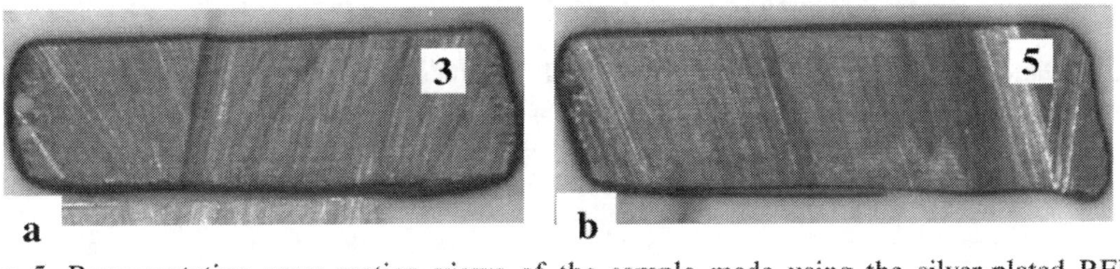

Figure 5. Representative cross-section views of the sample made using the silver-plated REBCO conductors. a) Section at sample location 3, and b) section at sample location 5.

4. Critical current measurements

Figure 6a shows the minimum excitation sequence needed to determine the critical current of our slit tape conductor samples. The data trace shows the magnetic flux density measured on the surface, near the center of the sample pole piece vs. the excitation current applied to the background field magnet for the uninsulated sample fabricated using the silver-plated REBCO tapes. The thin, solid black line shows the initial current upramp, with the room temperature (RT) sample installed in the bore of the magnet. The sample was then cooled to 77 K by liquid nitrogen transfer into the sample cryostat. The magnet current down ramp following cooldown is shown by the thick, dashed black line. The small vertical downward drop at zero external magnet current is attributed to flux creep in the sample as the induced current in the sample decreases from just over critical to below critical current. The thick dotted line shows a current upramp towards the end of the test sequence during which the liquid nitrogen is allowed to boil away at constant background field. The vertical upward portion of this curve is due to quench of the sample current as the sample temperature increases due to heat leak into the cryostat. The solid black curve shows a set of external magnet current sweeps following sample quench. The trace shows negligible change in the Hall probe calibration and offset with temperature and minimal magnetic hysteresis in the background magnetic field.

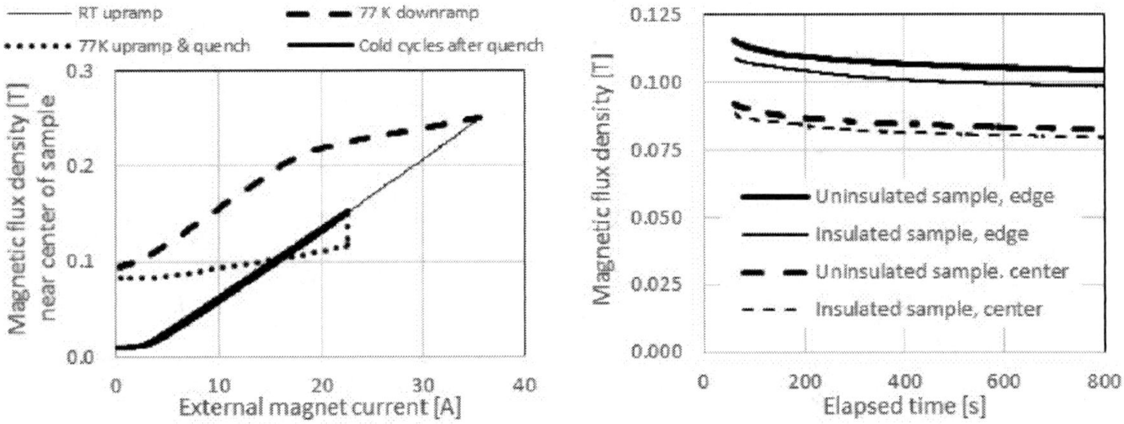

Figure 6. a) Flux-trapping test sequence, and b) comparison of flux creep portion of the test sequence before and after polyimide tape wrapping and epoxy potting of the silver-plated conductor stack.

Figure 6b compares the magnetic flux density trapped by the sample fabricated using 29 silver-plated REBCO conductor before and after application of the cable insulation. The vertical axis shows the magnetic flux density measured by the Hall probes mounted near the center and near the edge of the sample pole piece vs elapsed time following the end of the external magnet current downramp with sample immersed in liquid nitrogen. The thick lines show measurements for the uninsulated sample, while the thin lines show measurements for the insulated sample. The solid lines show measurements taken near the edge of the pole piece, while the dashed lines show measurements taken near the center. The slight reduction in the Hall probe readings for the insulated sample are due to the introduction of a 2.5 mm gap (corresponding to the silicone gasket) between the pole pieces during application of the cable insulation. The COMSOL models of the test arrangement indicate that all asymptotic Hall probe readings in Figure 6b are consistent with a sample current of 1720±20 A, compared with the estimated sample current of roughly 1830 A based on the magnetic field distribution across the cable cross-section near the ends of the slits, which is slightly lower than that near the center of the sample loop. The agreement is reasonable given inherent uncertainties in the COMSOL modeling, the assumed uniformity of critical current density across the width of each REBCO tape, and the position of the laser slitting line relative to the REBCO tape centerlines.

By comparison, results for the earlier sample, fabricated using 20 copper-clad REBCO tapes, were consistent with a sample current of 1150±10 A prior to insulation, which was close to the anticipated value, and a current of 940±10 A during retest following insulation. These results are for the poorly aligned sample that showed significant resin intrusion between the conductor tapes in Figure 4c.

5. Summary

We have tested a method for electrically insulating and epoxy impregnating high-current-capacity cables formed from REBCO tapes, using radiation-resistant materials suitable for use in large-scale superconducting magnet applications. The method combines a heat shrinkable polyimide tape, which acts as a barrier insulation, followed by impregnation with a low shrinkage resin, CTD-101G. Tight wrapping, followed by shrinkage of the polyimide tape effectively compacts the cable stack, potentially enhancing tape-to-tape current transfer, while sealing the cable stack perimeter to prevent resin intrusion between tapes during potting. The method was demonstrated using a series of small, slit-tape conductor samples. The results for the cable sample fabricated using silver-plated REBCO conductors shows that, when properly applied, the proposed insulation scheme effectively prevents critical current degradation during cooldown and subsequent operation.

Acknowledgments
This work was supported in part under the US Department of Energy, Office of Fusion Energy Science SBIR Phase I Grant DE-SC0011862.

References

[1] Bottura L, de Rijk G, Rossi L and Todesco E 2012 *IEEE Trans. Appl. Superconductivity* **22** 4002008
[2] Sorbom B N *et al* 2014 *arXiv:*1409.3540v1
[3] van der Laan D C, Lu X F and Goodrich L F 2011 *Superconductor Sci. Tech.* **24** 042001
[4] Takayasu M, Chiesa L, Bromberg L and Minervini J V 2012 *Superconductor Sci. Tech.* **25** 014011
[5] Goldacker W *et al* 2006 *J. Phys. Conf. Ser.* **43** 901
[6] Munshi N A *et al* 2013 *IEEE Trans. Appl. Superconductivity* **23** 7700104
[7] Miyazaki H, Iwai S, Tosaka T, Tasaki K, and Ishii Y 2014 *IEEE Trans. Appl. Supercond.* **24** 4600905
[8] Ekin J 2006 *Experimental techniques for low-temperature mesaurements* (Oxford: Oxford University Press)
[9] Iwasa Y 2009 *Case studies in superconducting magnets* (New York: Springer Science + Business Media)
[10] Takematsu T *et al* 2010 *Phys. C: Superconductivity* **470** 674
[11] Barth C *et al* 2013 *Superconductor Sci. Tech.* **26.5** 055007
[12] Otten S et al 2015 *Superconductor Sci. Tech.* **28** 065014
[13] Lee S *et al* 2013 *IEEE Trans. Appl. Superconductivity* **23** 4601305
Levin G A *et al* 2008 *Appl. Phys. Lett.* **93** 062504
[14] COMSOL multiphysics software.
[15] Takayasu M, Chiesa L and Minervini J V 2015 *IEEE Trans. Appl. Superconductivity* **25** 5400305
[16] Yanagisawa Y *et al* 2012 *Phys. C, Supercond.* **476** 19
Sato K *et al* 2013 *IEEE Trans. Appl. Superconductivity* **23** 4602504
[17] Hilton D K, Dalban-Canassy M, Weijers H W, Trociewitz U P and Larbalestier D C 2012 *Patent US2012142539-A1*
[18] Sundaram A 16 Jan. 2015 *personal correspondence*

Tensile strengths of polyamide based 3D printed polymers in liquid nitrogen

P Cruz[1], E D Shoemake[2], P Adam[2], and J Leachman[2]

[1]Gonzaga University, School of Engineering and Applied Science, Spokane, WA 99258 USA
[2]Hydrogen Properties for Energy Research (HYPER) Laboratory, Washington State University, Pullman, WA 99164-2920 USA

Abstract. Advances in additive manufacturing technology have made 3D printing a viable solution for many industries, allowing for the manufacture of designs that could not be made through traditional subtractive methods. Applicability of additive manufacturing in cryogenic applications is hindered, however, by a lack of accurate material properties information. Nylon is available for printing using fused deposition modeling (FDM) and selective laser sintering (SLS). We selected 5 SLS (DuraForm® EX, DuraForm® HST, DuraForm® PA, PA 640-GSL, and PA 840-GSL) and 2 FDM (Nylon 12, ULTEM) nylon variants based on the bulk material properties and printed properties at room temperature. Tensile tests were performed on five samples of each material while immersed in liquid nitrogen at approximately 77 Kelvin. Samples were tested in XY and, where available, Z printing directions to determine influence on material properties. Results show typical SLS and FDM nylon ultimate strength retention at 77 K, when compared to (extruded or molded) nylon ultimate strength.

1. Introduction

3D printing offers a new method of manufacturing thermoset and thermoplastic polymers. Deviations from well-known bulk material properties caused by this new method have yet to be extensively characterized at cryogenic temperatures, preventing adoption by the cryogenic industry. Polymers offer an alternative to conventional metal components due to lower thermal conductivity, density, and cost – ideal for lightweight insulating designs at a variety of temperatures. In many cases, bulk strength of these polymers increases as temperature drops to cryogenic levels [1,2], further improving their usefulness. However, very little material property information is available for 3D-printed materials at cryogenic temperatures.

Material, printing method, and print orientation contribute significantly to strength properties of printed plastics and are examined in this study. Strength of printed specimens is dependent on material and print orientation [3] and ultimately dictate the overall performance of a printed design. Two primary directions were studied for each material: (1) XZ plane and (2) ZX plane for FDM samples; (1) XY plane and (2) ZX plane for SLS samples. We anticipated that the anisotropic properties are significant in determining how materials react to loads at low temperatures.

2. 3D-printing technologies examined

IOP Conf. Series: Materials Science and Engineering **102** (2015) 012020 doi:10.1088/1757-899X/102/1/012020

2.1. *Fused deposition modelling (FDM)*

FDM is one of the most popular methods used today – especially in unmanned aerial vehicle (UAV) and space applications. A model is fabricated by extrusion of a thermoplastic material in a semi-molten phase onto a stage one layer at a time. At times, FDM models contain internal defects caused by the filament diameter and density affecting how the thermoplastic is extruded from the nozzle. The advantage of FDM lies in the ability to design a model using multiple materials as well as an ability to expand beyond conventional materials such as polycarbonate (PC) or acrylonitrile butadiene styrene (ABS) to glass reinforced polymers, metals, and ceramics [4].

2.2. *Selective laser sintering (SLS)*

SLS is a powder-based form of fabrication that has a wide array of engineering-grade materials available unique to the SLS printing method. SLS uses a high power laser to sinter polymer powders rather than using a liquid binder. Common materials are PC, polyvinyl chloride (PVC), ABS, nylon, resin, and polyester. Unlike FDM, SLS does not require scaffold support during the forming process, significantly saving on material cost; however, SLS models often suffer shrinkage and deformation from thermal expansion and cooling [4].

3. Materials and Methods

Stratasys Ltd. provided tensile specimens of FDM-printed Nylon-12 and ULTEM. Paramount Industries provided SLS-printed Duraform EX, Duraform HST, Duraform PA, carbon-glass reinforced nylon-12, and nylon-11. All specimens were manufactured according to ASTM D638 – Standard Test Methods for Tensile Properties of Plastics, Type I specimen. Five specimens were 3D printed for each test. FDM samples were tested in XZ and ZX orientations based off of the data we had from 3D printed material manufacturer specifications at room temperature. SLS materials were tested in XY and, where available, ZX orientations. KFH-6-350-C1-11L1M2R pre-wired Omega strain gauges were selected based on ASTM D638 recommendations and attached using cyanoacrylate. Strain gauges were set up using a type 1 quarter bridge with a thermal strain compensating dummy gauge fixed to a non-loaded segment in the same material immersed in liquid nitrogen. Strain from a quarter bridge arrangement is determined by

$$\epsilon = \frac{-4V_r}{GF(1+2V_r)}\left(1 + \frac{R_L}{R_g}\right) \quad (1)$$

Where V_r is the ratio of measured voltage output to excitation voltage, GF is the gauge factor specified by the strain gauge manufacturer, and R_L and R_g are the lead wire and strain gauge resistances, respectively. The gauge factor for the KFH-6-350-C1-11L1M2 is 2.04 and the ratio of $R_L/R_g = 0.001$. Elastic modulus was determined from the slope of the stress vs strain relationship in one direction only, not using an hysteresis curve as the samples were tested until failure. ASTM D638 does not specify use of an hysteresis curve when estimating elastic modulus. Stress was computed from the measured load divided by the measured cross-sectional area of each specimen. Strain was calculated using the measured voltage per equation 1. The average measured cross sectional areas of the specimens is provided in Table 1.

Table 1. Average cross sectional area of virgin specimens.

Material	A_0, 10^{-5} m^2
FDM Nylon-12	3.43
FDM ULTEM	3.46
SLS Duraform EX	3.28
SLS Duraform HST	3.33
SLS Duraform PA	2.93

SLS GFR N12	4.01
SLS GFR N11	2.84

Specimens were mounted in 350 mL disposable cups prior to testing. Slits made for mounting were sealed using Devcon 5-minute epoxy. In order to prevent yielding at the tabbed regions, a funnel system was rigged to provide a constant stream of LN2 at both the top and bottom tabs (Figure 1). Tensile test measurements were conducted using a 4.45 kN load cell on a 49 kN Instron load frame at a constant-crosshead speed of 2.0×10^{-3} cm/s. LN2 was added until thermal stress reached steady state then the specimens were axially loaded while LN2 was continually added to keep the specimen submerged until failure. Liquid nitrogen was also ported to the specimen clamps in an attempt to maintain uniform material temperature. Temperature sensors were not mounted on the specimen so we could not record the precise temperature.

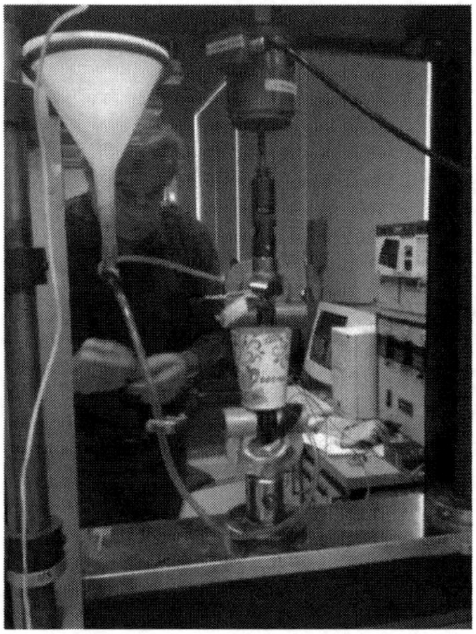

Figure 1. Mounted specimen prepared for tensile testing.

4. Results

Ultimate tensile strength and modulus are reported by specimen print orientation. The 77K ultimate tensile stress values increased for all plastics except for FDM Nylon-12 which decreased by 0.4% in the XY direction and decreased by 21% in the Z direction (figure 2).

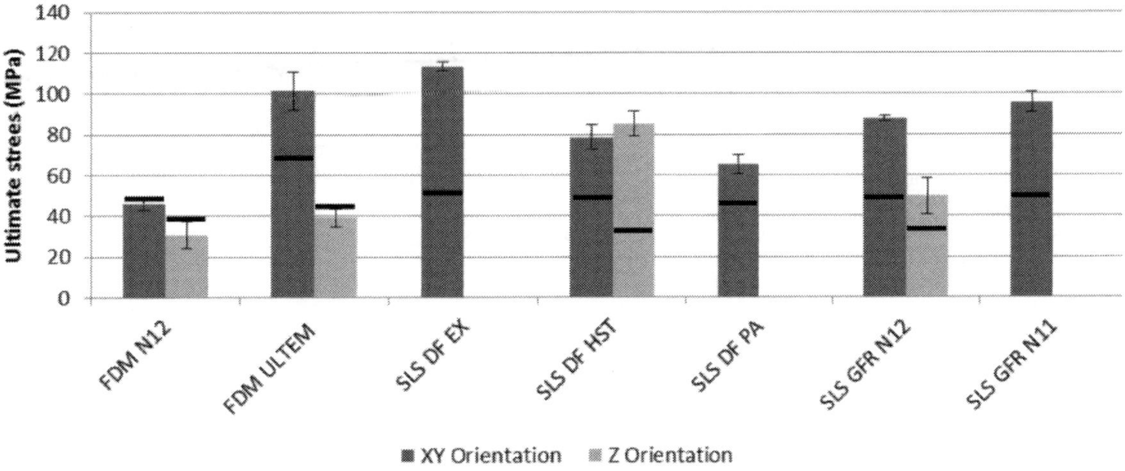

Figure 2. 77K mean ultimate stress and standard error. Black bars represent 3D printed material manufacturer's room temperature values.

Modulus increased for all plastics at 77 K, as shown in figure 3. Change in modulus at 77 K from room temperature is shown in Table 3.

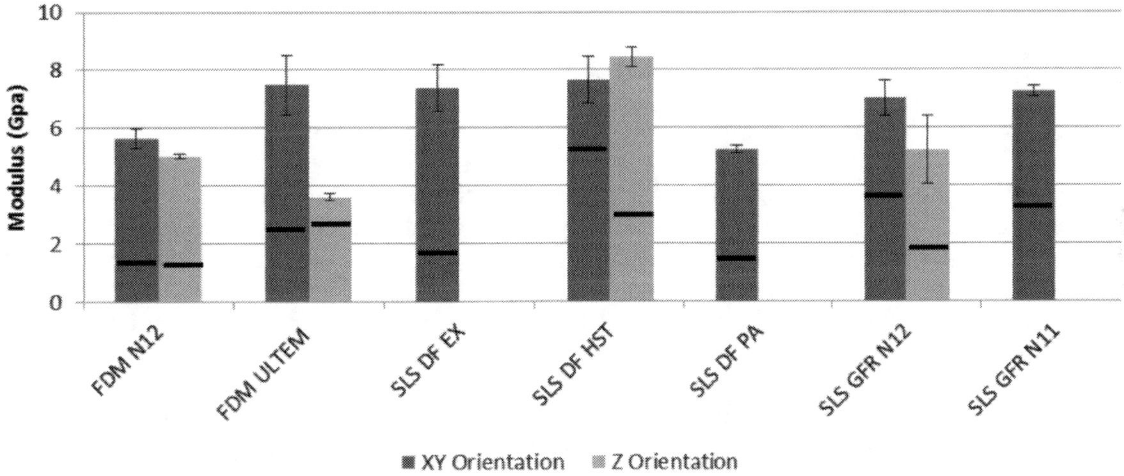

Figure 3. 77K mean modulus and standard error. Black bars represent material manufacturer's room temperature values.

Table 2. Ultimate stress and elastic modulus at 77K for XY/XZ and ZX print directions. Material manufacturer specification values are given in parentheses for comparison.

Process	Material	Orientation	Temperature (K)	Ultimate Stress (MPa)	Elastic Modulus (GPa)
Fused Deposition Modeling	N12	XZ	(295)	(46)	(1.28)
			77	45.8	5.63
		ZX	(295)	(38.5)	(1.14)
			77	30.6	5.02
	ULTEM	XZ	(295)	(69)	(2.15)
			77	101	7.48
		ZX	(295)	(42)	(2.27)
			77	39.1	3.60
Selective Laser Sintering	DF EX	XY	(295)	(48)	(1.52)
			77	113	7.37
		ZX	(295)	-[a]	-[a]
			77	-[a]	-[a]
	DF HST	XY	(295)	(50)	(5.73)
			77	78.6	7.64
		ZX	(295)	(33)	(3.0)
			77	85.3	8.45
	DF PA	XY	(295)	(43)	(1.59)
			77	65.3	5.25
		ZX	(295)	-[a]	-[a]
			77	-[a]	-[a]
	GFR N11	XY	(295)	(48)	(3.38)
			77	95.7	7.26
		ZX	(295)	-[b]	-[b]
			77	-[b]	-[b]
	GFR N12	XY	(295)	(49)	(3.82)
			77	87.7	7.01
		ZX	(295)	(33)	(1.95)
			77	49.5	5.23

[a] A ZX print direction was not available from the manufacturer.
[b] Samples weren't available for testing.

In general, the ultimate strength of 3D printed polymers increases at 77 K. The ultimate strength of material specimens printed using fused deposition modelling decreased for Nylon-12 in both the XY and Z print directions and decreased for Ultem in the Z print direction. Reduced mechanical properties in the Z print direction may be attributable to weaker inter-layer bond strength between successive layers deposited while printing that are normal to the tensile load. The reduction in ultimate strength for Nylon-12 in the XY print direction, parallel to the tensile load is not clear at this scale of testing and would require assessing the micro structure of FDM plastic to determine the micro-mechanical properties. The average ultimate stress of all materials tested is greater in the XY print direction than in the Z print direction at room temperature. The relationship holds at cryogenic temperature for all materials tested here except for SLS printed Duraform HST, in which the ultimate stress is greater in the Z print direction than it is in the XY print direction. The mean for each print direction is on the standard error bound for the opposite print direction indicating negligible difference in ultimate strength rating between the two orientations.

The mean ultimate stress values for SLS printed plastics (XY orientation) vary between 43 and 50 MPa at room temperature; a range of 7 MPa. At cryogenic temperature, ultimate stress varies between 65 and 113 MPa; a range of 48 MPa that represents a large increase in variability between the plastics at cryogenic temperatures. Duraform HST, glass fiber reinforced Nylon-11 and glass fiber reinforced Nylon-12 are considered fibre reinforced composites by their respective manufacturers. Glass fiber reinforced Nylon-12 is filled with carbon fibers and glass spheres to increase mechanical strength while reducing mass. It is not clear which component of the composite material is responsible for the increase in mechanical strength or why the differences are amplified at cryogenic temperatures.

The modulus of all 3D printed polymers increased at 77 K. At room temperature, the modulus for plastics printed in the XY direction is typically greater than the modulus for plastics printed in the Z direction except for Ultem.

5. Conclusions

Additive manufacturing opens a new pathway for creating complex geometry that would otherwise be too expensive or impossible using traditional manufacturing methods. Devices that operate at cryogenic temperatures and require the low thermal conductivity and high strength of engineering polymers can benefit from the advantages in additive manufacturing. We conducted tensile strength measurements at cryogenic temperatures of currently available engineering plastics that are being used in fused deposition modelling and selective laser sintering manufacturing methods. In general, the strength of the polymers we tested at 77 K that were manufactured using SLS increased in ultimate strength by 35.8% to 175%, while the modulus increased by 29.4% to 228% when compared to room temperature values. The ultimate strength of Duraform EX and Ultem decreased in 3 out of 4 cases by 0.4% to 20.6% while the modulus increased from 58.5% to 341% between room temperature and 77K. The general gains in strength and stiffness indicate 3D printed plastics have potential for use in cryogenic environments, although additional fatigue, shear strength, and notch sensitivity studies should be conducted prior to use in critical applications.

6. Acknowledgements

We would like to acknowledge the assistance of Robert Lentz in guiding the instrumentation process and for the use of the Instron load frame. Funding for this work was provided by the Washington Joint Center for Aerospace Technology and Innovation (JCATI) and the National Science Foundation (NSF) Research Experiences for Undergraduates (REU) grant # DMR 1062898.

References

[1] Perepechko, I. I. 1980 *Low temperature properties of polymers.*
[2] Hartwig G. 1997 *Polymer properties at room and cryogenic temperature.*
[3] Vega, V., et al., *The effect of layer orientation on the mechanical properties and microstructure of a polymer.* Journal of materials engineering and performance, 2011. **20**(6): p. 978-988.
[4] Gross, B.C., et al., *Evaluation of 3D Printing and Its Potential Impact on Biotechnology and the Chemical Sciences.* Analytical Chemistry, 2014. **86**(7): p. 3240-3253.

IOP Conf. Series: Materials Science and Engineering **102** (2015) 012021 doi:10.1088/1757-899X/102/1/012021

Thermal and mechanical properties of advanced impregnation materials for HTS cables and coils

N Bagrets, S Otten, K.–P. Weiss, A. Kario, W. Goldacker

Institute for Technical Physics, Karlsruhe Institute of Technology, Germany
E-mail: nadezda.bagrets@kit.edu

Abstract. In the growing field of high-temperature superconducting (HTS) applications, finding an appropriate impregnation material for cables and coils remains a challenging task. In HTS cables and coils, tapes have to be able to withstand mechanical loads during operation. Impregnation is playing a role as mechanical stabilization. However, material properties usually change significantly when going to low temperatures which can decrease performance of superconducting devices. For example, a large mismatch in thermal expansion between a conductor and impregnation material at low temperatures can lead to delamination and to degradation of the critical current. Impregnation materials can insulate tapes thermally which can lead to damage of the superconducting device in case of quench. Thus, thermal conductivity is an important property which is responsible for the temperature distribution in a superconducting cable or in a coil. Due to Lorentz forces acting on structural materials in a superconducting device, the mechanical properties of these materials should be investigated at operating temperatures of this device. Therefore, it is important to identify an advanced impregnation material meeting all specific requirements. In this paper, thermal and mechanical properties of impregnation material candidates with added fillers are presented in a temperature range from 300 K to 4 K.

1. Introduction

Impregnation of HTS devices is applied for mechanical reinforcement against Lorentz forces or to facilitate heat conduction in conduction–cooled coils. Common impregnation materials for superconducting as well as resistive coils are epoxy resins. Early attempts to impregnate HTS devices with such resins, however, showed that the thermal expansion mismatch can lead to damage of the conductor [1]. Therefore, alternative impregnation materials including epoxy resins with filler particles [2], paraffin [3], or introducing barrier between conductor and epoxy resin [4] are considered and applied. Filler particles decrease the thermal expansion mismatch with the conductor; beeswax, paraffin, or epoxy with shrink tube as barrier do not bond to the conductor and, in this way, prevent transmission of thermal stresses. The main aim of this work is characterisation and analysis of several commercially available resins for Roebel cable impregnation. Recently, it was shown that transverse mechanical loads lead to degradation of critical current of the Roebel cable due to its complex architecture [5]. However, such transverse stresses are unavoidable in superconducting magnets, therefore appropriate impregnation of the cable is needed.

In this work, the low–temperature properties of several commercially available filled epoxy resins are investigated. The thermal conductivity (TC) and thermal expansion (TE) are determined as a function of temperature in the range from 4 K up to 300 K. The tensile stress–strain behaviour is measured at room temperature, in liquid nitrogen ($T = 77$ K) and in liquid helium ($T = 4.2$ K).

2. Samples and measurement procedure

2.1 Materials description

Two groups of epoxies – candidates for impregnation of cables – were investigated: electrically conductive and insulating. Silver and graphite–filled epoxies (Duralco 125/127) were purchased from Polytec. Insulating resins were offered for testing by Huntsman Corporation. These are Araldite CY5538 with hardener HY5571, and Araldite CW5730N with hardener Aradur HY5731 and with $Al(OH)_3$ as a filler. Araldite CY5538 with hardener HY5571 is supplied unfilled. As filler in different mixing ratios (50 and 60 %), fused silica flour Silbond FW600 EST with a median grain size of 4μm was used. Araldite CW5730N is pre–filled with 56 wt% aluminium hydroxide. An overview of the materials tested is shown in table 1.

The samples for the thermal expansion measurements were prepared in a Teflon mould according to the instructions of the manufacturer. The resulting samples had a size of 60 mm x 10 mm x 5 mm. For TC measurements, cylindrical samples of 6mm diameter and 2–3 mm length were used. These were cut from the larger TE samples.

The samples for mechanical test were shaped according to the DIN EN ISO 527–2 norm. Those samples were machined from a 30 mm x 50 mm x 90 mm block. In the case of silica–filled resin, preparing such a large sample could result in agglomeration of particles leading to an inhomogeneous sample. To avoid this, the mould was rotated continuously around its horizontal axis during the curing process.

Table 1. Overview of tested materials.

	Viscosity [Pa.s]/ Processing temperature [°C]	Filler	Filling ratio [wt%]	Product name
Electrically conductive fillers				
epoxy+silver	20/20	Silver	60-80	Duralco 125
epoxy+graphite	50/20	Graphite	50-60	Duralco 127
Electrically insulating fillers				
Epoxy	<0.2/40			Araldite MY750/ Aradur HY5922
epoxy+silica 60	<4.5/80	Silbond FW600 EST	60	Araldite CY5538/ Aradur HY5571-1
epoxy+silica 50	<5/80	Silbond FW600 EST	50	Araldite CY5537/ Aradur HY5571s-1
epoxy+$Al(OH)_3$	0.7/60	$Al(OH)_3$	56	Araldite CW5730N/ Aradur HY5731

2.2 Measurement methods

The thermal conductivity is measured with the axial heat flow method within the Physical Property Measurement System of Quantum Design using a Thermal Transport Option in the steady–state measurement mode [6]. The thermal conductivity k is defined as

$$k = \frac{P\Delta x}{A\Delta T} \qquad (1)$$

where P is the heat power, A is the cross–sectional area, Δx is the distance between the two measurement points with the temperature gradient ΔT.

Thermal expansion measurements were performed in the temperature range from 4.5 K up to 290 K in a glass–fibre reinforced plastic (GFRP) l^4He–bath cryostat. For sample deformation measurements clip-on extensometers are used [7]. The samples with extensometers mounted on it is then placed in a ^{4}He-bath cryostat and cooled down in a flow of He–gas. Temperature sensor was placed directly on the sample. At the end, a moderate amount of liquid He is filled. The measurement is performed when all liquid ^{4}He is evaporated and the temperature in the cryostat slowly rises due to the non-ideal insulating vacuum. The warm-up takes approximately 12 hours for the whole temperature range from 4.5 K to 290 K at an average rate of < 0.5 K/min.

For load-displacement measurements, a 100 kN tensile testing machine was used. Tests were performed according to DIN EN ISO 527–2. For measurements of the elongation, a system of extensometers was used, which was mounted directly on the sample. The start length is defined as the distance between the tips of extensometers. Further, the elongation of the sample results in a deformation of the extensometer frames and change in output voltage signal, which provides the real elongation via a calibration factor. The strain is then evaluated as

$$\varepsilon = \Delta L/L_0, \qquad (2)$$

where ε is a strain, ΔL is an elongation, and L_0 is a start length.

3. Results

As it was mentioned above, thermal expansion is an important property for impregnation of cables and coils since it determines the level of thermal stresses between epoxy and *RE*BCO tapes. It differs by up to two times for different epoxies and fillers. TE in temperature range from RT down to 4.5 K is shown in Table 2. TE of *RE*BCO tape, epoxy with alumina (Stycast® 2850FT blue) [8] are shown for comparison.

Table 2. TE of different resins in temperature range from 4.5 K up to RT.

Material	TE RT-77 K %	TE RT-4.5 K %
epoxy+silver	-0.96	-1.04
epoxy+graphite	-0.53	-0.58
Epoxy	-1.04	-1.15
epoxy+silica 60	-0.51	-0.60
epoxy+silica 50	-0.61	-0.67
epoxy+$Al(OH)_3$	-1.02	-1.11
Stycast® 2850FT	-0.43	-0.49
*RE*BCO tape	-0.25	-0.26

The TE of unfilled epoxy is several times larger than that of *RE*BCO tape. Pure epoxy has a thermal expansion of -1.15 % going from RT to 4.5 K. Since filler materials themselves have lower TE, all fillers decrease the thermal expansion as it was expected. Epoxy with silver and epoxy with $Al(OH)_3$ filler shows the thermal expansion of about -1% going from RT to 4.5 K. Effect of adding fillers is much more pronounced for epoxies with silica and graphite – their TE is about -0.5 – -0.7 %, which is much closer to TE of a *RE*BCO tape.

TC measurements results are shown in Fig. 1. Thermal conductivity of epoxies with good conductive fillers (aluminium, graphite, silver) is much higher than that of pure epoxy resin. However, effect of fillers is much more pronounced for RT and 77 K than for 4 K. At 4 K thermal conductivity of all resins is quite low – around 0.05 W/Km (except of epoxy+silver, which TC is approx. three times higher) – and the effect of fillers doesn't show any clear trend.

Pure epoxy has lowest thermal conductivity. Adding non-metallic fillers increases TC. Adding silica fillers doesn´t show that large impact on TC, whereas graphite filler increases TC appreciably. Epoxy with silver shows highest TC values, due to high thermal conductivity of silver.

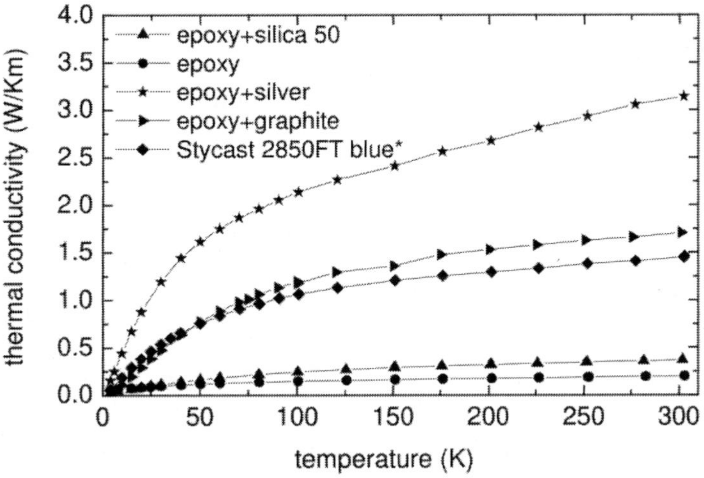

Fig.1 Thermal conductivity as a function of temperature for different epoxies with fillers TC of *Stycast® 2850TF blue [8] is shown as well for comparison.

For impregnation of Roebel cables and coils epoxies with fillers with possibly lower TE und higher TC are desired. However not all materials described above can be recommended for this application. One reason is high viscosity of electrically conductive resins (see table 1) that impedes epoxy flow into open spaces within the cable. Conductive resins are not needed for impregnation of coils, however they can be considered for other HTS applications where electrical connection is desired as moderate coupling of strands in cables.

Taking into account TE, TC and viscosity two epoxy's were chosen for mechanical tests pure epoxy (Araldite MY750/HY5922) and epoxy+silica 50 (Araldite CY5537/HY5571–1 with Silbond FW600 EST).

In Table 3 the main mechanical properties – Young´s modulus, proof strength $R_{p0.1}$, and ultimate strength – are summarized for all samples at RT, 77 K, and 4.2 K. The proof strength was evaluated using the offset method with 0.1 % of strain.

Table 3. Mechanical properties of epoxy and epoxy with filler at different temperatures.

Material	Temperature K	Young´s modulus GPa	$R_{p0.1}$ MPa	Ultimate strength MPa
Epoxy	RT	3.06	38	57.6
Epoxy	77	6.67	125	177
Epoxy	4.2	7.28	156	158
epoxy+silica 50	RT	6.91	48	66.4

epoxy+silica 50	77	13.92	-	149
epoxy+silica 50	4.2	15.33	-	101

Adding silica filler to epoxy increases Young´s modulus approx. twice. The data obtained for epoxies correlate well with data given in literature [9]. Young´s moduli for unfilled epoxy are very similar to that reported in [9], whereas for filled epoxy we got higher values as reported in [9] for other commercially available filled resins.

Ultimate strength is higher for unfilled resin than that filled for at 77 K, and 4 K. At room temperature ultimate strength is slightly higher for filled resin. Ultimate strength shows the same trend and similar values for filled and unfilled resin, as it was reported in [9]. Ultimate strength is higher at 77 K than that at RT, and it is lower for 4 K than that for 77 K.

4. Conclusions

The thermal expansion, thermal conductivity and mechanical behaviour under tensile stress were analysed for several filled epoxy resins in the temperature range 4.5–293 K. It was found that all tested fillers (silver, graphite, fused silica and aluminium hydroxide) lead to a significant reduction of the thermal expansion, which will decrease thermal stresses when cooling down in impregnated devices. Filler particles, especially silver, increase thermal conductivity. However, this increase is not pronounced at liquid helium temperature. Fused silica filled resin showed an increased Young's modulus and decreased ultimate strength in the entire temperature ranges. The first impregnation result of Roebel cable with fused silica filled resin was recently shown [10]. Sufficient improvement in stability of cable against transversal stress was achieved, however still some degradation of critical current was reported. Trials on cable impregnation avoiding current reduction are ongoing.

The resins investigated were qualified for cryogenic usage; the needs of the specific HTS applications have to refer to the individual enhanced properties, where desired strength, heat conduction and thermal expansion play a role.

Acknowledgments

We would like to thank Huntsman Corporation for providing epoxy resins and Quarzwerke GmbH for providing the silica flour.

[1] Takematzu T, Hu R, Takao T, Yanagisawa Y, Nakagome H, Ugliette D, Kiyoshi T, Takahashi M, Maeda H 2010 *Physica C.* **470.1718** 674

[2] Barth C, Bagrets N, Weiss K, Bayer C, Bast T 2013 *Superconductor Science and Technology.* **26.5** 055007

[3] Matsumoto S, Kiyoshi T, Otsuka A, Hamada M, Maeda H, Yanagisawa Y, Nakagome Y, suematsu H 2012 *Superconductor Science and Technology.* **25.2** 025017

[4] Trociewitz U P, Dalban-Canassy M, Hannion M, Hilton D K, Jaroszynski J, Noyes P, Viouchkov Y, Weijers H W and Larbalestier D C 2011 *Appl. Phys. Lett.* **99** 202506

[5] Uglietti D, Wesche R and Bruzzone P 2013 *Superconductor Science and Technology.* **26** 074002

[6] PPMS - Thermal Transport Application Note [Online]. Available: www.lot.oriel.com/uk/en/home/ppms, 2011.

[7] Nyilas A. *Conf. Proc. 824 Advances in Cryogenic Engineering: Transactions of the International Cryogenic Materials Conference-ICMC.* **52,** 27.

[8] Bagrets N, Goldacker W, Schlachter S I, Barth C, Weiss K. *Cryogenics.* **61** 8

[9] Reed R P, Walsh R P, *Conf. Proc. Advances in Cryogenic Engineering: Transactions of the International Cryogenic Materials Conference-ICMC.* **40,** pp. 1129-1136.

[10] Otten S, Dhallé M, Gao P, Wessel W, Kario A, Kling A, Golgacker W 2015, *Superconductor Science and Technology.* **28** 065014

IOP Conf. Series: Materials Science and Engineering **102** (2015) 012022 doi:10.1088/1757-899X/102/1/012022

Thermal and mechanical properties of selected 3D printed thermoplastics in the cryogenic temperature regime

K-P Weiss[1], N Bagrets[1], C Lange[1], W Goldacker[1], J Wohlgemuth[2]
[1]Institute for Technical Physics, Karlsruher Institute of Technology, Hermann von Helmholtz Platz 1, 76344 Eggenstein-Leopoldshafen, Germany
[2]Institute of Functional Interfaces, Karlsruher Institute of Technology, Hermann von Helmholtz Platz 1, 76344 Eggenstein-Leopoldshafen, Germany

E-mail: klaus.weiss@kit.edu

Abstract. Insulating materials for use in cryogenic boundary conditions are still limited to a proved selection as Polyamid, Glasfiber reinforced resins, PEEK, Vespel etc. These materials are usually formed to parts by mechanical machining or sometimes by cast methods. Shaping complex geometries in one piece is limited. Innovative 3D printing is now an upcoming revolutionary technology to construct functional parts from a couple of thermoplastic materials as ABS, Nylon and others which possess quite good mechanical stability and allow realizing very complex shapes with very subtle details. Even a wide range of material mixtures is an option and thermal treatments can be used to finish the material structure for higher performance. The use of such materials in cryogenic environment is very attractive but so far poor experience exists. In this paper, first investigations of the thermal conductivity, expansion and mechanical strength are presented for a few selected commercial 3D material samples to evaluate their application prospects in the cryogenic temperature regime.

1. Introduction

The application of High Temperature Superconductors is actually expanding to a couple of new fields, in particular space and airborne applications. The design and development of future low or zero emission airplanes is a challenge which requires the integration of the technology of superconductivity, which will be applied in different parts of the vehicle as power generator, electrical combustion drives and transfer lines depending on the philosophy of the approach. The advantage of the high energy density and low losses of superconducting wires and cables allows very compact designs, meets however the disadvantage of a necessary thermal insulation by a containment with vacuum shield, a cryostat, to provide the operation temperature of superconductors and to restrict thermal losses. Such systems are not only expanding the required space but also contribute to the payload of the airplane or space vehicle. Ultra-low weight solutions for superconducting component are not developed so far since they need to use low weight materials, in particular plastics, resins and polymers. Only a limited number of such materials, Polyamides (PA), PEEK, PU etc. are qualified for cryogenic temperatures so far and used. But a large number of other candidates have a good potential and need to be investigated. Plastics, resins and polymers have the disadvantage of a high thermal contraction compared to metals and superconductors, which leads in composites to a thermal mismatch upon cooling to cryogenic temperatures. Dispersion and glass-fiber reinforced are

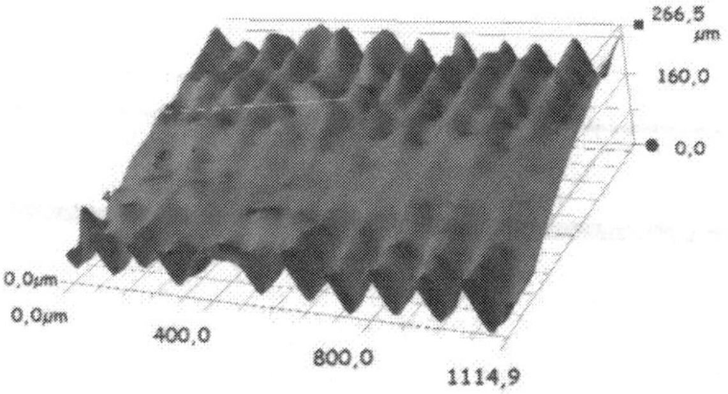

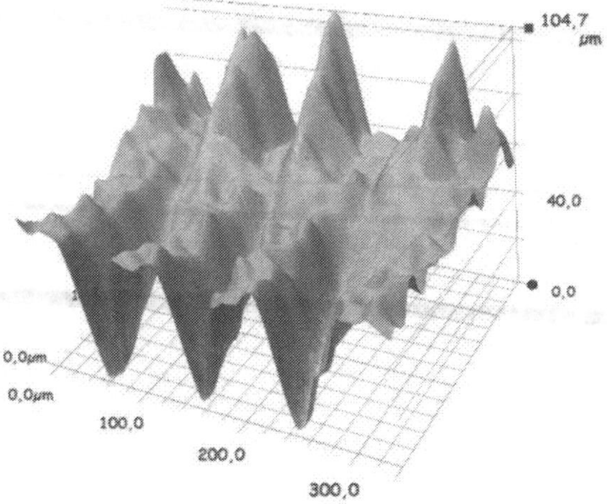

Figure 1. Structure of ABS to investigate the printed layer thickness using the surface profile obtained by optical microscope (left 300x and right 1000x magnification).

materials used to adjust the thermal contraction to those of metals, alloys and composites and to increase the mechanical performance. 3D-printed thermoplastics are so far rarely considered for such cryogenic applications but are promising candidates for complex cryogenic parts and constructions. Rare or no material data on mechanical and thermal properties however are available for this temperature regime. It is the goal of this work to start first investigations of the mechanical properties (tensile stress and compression), the thermal expansion and the thermal conduction on some examples of 3D printed materials to get first data at temperatures in the range of 4-300K.

2. 3D Printing methods of plastics

In this paper the choice of materials was limited to the used printers and therefore to ABS and the photopolymer VeroWhite fullcure 830™, an advanced plastics developed by Stratasys. Two 3D printer with different print technology were used for manufacturing the test specimen. The main difference between the methods is the resolution, layer thickness of the pixels, the densification of the material and the surface and geometry accuracy.

2.1. Fused Deposition Modeling

The first device was a modified Ultimaker-Original for Fused Deposition Modeling, which is an easy

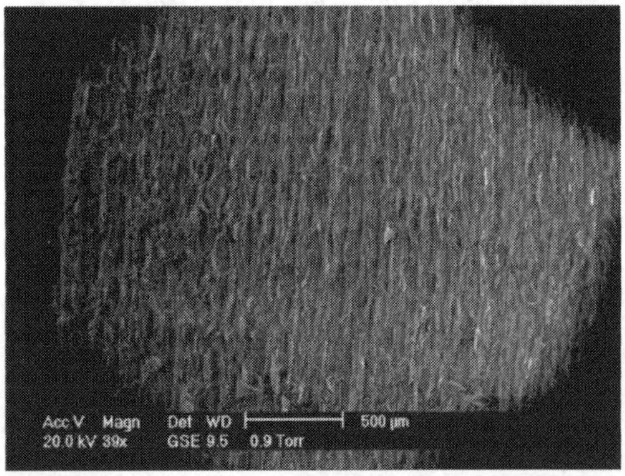

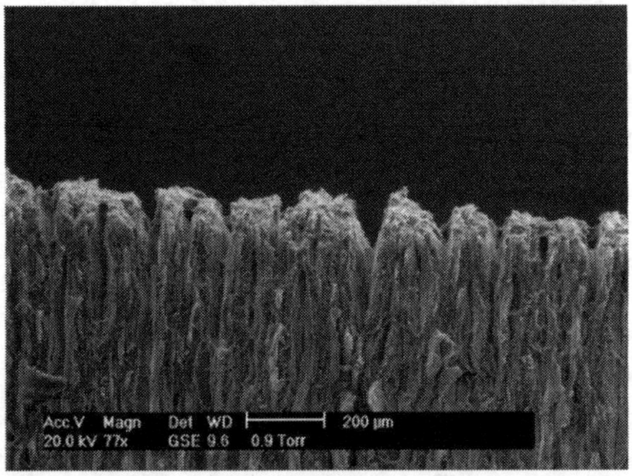

Figure 2. Structure of Polyjet-Modeling material VeroWhite™ to visualize the thickness of the printed layers (pictures by ESEM).

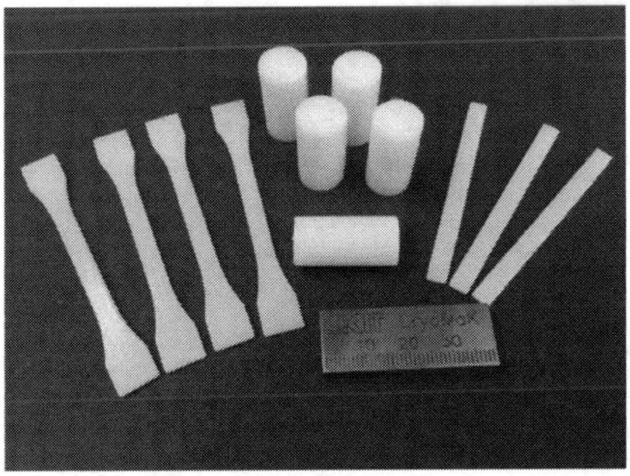

Figure 3. Specimen geometries manufactured by 3D printing method. Shown specimen for tensile (left), compression (middle), and thermal expansion (right) measurement.

and low cost method. The resolution range and sheet thickness is about 100 μm. It allows melting of e.g. ABS (Acrylnitril-Butadien-Styrol) or PLA (Polylactacide) material at ~245°C by depositing on glass plate with a temperature of ~100°C. For this work ABS was chosen. In Fig.1 the layer structure of about 100 μm thickness is visible. The surface has evenly distributed grooves along the deposited sheets with approximately 50 μm depth.

2.2. Polyjet-Modeling

The second approach was Polyjet-Modeling using a Stratasys EDEN 260V device with a resolution in the x,y-plane of ~42 μm with a possible sheet thickness of ~17 μm. The temperature of the system during printing was ~50-60°C. The deposited material is flattened by a roller followed by hardening of the acrylate photopolymer (see datasheet VeroWhite Fullcure830™ [4]) resin performed by UV radiation. This procedure will lead to a more dense body, leak tight regarding liquids [7]. A separate support material can be included into the printed model to create voids. That material can be removed by rinsing with sodium hydroxide solution. Pictures of the surfaces are shown in Fig.2 indicating again the layered structure, indicating a comparable sheet thickness of about 100 μm.

3. Material Properties

To allow an assessment of the cryogenic material properties it was decided to manufacture specimen geometries for thermal expansion, thermal conductivity, as well as mechanical tensile and compressive measurements. In Fig. 4 the shape of the produced specimen by 3D printing method is shown, the dimensions are identical for the ABS and the VeroWhite™ material.

3.1. Thermal expansion

Rectangular beams with a length of about 60 mm and a cross sectional area of about 5 x 2 mm^2 were used to measure the thermal expansion. Details regarding the measurement method can be found in [1, 2]. The contraction behaviour from 290 K down to 4.2 K is depicted in Fig. 5, the typical error is in the range of about 15 μm/m. As expected for such polymers the contraction is in the range of about 12500 μm/m and 15000 μm/m of VeroWhite™ and ABS, respectively. These results in this temperature range are comparable to other plastics like PEI (9100 μm/m), PEEK (10200 μm/m), or PA (13800 μm/m) [2]. On the contrary composite materials like G10 (glass-fibre reinforced epoxy, thermally matched to steel, copper or Nb_3Sn superconductors) exhibit a significant lower thermal expansion due to their glass-fibre addition (normal direction: 6000 μm/m and warp direction: 2400 μm/m) [3]. However, to avoid high mechanical stresses due to a significant mismatch of connected materials, special care needs to be taken.

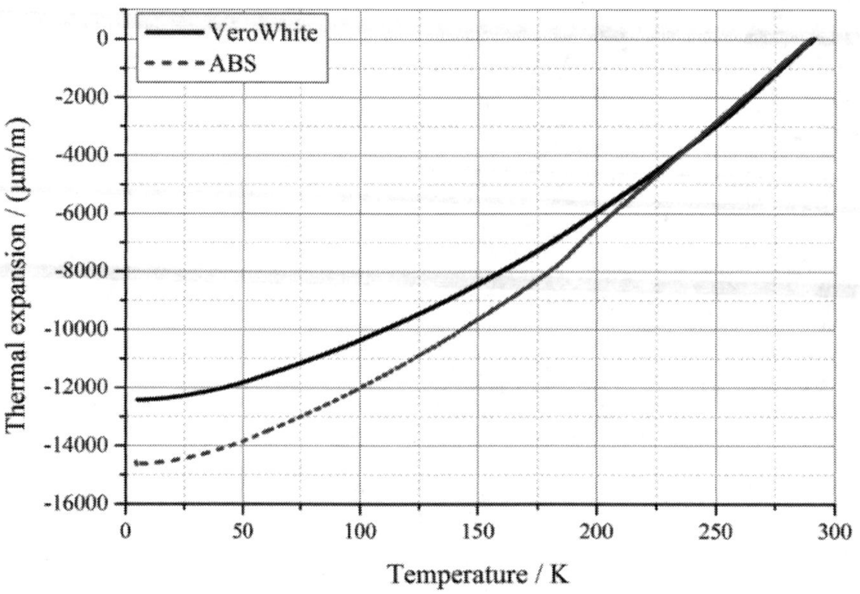

Figure 4. Thermal contraction between 290 K and 4.2 K of ABS and VeroWhite™.

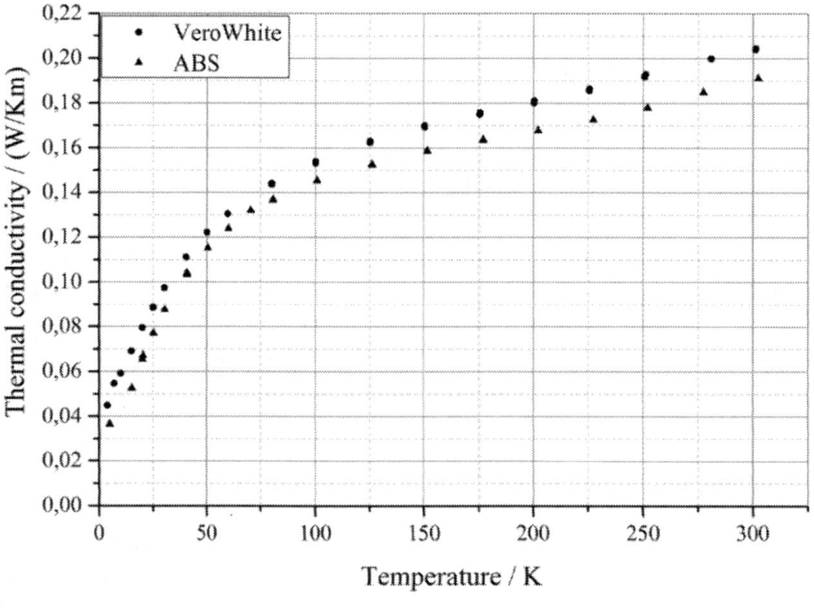

Figure 5. Thermal conductivity between 300K and 4.2 K of ABS and VeroWhite™.

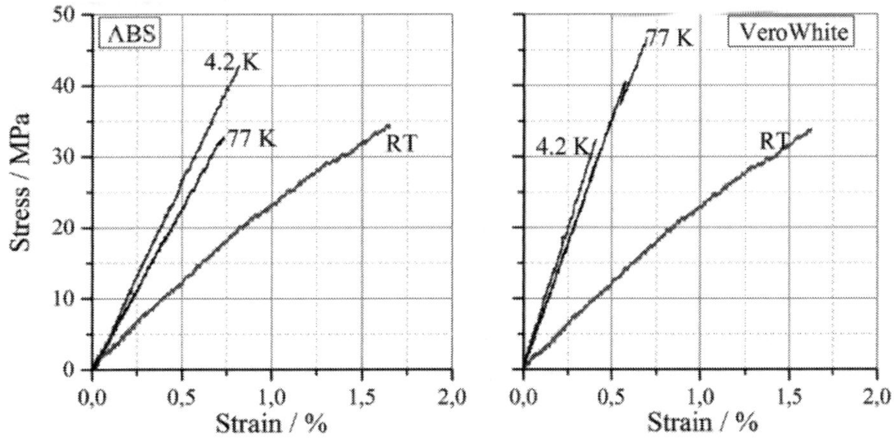

Figure 6. Tensile test results of ABS (left) and VeroWhite™ (right) at RT, 77 K and 4.2 K.

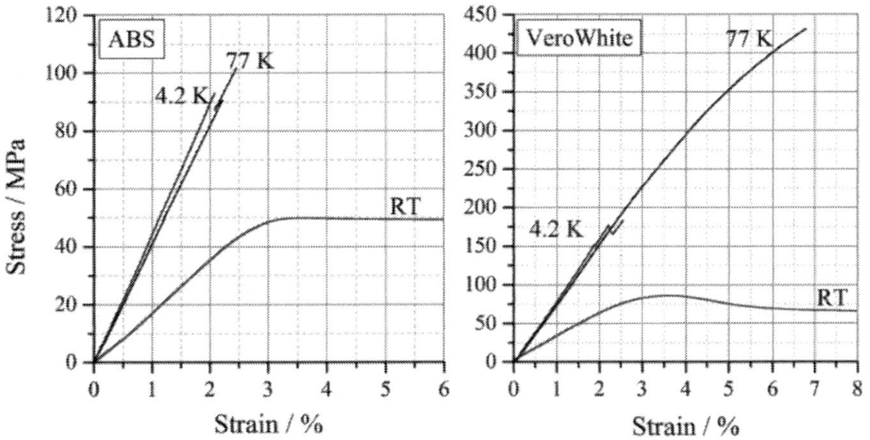

Figure 7. Compressive test results of ABS (left) and VeroWhite™ (right) at RT, 77 K and 4.2 K.

3.2. Thermal conductivity

To investigate the thermal conductivity a *Physical Properties Measurement System PPMS* was utilized [1, 2]. Circular sheets with a diameter of 6 mm and a thickness of up to 2 mm were used to determine the thermal conductivity applying the two-point measurement arrangement. The typical deviation is below 0.01 W/(K m) [2]. In Fig.6 the measurement results are summarized. The thermal conductivity at room temperature is in the range of 0.2 W/(K m) and decreases for both materials to approximately 0.04 W(K m). These low values are expected for such insulating materials. Similar values can be found for pure Araldit or Polyimid (PI, e.g. Kapton) [2]. Compared to G10 composites at 300 K (normal direction: 1 W/(K m) and warp direction: 0.7 W/(K m)) where the glass filler material increases the thermal conductivity slightly [2]. These results encourage the possibility to use such materials as thermal insulators for cryogenic applications.

Table 1. Mechanical material properties tested at RT, 77 K and 4.2 K.

Material	Temp.	Tensile Youngs Modulus, GPa	Tensile Yield strength, MPa	Tensile Ultimate Strength, MPa	Compressive Modulus, GPa	Compressive Yield Strength, MPa	Compressive Ultimate Strength, MPa
Polyjet-Modeling							
VeroWhite	RT	2.497 (2.5)[a]	32	36 (50)[a]	3.108	81	86
	77K	6.850	-	46	7.813	255	431
	4K	8.876	-	32	7.752	-	183
Fused Deposition Modeling							
ABS	RT	2.583 (2.4)[b]	33	35 (45)[b]	1.591	49	50
	77K	4.476	-	32	4.143	-	101
	4K	5.570	-	43	4.397	-	93

Data at RT from datasheet [a][5], [b][6]

3.3. Mechanical Properties

Beside the thermal properties mechanical characteristics of such materials are of interest. Therefore tensile test specimen were printed and tested according to DIN EN ISO 527-2. Compressive tests were performed on cylindrical 12.7 mm diameter and 25.4 mm high specimen following ASTM D695.

The tests were performed at room temperature of about 293 K (RT), liquid nitrogen temperature (77 K) and immersed in liquid helium (4.2 K). The elastic Young's modulus was determined along with the ultimate strength, and if appropriate the yield strength at 0.2% offset is given. In Table 1 all results are listed. At RT the expected values given by the supplier is added in brackets for information. Taking into account the calculated uncertainty of the obtained values following [8, 9] and a coverage probability 95.45% ($k = 2$) the Young's modulus has $u_E = 1.011$ GPa, the compressive modulus about $u_C = 0.074$ GPa and the yield or ultimate strength $u_{YS/UTS} = 2.120$ MPa. However taking into account the structural integrity, e.g. porosity or interlayer bonding, the estimated uncertainty is significantly higher. Further investigations like reproducibility and micrographs of the cross section area can help to clarify this point. Still, the obtained values are remarkably close to the room temperature values of the supplier.

Looking at the results in Table 1 the typical increase of the Young's Modulus of such materials is found. However, the tensile tests give low strength values around 40 MPa. The largest difference is in the compressive strength, where the VeroWhite™ material is at least double in strength than the ABS for all temperatures measured. This might be attributed to the production route of the Polyjet-Modeling, as each layer of the VeroWhite™ material is slightly deformed by rollers during printing, leading to an increased network structural integrity (see Fig 2).

The same behaviour can be observed in artificially reinforced G10 material [3]. Comparing the obtained values form this work with measurements on G10 material the tensile behaviour normal to the glass-fibres at 4 K is in the same range of about 30 MPa. However due to the reinforcement of the epoxy by glass-fibres the structural integrity is superior in fibre direction (145 MPa). As already stated due to the net structural integrity of the glass fibres, the G10 material exhibits compressive strength values far above 200 MPa allowing the application of such materials for structural reinforcement at cryogenic temperatures.

4. Conclusion and outlook

First investigations on 3D-printed materials for cryogenic temperatures are already promising, although the material choice was limited to the availability of resources for this first step. Two different manufacturing routes (*Fused-Deposition-Modeling and Polyjet-Modeling*) were used to

allow the investigation of the advanced polymer VeroWhite™ with ABS as reference, being aware that ABS is inferior for cryogenic use.
Within this work the samples investigated were printed in the direction of the flat side. An open question is the existence of an anisotropy of properties with respect to the print direction (Fig. 2). Additionally, further tests are necessary to determine the uncertainty of the measured.
The thermal properties obtained show the typical behaviour of such materials leading to very good thermal insulation properties. However, due to the thermal expansion combined with moderate mechanical properties special care has to be taken to avoid failure due to stresses because of temperature gradients during cool down or mismatch of contraction as composite material.
Further investigations regarding the possibility to print composite structures to customize the material properties regarding the needs of specific applications have to be done.
In future the investigations will be extended to a couple of other 3D-printed materials, including investigations on high voltage insulation, thermal cycling, and permeability of gases.

5. References

[1] Bagrets N, Weiss E, Westenfelder S and Weiss K-P 2012 *IEEE Trans. Appl. Supercond.* **22** NO.3 9501204
[2] Barth C 2013 *KIT Scientific Publishing* ISBN 978-3-7315-0065-0 p.53-68
[3] Weiss K-P, Westenfelder S, Jung A, Bagrets N and Fietz W 2012 *AIP Conf. Proc.* **1435** 148
[4] http://isites.harvard.edu/fs/docs/icb.topic993207.files/Verowhite.pdf (23 June 2015)
[5] http://www.3dprinterscanada.com/assets/uploads/ProductBrochures/FullCurePDF3DPrinters Canada.pdf (23 June 2015)
[6] http://www.kern.de/cgi-bin/riweta.cgi?nr=2110&lng=1 (23 June 2015)
[7] Hübner J, Brakowski R, Wohlgemuth J, Brenner-Weiß G and Franzreb M 2015 *Engineering in Life Sciences* DOI: 10.1002/elsc.201400171
[8] Urbach E, Bagrets N, Weiss K.-P 2013 *Fusion Engineering and Design* **88** 2631
[9] ISO/IEC 98-3 Guide, Uncertainty of measurement. Part 3: Guide to the expression of uncertainty in measurement (1995)

ANALYTICAL INVESTIGATION IN BENDING CHARACTERISTIC OF TWISTED STACKED-TAPE CABLE CONDUCTOR

Makoto Takayasu[1] and Luisa Chiesa[2]

[1] MIT Plasma Science and Fusion Center, Cambridge, MA 02139, U.S.A.

[2] Tufts University, Mechanical Engineering Department, Medford, MA 02155, U.S.A

takayasu@psfc.mit.edu

Abstract. An analytical model to evaluate bending strains of a twisted stack-tape cable (TSTC) conductor has been developed. Through a comparison with experimental results obtained for a soldered 32-tape YBCO TSTC conductor, it has been found that a Perfect-Slip Model (PSM) taking into account the slipping between tapes in a stacked-tape cable during bending gives much better estimation of the bending performance compared to a No-Slip Model (NSM). In the PSM case the tapes can slip so that the internal longitudinal axial strain can be released. The longitudinal strains of compression and tension regions along the tape are balanced in one twist-pitch and cancel out evenly in a long cable. Therefore, in a cable the strains due to bending can be minimized. This is an important advantage of a TSTC conductor. The effect of the cable diameter size on the bending strain is also expected to be minor, and all tapes composing a TSTC conductor have the same strain response under bending, therefore the cable critical current can be characterized from a single tape behaviour.

1. Introduction

The second generation High Temperature Superconductor (HTS) Rare-Earth Barium Copper Oxide (REBCO) tapes are very attractive to various applications of high field magnets and transmission power cables. Cabling methods for the HTS flat tapes cabling such as Roebel Assembled Coated Conductor (RACC) [1], Conductor-On-Round Core (CORC) [2], Twisted Stacked-Tape Cable (TSTC) [3] and a few other alternatives [4], [5] have been proposed and are being investigated for high current, high field applications.

We have been developing TSTC cabling method, which consists of stacking flat tapes and twisting them along the stack axis. This compact cabling technique using REBCO tapes is very useful for both high field magnet conductors and power transmission. TSTC conductors has been proposed and fabricated by several methods, including sheathing the stack of tapes with a copper tube and embedding the stack in single and multiple helical grooves formed in a circular rod [3], [6]. In real applications, the bendability of a REBCO tape cable is very important to effectively fabricate and transport a long cable and ultimately wind a magnet [5], [6]. We have experimentally examined bendability of a TSTC conductor [7]. Bending characteristic of various TSTC conductors is investigated by an analytical calculation method. Two models, No-Slip Model (NSM) and Perfect-Slip Model (PSM), are presented, and critical current degradation due to bending is discussed.

2. Analysis Modelling

When a TSTC conductor is circularly bent as shown in figure 1, the bending neutral axis of the cable is at the center of its axis, which is on the bending plane. Therefore the portions of the tape located on the outside of the cable neutral plane experiences tension, while the inner portions experience compression. Note that the cable neutral plane means a circular plane determined by the cable neutral axis, which is perpendicular to the cable bending plane.

Since the thickness of the coated conductor tape and the superconductor layer thickness are much smaller than the other dimensions (tape width, the cable diameter and the cable bending radius), the tape thickness effect on the bending strain is neglected in this calculation.

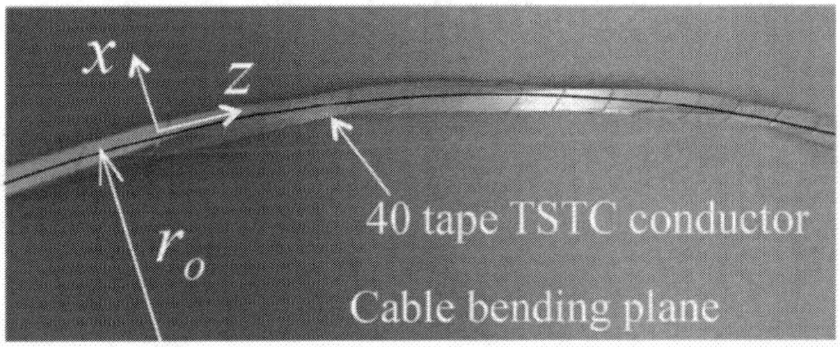

Figure 1. Illustration of a 40 tape Twisted Stacked-Tape Cable (TSTC) conductor bent with the bending radius r_o=500 mm.

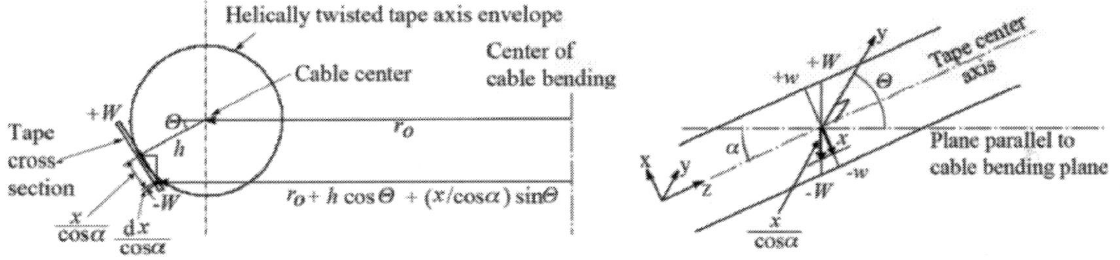

(a) (b)

Figure 2. Schematic illustrations showing a bending model of a TSTC conductor: (a) a cross-section view of a tape in a TSTC conductor bent with the radius r_o, and (b) an enlarged view of a part of a twisted tape.

The distance between the tape center axis and the cable center axis is defined by h (named the tape off-center distance in this manuscript), as shown in figure 2(a). Figure 2(b) illustrates an enlarged view of a segment of a twisted tape, which is one of the tapes composing the stacked-tape cable. A segment on the tape is denoted by x in the tape width direction and z in the tape center axis, as shown in the figure. The axial bending strain ε_b of an element ($dx \cdot dz$) on a tape due to a twisted-tape cable bending is given for the bending radius r_o as

$$\varepsilon_b = \frac{\left(\frac{x}{\cos\alpha}\right)\sin\theta + h\cos\theta}{r_o}\cos\alpha \tag{1}$$

$$= \frac{x\sin\theta + h\cos\theta\cos\alpha}{r_o} \tag{2}$$

where Θ is the twist angle, and α is the cable twist angle of a TSTC conductor which are given by the tape off-center distance h and the twist-pitch L_p as follows,

$$\alpha = \tan^{-1}(2\pi h/L_p) \tag{3}$$

$$\theta \equiv \frac{2\pi z \cos\alpha}{L_p} \approx \frac{2\pi z}{L_p} \tag{4}$$

It is noted that in the present analysis of the bending strain of a TSTC conductor composed of twisted tapes, the strains induced in the tapes during the twisting of the stack are neglected, as they are negligible compared with the bending strain. Therefore we calculate only the additional strain due to bending. It was shown that the twisting strain effect on the critical current for a TSTC conductor of 4 mm REBCO tapes for twist-pitches longer than 200 mm is much smaller than the bending strains discussed here [7].

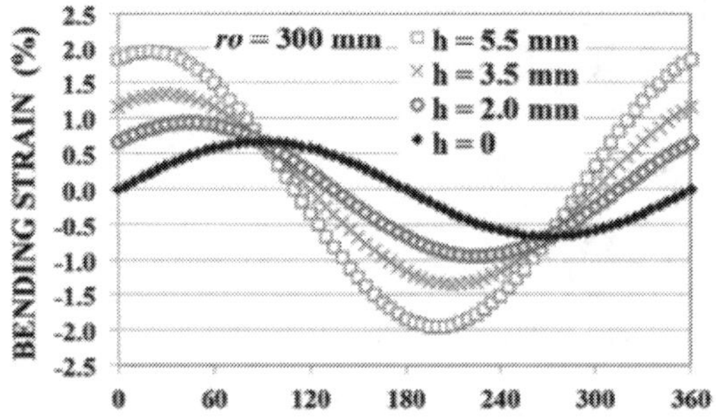

(a)

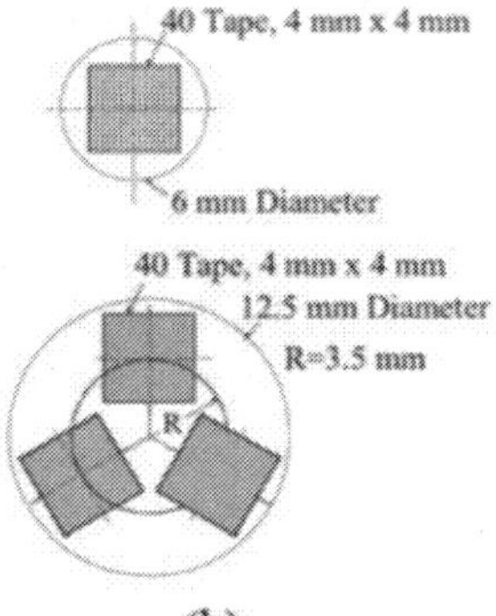

(b)

Figure 3. (a) The bending strain distribution at $x = +w$ for NSM along a twisted tape of 4 mm width and 200 mm twist-pitch bent with the radius of r_o=300 mm with various tape off-center distance h values up to 5.5 mm. (b) The insert figures show a single stack cable with $0 \le h \le 2$ mm and three-stack cable with $1.5 \le h \le 5.5$ mm.

Figure 3 shows the bending strains at $x = +w$ for one of the 4 mm tapes of a twisted stacked-tape cable with a twist-pitch of 200 mm. The bending strain is evaluated using equation (2) as a function of the twist angle Θ with varios values of the tape off-center distance h. Figure 3(b) shows two TSTC conductors where the tapes have various h-values. Distributions of the bending strain ε_b significantly change with the tape off-center distance h. The second term, $h \cdot \cos\alpha \cdot \cos\Theta$, in the numerator of equation (2) contributes the internal longitudinal strain along a tape. As seen in figure 3, the larger the tape off-center distance h is, the larger are the bending strains.

We consider two limiting cases to understand the bending behaviour of a TSTC conductor: No-Slip Model (NSM) and Perfect-Slip Model (PSM). In the case of the No-Slip Model, the tapes in the stacked-tape cable cannot slip on one-another and behave like a solid block during bending due to high friction between the tapes. On the other hand, in the case of the Perfect-Slip Model the tapes can slip freely as friction is assumed to be zero between the tapes.

In the case of PSM, equation (2) can be simplified. In the right hand of equation (2) the first term of x reflects a strain due to the tape width, and the second tem of h component induces a longitudinal strain along a tape. If the tapes can freely slip in a stack during bending, the longitudinal strain can be released by slipping. In a TSTC conductor tapes are twisted, therefore compressive and tensional stresses in a full twist-pitch balance and cancel out over the short length of one twist-pitch even for a long cable. Therefore the longitudinal strain due to the second term of equation (2) can be disregarded. In this case the bending strain ε_b from equation (2) reduces to

$$\varepsilon_b = \frac{x \sin\theta}{r_o} \tag{5}$$

This equation is independent of the tape off-center distance h.

The strain distributions along both tape edges ($x = \pm w$) for both NSM and PSM were calculated from equations (2) and (5), respectively, and are shown in figure 4. As expected, in the case of PSM described by equation (5), all stacked tapes composing a cable experience the same strain distribution, as the strain does not depend on the tape off-center distance h. The neutral axis of each tape is the center axis of the tape. The distribution of the bending strains ε_b along the width is symmetrical with equal amounts of compression and tension, as seen in figure 4(b).

Bending strains for various bending radiuses for both NSM with the tape off-center distance h=2 mm and PSM obtained from equations (2) and (5), respectively, are shown in figure 5. These figures show the bending strains at the tape edge $x = +w$. The peak strains exceed 2%, and they become very large in the NSM case. The large strain resulted from those bending values will be used for the critical current evaluation in the following calculations. We should notice that the large strains will cause the yielding of the material and permanent degradations of the critical current. Including those phenomena in the models discussed in this paper is beyond the scope of this work.

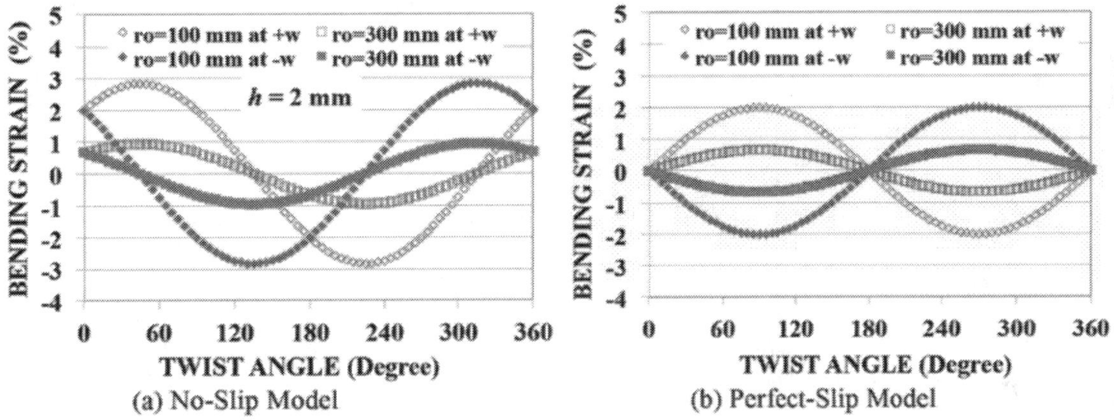

Figure 4. Bending strain distributions for a tape of 4 mm width and the twisted pitch 200 mm along a tape as a function of the twist angle Θ. The curves show the strains at the edges of $x = \pm w$ for the bending radius r_o=100 mm and 300 mm. (a) No-Slip Model (NSM) with h=2 mm, and (b) Perfect-Slip Model (PSM).

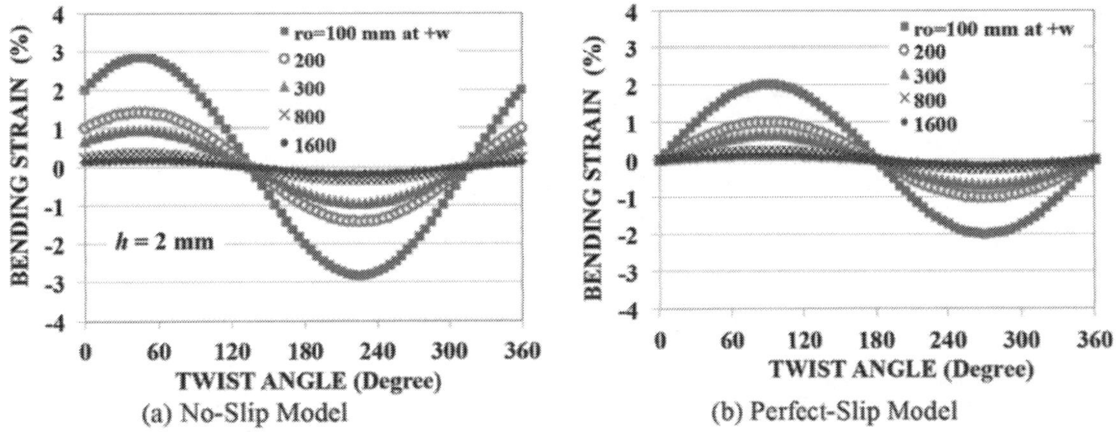

Figure 5. Bending strains at the tape edge x=+w for (a) No-Slip Model (NFM) with h= 2 mm and (b) Perfect-Slip Model (PSM) calculated for various bending radiuses r_o. The tape width, $2w$, and twist-pitch, L_p, are 4 mm and 200 mm, respectively.

3. Critical Current

The critical current locally varies with the bending strains at locations given with x in the width and z or the twist angle Θ along the length. To simplify the calculation of the critical current of the twisted bent tape, first the critical current is averaged over the width between $x = +w$ and $x = -w$ in order to evaluate the strain distribution effect over the width. The average critical current $I_{c\Theta}$ over the width is given by integration,

$$I_{c\theta} = \frac{1}{2w}\int_{-w}^{w} I_c\,(\varepsilon_{b\theta},\ x)\ dx \tag{6}$$

The critical current as a function of the strain was approximated using reported data, which was given by the following polynomial equation [3], [8],

$$\begin{aligned} {}^{I_c}/_{I_{co}} &= 0.057713\times10^{12}\,\varepsilon_{b\theta}{}^{6} - 0.03979215\times10^{10}\,\varepsilon_{b\theta}{}^{5} - 0.2090279\times10^{8}\,\varepsilon_{b\theta}{}^{4} \\ &+ 0.02385557\times10^{6}\,\varepsilon_{b\theta}{}^{3} - 0.1668065\times10^{4}\,\varepsilon_{b\theta}{}^{2} - 0.003\,662115\times10^{2}\,\varepsilon_{b\theta} + 1.0 \end{aligned} \tag{7}$$

This equation was obtained by curve fitting based on the experimental data of the strain ranges between about -1.0% and +0.52%, however in this analysis the critical current was evaluated using the equation for strain ranges beyond those values.

Equation (6) was solved by a numerical method using the Gaussian integration method of order 40 with Microsoft Excel®. The strain $\varepsilon_{b\Theta}$ is given by equation (2) for NSM and equation (5) for PSM.

The critical current evaluated with equation (6) is given as a function of the twist angle Θ along the tape length. That is, the critical current varies along the length. When the tape current is given, the voltage distribution during a transition from superconducting to normal state varies as shown in figure 6. Here, the electric field distribution calculated for a given strain using equation (5) (PSM case) and $E/E_c=(I/I_c)^n$ with n=25 is shown as a function of the twist angle Θ along the length. The critical current of a tape in the stacked tape cable is obtained from the criterion of E_c (100 μV/m) for the integrated total voltage of one full twist-pitch length L_p. Similar curves can be obtained for the NSM case and various h values.

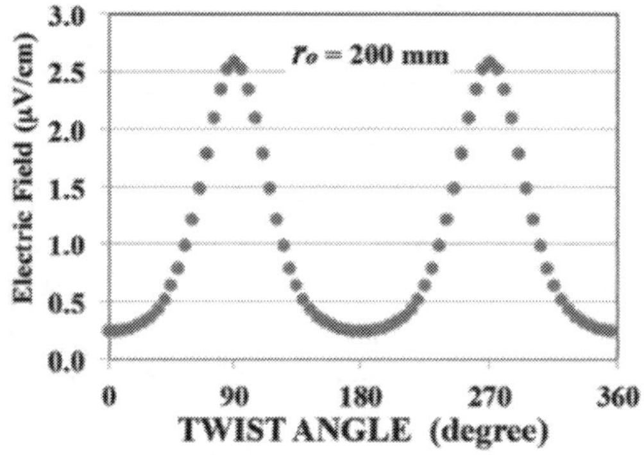

Figure 6. Electric-field distribution calculated from bending strain distribution along the tape of one twist pitch when the total voltage of one twist pitch reached the critical current criterion of E_c. Tape width $2w$=4 mm, twist-pitch L_p=200 mm, h=0, n=25, r_o=200 mm.

The total voltage of one twist-pitch at the tape current I_o is given by

$$V = \int_0^{L_p} \Delta V\ dz = \int_0^{L_p} E_c\left(\frac{I_o}{I_{c\theta}}\right)^n dz = E_c\, I_o{}^n \int_0^{L_p} I_{c\theta}{}^{-n}\ dz \tag{8}$$

On the other hand the voltage V of one twist-pitch L_p is given for a criterion E_c by V= $E_c\ L_p$. Using equations (8) and (3), the tape critical current $I_{c\ Tape}$ for a twisted bent tape is given by

IOP Conf. Series: Materials Science and Engineering **102** (2015) 012023 doi:10.1088/1757-899X/102/1/012023

$$I_{c\,Tape} = \sqrt[n]{\frac{L_p}{\int_0^{L_p} I_{c\Theta}{}^{-n}\,dl}} = \sqrt[n]{\frac{2\pi}{\int_0^{2\pi} I_{c\Theta}{}^{-n}\,d\Theta}} \tag{9}$$

The tape critical current was evaluated with Θ=5 degree step analysis using Microsoft Excel®. In this case equation (9) is reduced to

$$I_{c\,Tape} = \sqrt[n]{\frac{L_p}{\int_0^{L_p} I_{c\Theta}{}^{-n}\,dl}} = \sqrt[n]{\frac{72}{\sum I_{c\Theta}{}^{-n}}} \tag{10}$$

The critical current of a tape was analysed using equation (10). Cable critical current of a cable composing of k tapes can be obtained from $k\,I_{c\,Tape}$ for PSM as all the tapes experience the same strain during bending.

4. Critical Current Results Analysed and Comparison with Experimental Results

Figure 7 shows the critical current characteristics with various values of the tape off-center distances h as a function of the bending diameter. They are obtained for the tape width $2w$=4 mm, the twist pitch L_p=200 mm and n=25 using equation (10). The critical currents degrade significantly as the tape off-center distance h increases. The critical current for the PSM case corresponds to the case of h=0 in figure 7.

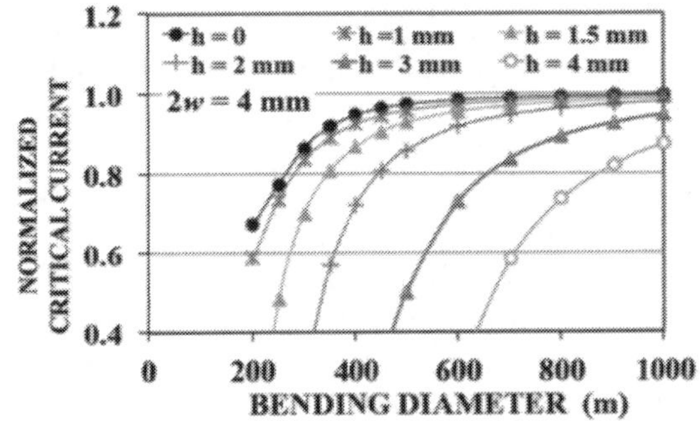

Figure 7. Normalized critical current in NSM for various values of h as a function of the bending diameter. Tape width $2w$=4 mm, twist-pitch L_p=200 mm, and n=25. The solid circles of h=0 also apply to the critical current in PSM (The same solid circle curve is shown in figure 8.).

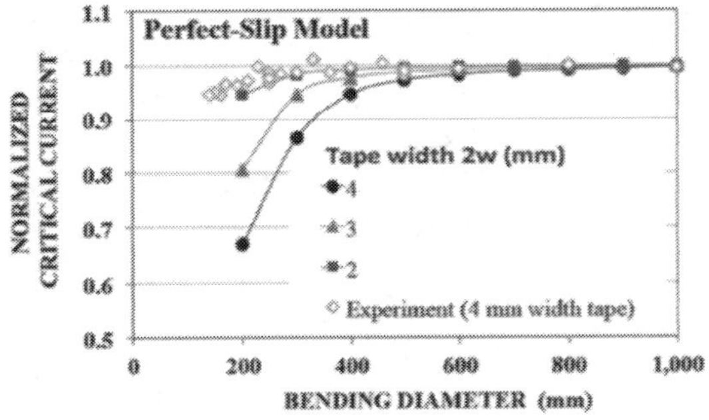

Figure 8. Normalized critical current for PSM with various tape width $2w$ as a function of the bending diameter, comparing with an experimental result of 4 mm width, 32-tape soldered YBCO TSTC conductor. Twist-pitch L_p=200 mm, and n=25.

Figure 8 shows the critical currents in PSM (h=0) for 2 mm, 3 mm and 4 mm width tapes with experimental data obtained for a 32-tape, 4 mm width cable made of SuperPower YBCO tapes. The cable was soldered before bending in a straight condition [7]. A narrower width tape has smaller degradation due to bending. The calculated critical currents for $2w$=4mm degrade much more with the bending diameter than those obtained experimentally. The experimental result for 4 mm tape was similar to the calculation value for 2 mm width. The experimental work, as mentioned in [7], was performed to wind a soldered 32-tape TSTC conductor on the side surface of a disk of a given diameter without a groove. The cable was not forced to make a very tight winding with the disk, so that the overall cable bending diameter was correctly achieved. This caused some local waving of the cable due to the nature of a twisted cable made of stacked flat-tapes. Such short range waving of the cable was observed in a few places. This might be a reason why the experimental result showed less critical current degradation than the calculated one (locally the tapes experienced less bending). Overall this result together with what observed in figure 7 (NSM vs. PSM) suggests that the model analysis of Perfect-Slip Model (PSM) can give a good evaluation for bending characteristics of a TSTC conductor and it is more appropriate in describing the bending behaviour of a TSTC compared to the No-Slip Model (NSM).

5. Discussion

The maximum value of the bending strain, ε_{bMax} for NSM is given from equation (2) as

$$\varepsilon_{bMax} = \frac{\sqrt{w^2+(h\cos\alpha)^2}}{r_o} \tag{11}$$

This maximum value is established at the hard bending (in-plane bending) [9], [10] where a flat tape bends in the tape plane. For a TSTC conductor the maximum value of h will be approximately the size of the cable radius. Bending strain increases with h. However, our experiments of the critical current tests for TSTC conductors did not show the significant degradations expected from the strain evaluated with equation (11).

In the right end side of equation (11) the first term of w reflects a strain due to the tape width, and the second term of h component induces a longitudinal strain. Given the experimental observations (lower degradations than what evaluated with equation (11)) we can conclude that the tapes in a TSTC conductor are not rigidly held and that the PSM case could be acceptable and more appropriate in describing the behaviour of a TSTC conductor under bending. Therefore it is reasonable to say that the slipping of 2G stacked tapes in a TSTC conductor can occur, and the tapes can freely slip in a stack during bending. The stacked tapes are twisted, therefore in each half cycle of the twisting the strains change from tension to compression. It follows that the internal longitudinal strains given by the h term in the bending strain equation can be released in a full twist-pitch even for a long cable. This is an important and beneficial consequence of the twisting of a TSTC conductor.

In the slipping condition (PSM) the bending strain of equation (2) can be reduced to equation (5), which becomes independent of the tape off-center distance h. All tapes of a TSTC conductor experience the same bending strain. That is, a cable critical current can be obtained from multiplying the number of tapes with the single tape critical current.

Based on the Perfect-Slip Model (PSM), the critical current of a twisted tape was evaluated for one full twist-pitch length and compared with experimental results. The agreement was not good as seen in figure 8, however it was much better than what estimated with the No-Slip Model (NSM) as seen in figure 7. The calculated result for PSM showed more degradations than the experimental results. This could be due to the soft bending applied in the experiment to respect the natural irregularity during cable bending. Taking into account this observation the analytical results seem to show a fairly good agreement with the experimental results.

In the present work the tape width and the twist pitch were 4 mm and 200 mm, respectively. As seen in equation (5) for the PSM case, the bending strain is proportional to the width. A narrower tape gives less degradation for bending. The twist-pitch L_p in equation (3) affects cosα in the strain of equation (2), however its effect on the bending strain is minor.

6. Conclusions

In order to analyse bending characteristics of a HTS Twisted Stacked-Tape Cable (TSTC) conductor No-Slip Model (NSM) and Perfect-Slip Model (PSM) have been developed. Comparing with experimental results, it was found that the PSM case fits the experimental results much better than the NSM case.

In the NSM case, 2G tapes are rigidly stacked and they do not slip between tapes. On the other hand in the PSM case the tapes can slip so that the internal axial strain can be released. The longitudinal strains of compression and tension regions along the tape are balanced in one twist-pitch and cancel out even for a long cable. Therefore the strains due to a cable bending can be minimized. Additional strains over the width are induced during the bending of the cable. This strain causes the critical current degradation due to bending. The bending strain effect on the critical current of a TSTC conductor was evaluated for various bending diameters.

The Perfect-Slip Model analysis has shown that the tape distance (the tape off-center distance h) from the cable center does not affect the magnitude of the strain due to bending, since a tape can slip and release the longitudinal strain. The bending strains are induced equally into all stacked tapes in a TSTC conductor. Therefore the cable current can be characterized from the behavior of one single tape, and the total cable current can be obtained from a multiplication of the current of one tape by the number of tapes. Further investigation to explain the discrepancy between experimental results and the model analysis will be necessary using both experiments of various TSTC conductors and Finite Element Analysis.

Acknowledgements
This work supported by the U. S. Department of Energy, Office of Fusion Energy Science under Grants: DE-FC02-93ER54186.

References

[1] Goldacker W, Frank A, Kudymow A, Heller R, Kling A, Terzieva S and Schmidt C 2009 Improvement of superconducting properties in ROEBEL assembled coated conductors (RACC) *IEEE Trans. Appl. Supercond.* **19**, 3098–3101

[2] van der Laan D C, Noyes P D, Miller G E, Weijers H W and Willering G P 2013 Characterization of a high-temperature superconducting conductor on round core cables in magnetic fields up to 20 T *Supercond. Sci. Technol.* **26** 045005

[3] Takayasu M, Chiesa L, Bromberg L and Minervini J V 2012 HTS twisted stacked-tape cable conductor *Supercond. Sci. Technol.* **25**, 014011

[4] Celentano G, De Marzi G, Fabbri F, Muzzi L, Tomassetti G, Anemona A, Chiarelli S, Seri M, Bragagni A and della Corte A 2014 Design of an Industrially feasible twisted-stacked HTS cable-inconduit conductor for fusion application *IEEE Trans. Appl. Superconduct* **24** 4601805

[5] Uglietti D, Wesche R and Bruzzone P Design and strand tests of a fusion cable composed of coated conductor tapes *IEEE Trans. Appl. Superconduct* **24** 4800704

[6] Takayasu M, Minervini J V, Bromberg L, Rudziak M K and Wong T 2012 Investigation of twisted stacked-tape cable conductor *Adv. Cryo. Eng.* **58** Plenum, N.Y. 273-280

[7] Takayasu M, Mangiarotti F J, Chiesa L, Bromberg L, and Minervini J V 2013 Conductor characterization of YBCO twisted stacked-tape cables *IEEE Trans. Appl. Superconduct* **23** 4800104

[8] Data of 2G Strain effect measured by Danko van der Laan, NIST, available at http://superpower-inc.com, SuperPower technical document in 2008: Xie Y Y 2008 Recent results in 2G HTS wire technology development at SuperPower Korea Superconductivity Society Annual Meeting

[9] van der Laan D C, and Ekin J W 2008 Dependence of the critical current of $YBa_2Cu_3O_{7-\delta}$ coated conductors on in-plane bending *Supercond. Sci. Technol.* **21** 115002

[10] Markiewicz W D, and Swenson C A 2010 Winding strain analysis for YBCO coated conductor *Supercond. Sci. Technol.* **23** 045017

IOP Conf. Series: Materials Science and Engineering **102** (2015) 012024 doi:10.1088/1757-899X/102/1/012024

Critical Current Properties of HTS Twisted Stacked-Tape Cable in Subcooled- and Pressurized-Liquid Nitrogen

M Tomita[1], K Suzuki[1], Y Fukumoto[1], A Ishihara[1], T Akasaka[1], Y Kobayashi[1], A Maeda[1], and M Takayasu[2]
1)Applied Superconductivity, Materials Technology Division, Railway Technical Research Institute, 2-8-38 Hikari-cho, Kokubunji-shi, Tokyo 185-8540, Japan
2) Plasma Science and Fusion Center, Massachusetts Institute of Technology, Cambridge MA 02139 USA

E-mail: tomita@rtri.or.jp

Abstract. A 2 m length Twisted Stacked-Tape Cable (TSTC) conductor which was fabricated by 32-YBCO-tapes (4 mm width) with a 200 mm twist pitch was investigated at various temperatures near 77 K in subcooled- and pressurized-liquid nitrogen. The critical current of the TSTC cable which was 1.45 kA at 77 K measured from 64 K to 85 K by controlling the equilibrium vapor pressure of nitrogen bath and were varied from 3.65 kA at 64 K to 0.42 kA at 85 K. The temperature dependence of cables' critical current agrees with that of the 4 mm width YBCO tape. These results are encouraging for applications of a compact Twisted Stacked-Tape Cable application in railway systems.

1. Introduction

Railway Technical Research Institute (RTRI) in Japan has developed superconducting cable applications for railway systems. The aim of this development has been to investigate a new system to feed electric power to electric trains by using high temperature superconducting (HTS) cables to replace copper cables. Introduction of superconducting cables in railways could significantly reduce the voltage drop in feeder. Therefore, DC HTS electrification has various advantages, such as the reduction of transmission loss, the increase of regeneration efficiency, and fewer substations. More specifically, the HTS feeder cable could result in significant energy savings, while improving the power handling capacities of feeder cables and railway substations [1].

At present, we are developing a power testing system using a conventional superconducting power transmission type cable with helically wound HTS tapes on a cylindrical former, operated at 77 K in liquid nitrogen [1]. However the current density and cooling of the power cable are not sufficient especially for the catenary lines above ground as a feeder. From a practical application perspective in railway systems, we have investigated a compact superconducting cable system based on a TSTC conductor of REBCO HTS tapes [2].

2. Experimental details

The TSTC cable is typically made from twisting freestanding flat tapes torsionally along the axis of stack. A 2 m long, 32-YBCO-tape (4 mm width) TSTC cable with a 200 mm twist pitch was tested at various temperatures from 64 K to 85 K in subcooled- and pressurized-liquid nitrogen using a variable temperature power cable test system.

2.1. TSTC conductor

To investigate of a TSTC conductor performance for electric railway applications, an existing 32-tape YBCO cable of about 2 m length was used. The cable is composed of 32-YBCO tapes made by SuperPower Inc. The YBCO HTS tapes were 4 mm width and 0.1 mm thick (SCS4050-AP, 2010), and the average critical current was 85 A at 77 K in self-field. The tapes were stacked between copper strips 0.82 mm thick and 4.8 mm width, and twisted together with a twist pitch 200 mm. This TSTC conductor shown in Figure 1 was originally fabricated and used for a bending [2], [3].

2.2. Cable Termination

The YBCO layer of a HTS tape is deposited on a base substrate, therefore it has an asymmetric conductivity and the electrical conductivity of the YBCO layer side (conductive side) with the silver layer is better than that of substrate side. Given this asymmetry in electrical conductivity, to fabricate a termination of YBCO tapes, the conductive side must be properly soldered or connected to an electric conductor of copper. The YBCO-BSCCO termination method developed earlier was used [2]. To assemble the YBCO-BSCCO terminator, 32 YBCO tapes were sandwiched between the BSCCO tapes after cleaning all contact surfaces, and then they were clamped with 6.4 mm thick, 70 mm length stainless steel plate without soldering. The contact pressure of about 110 MPa between YBCO and BSCCO tapes was applied by fourteen 6–32 screws with a known torque measured with a torquemeter. The copper tube of the terminator was clamped into a copper block to be connected to a current lead of the cryostat. The total length of the cable which was encapsulated with a glass sleeve for electrical insulation was 2335 mm including the terminations (about 222 mm each end) which could fit in the test cryostat.

2.3. Test cryostat system

Figure 2 shows a cryostat used for a TSTC conductor test [4]. A 2 m cable can be mounted in the horizontal corrugated pipe between the current lead towers. Test sample current-terminations were connected to copper terminators with mechanical clamping at the bottom of the current leads. Side flange on each current lead tower can be opened to mount a sample and to make connections of current leads. The temperature of operation during the test whose values were determined using platinum resistance temperature sensors mounted at the center of the conductor as shown in Figure 1 and installed near the bottom of each current lead was controlled with nitrogen vapour pressure (0.0146-0.2290 MPa) by adjusting the inner cryostat pressure using a vacuum pump or nitrogen gas cylinder with pressure controlled valves shown in Figure 2. The TSTC conductor was inserted horizontally into the cryostat from the side flange windows. An overall view of the cryostat is shown in Figure 2.

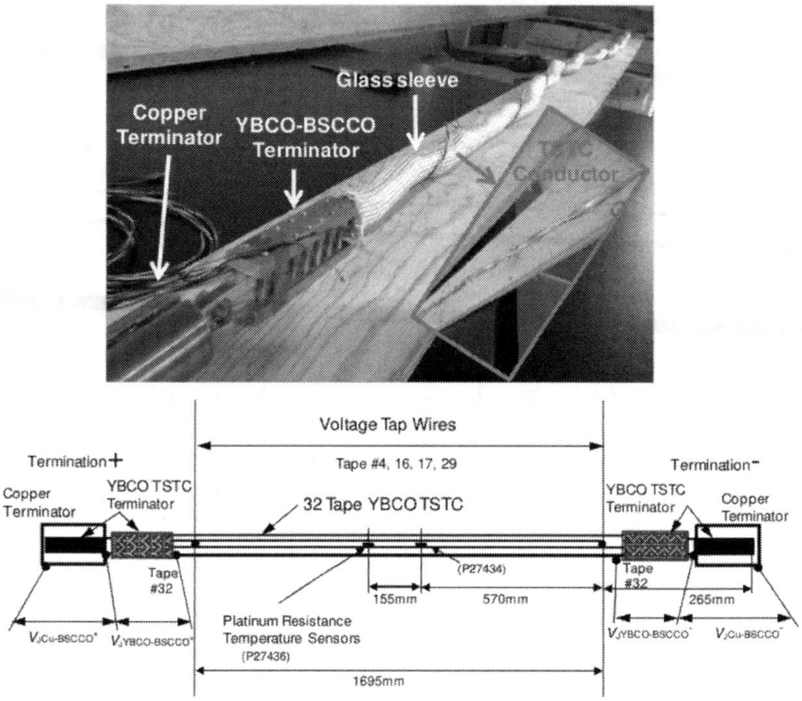

Figure 1. 32-tape YBCO TSTC conductor encapsulated with a glass sleeve for insulation before mounting in the cryostat with instrumentation wires: four voltage tap wires on Tapes #4, #16, #17 and #29, and two termination resistance measurements at each termination, and two platinum resistance temperature sensors on the conductor.

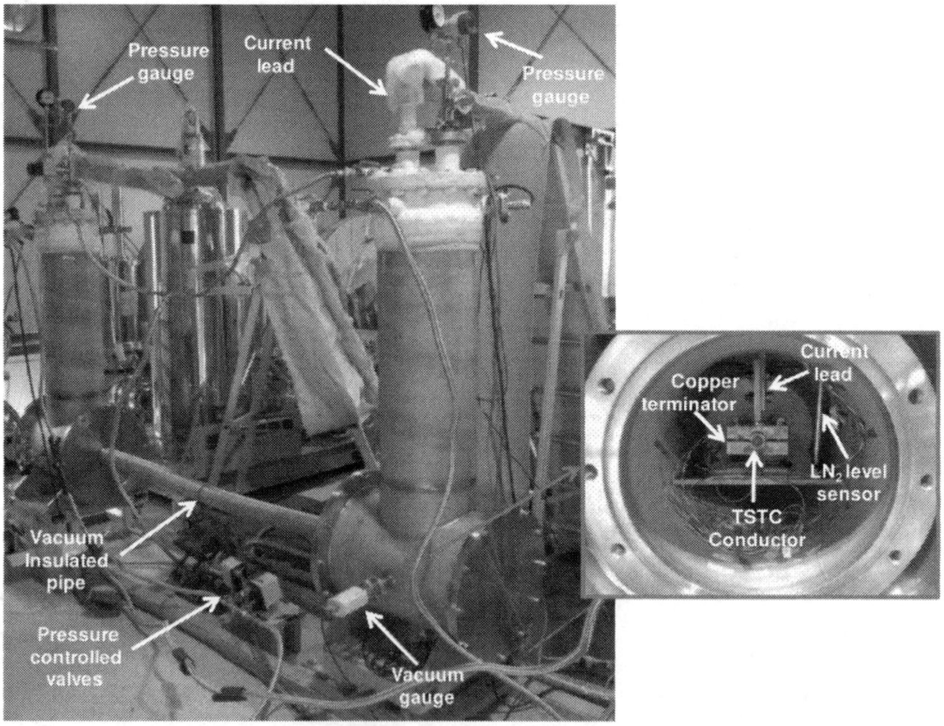

Figure 2. Test cryostat.

3. Results and discussion

3.1. Test Results of 32-Tape TSTC conductor

DC current of the maximum capacity of 7200 A was supplied. For the critical current measurement, four pairs of voltage tapes mounted on four tapes selected in the 32-tape stack (#4, #16, #17 and #29), as shown in Figure 1. To measure the termination resistance, two pairs of voltage taps were provided at each termination: one to measure the YBCO-BSCCO joint resistance between YBCO cable and BSCCO tapes ($V_{jYBCO\text{-}BSCCO}$), and other to measure the contact resistance between the copper tube of YBCO-BSCCO terminator and the copper terminator ($V_{jCu\text{-}BSCCO}$). The termination resistances were mesured: $R_{jYBCO\text{-}BSCCO}$ (Positive side) 13 nΩ, $R_{jYBCO\text{-}BSCCO}$ (Negative side) 10-18 nΩ, $R_{jCu\text{-}BSCCO}$ (Positive side) 5.0 $\mu\Omega$, and $R_{jCu\text{-}BSCCO}$ (Negative side) 1.2 $\mu\Omega$. These joint resistances between YBCO cable and BSCCO tapes for both sides were as low as expected, however the contact resistance between the copper tube of YBCO-BSCCO terminator and the copper terminator were very high. The high resistances were resulted from the diameter mismatch between the copper tube diameter and the copper terminator. These resistances can be reduced by one order of magnitude using properly fitted adapters.

A 32-tape TSTC conductor was evaluated under various temperatures between 64 K and 85 K. During the test, the temperature was decreased first from 77 K to 64 K, and then next day it was raised from 77 K to 85 K by 1 K step. During the experiment the *V-I* curves were plotted and critical currents and *n*-values were analyzed. All the *V-I* curves, critical currents and *n*-values measured between 85 K and 64 K are plotted in figures 3 and 4.

The critical current was 1421 A at 77 K in self-field, and varied between 421 A at 85 K and 3620 A at 64 K, (I_c at 64 K is roughly 8.6 times the I_c at 85 K).

We have evaluated the performance of a HTS tape power conductor taking into account of the self field effect on the degradation. It was based on the analytical averaged self-field of a TSTC conductor and the "minimum" critical currents of a single tape as a function of the angle between the tape and the magnet field applied at low magnetic fields [5]. To evaluate the critical current temperature dependence of a 32-tape TSTC conductor, we need the field orientation dependence of the critical current at low fields and at various temperatures. Figure 5 shows estimated the "minimum" critical current of a single tape at various temperatures. Minimum critical current has been defined in reference [8] as the minimum critical current obtained from a field angle dependence between 0 degrees ($B_{//a\text{-axis}}$) and 90 degrees ($B_{//c\text{-axis}}$). In figure 5 the *I-B* curve at 77 K was measured [2], while the other curves were calculated from the field dependence at various temperature [6].

A load line given with 47.8 A at 100.2 mT in figure 5 was obtained by a self-field analysis of 32-tape TSTC conductor [2]. From the intersection points of this load line and the critical current curves the operation critical current can be obtained for a single tape. The cable critical current for the 32-tape cable in self-field can be calculated by multiplying by 32. Calculated cable critical currents of a 32-tape TSTC conductor are plotted with the measured results in figure 6. The calculated critical currents agree quite well with the experimental results. The maximum differences between them were about 15%. It is desired to measure critical current temperature dependences of the same tape as that of the cable under various fields up to 0.3 T in order to evaluate cable performance more accurately.

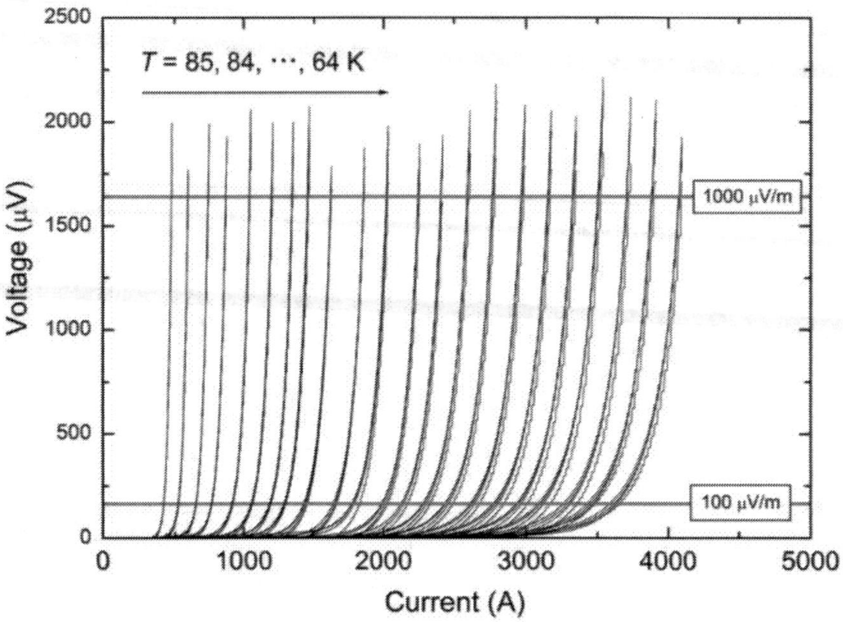

Figure 3. *V-I* curves measured for four voltage taps at 85 K to 64 K.

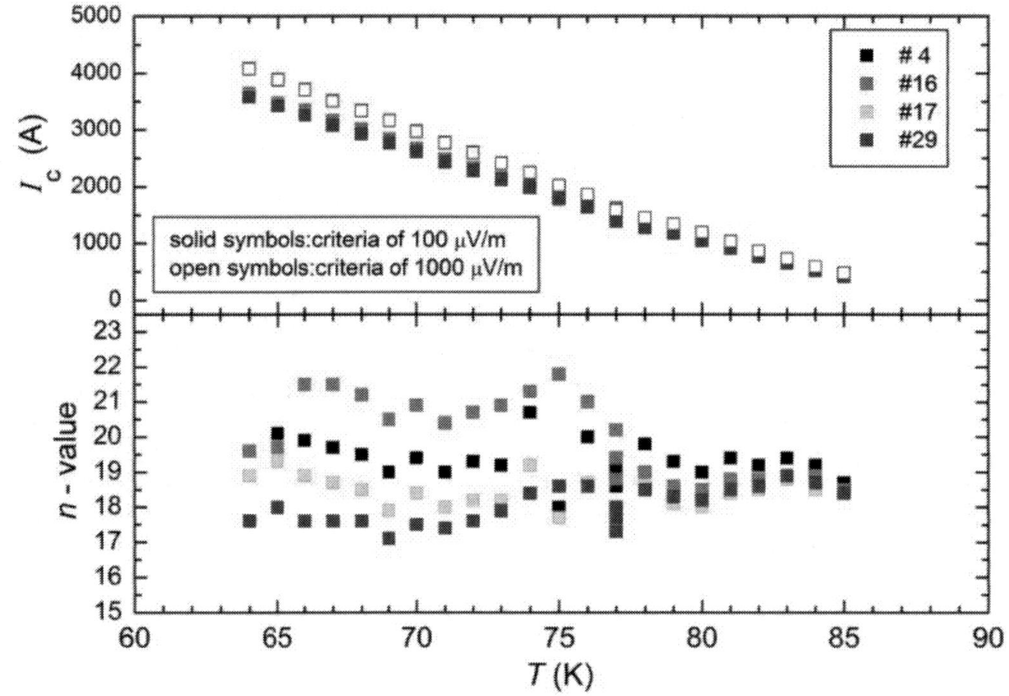

Figure 4. Test results of the critical currents and *n*-values for four voltage taps of Tapes #4, #16, #17 and #29 at 64 K to 85 K. The upper figure shows the critical current with both criteria of 100 μV/m (solid symbols) and 1000 μV/m (open symbols).

IOP Conf. Series: Materials Science and Engineering **102** (2015) 012024 doi:10.1088/1757-899X/102/1/012024

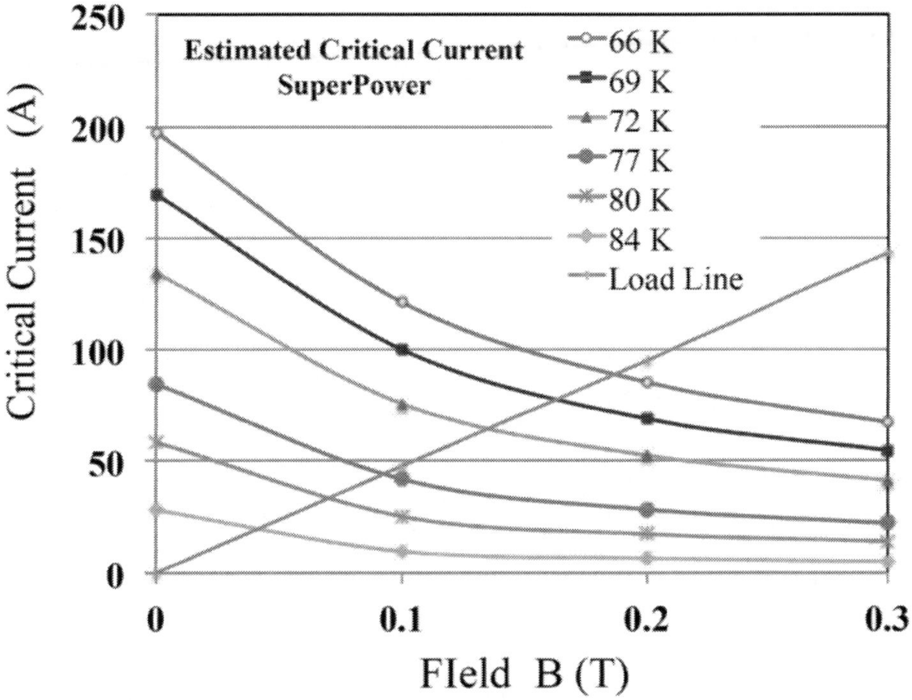

Figure 5. Estimated the minimum critical-current vs. field of a single tape at various temperatures with a load line of self-field for a 32-tape TSTC conductor. The intersection between the load line and the critical current curve gives the critical current of one tape in a cable.

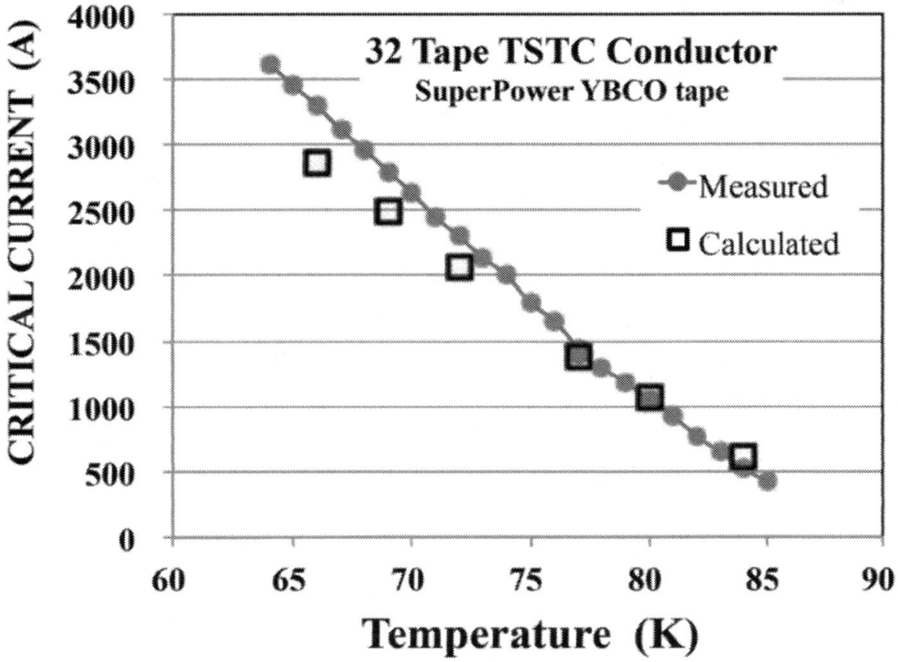

Figure 6. Calculated temperature dependence of the critical current for 32-tape TSTC conductor with the measured results.

4. Conclusions

A 32 tape YBCO TSTC cable has been successfully tested at various temperatures near 77 K. By decreasing operation equilibrium vapor pressure of liquid nitrogen the cable critical current was improved significantly (1.4 kA at 77 K and 3.6 kA at 64 K). The cable capacity can be increased roughly 2.6 times by decreasing the nitrogen vapor pressure.

The test results of critical currents in the TSTC cable were evaluated with our power cable design method developed for a TSTC conductor. The comparison between the measured and calculated results agrees in an error of less than 15% in the temperature range between 66 K and 84K, although the YBCO tape used for temperature test was not exactly same type.

In an electric railway application the power load fluctuates during the day, therefore it is really beneficial that the power cable capacity can be increased when the load increases by subcooling system. Compactness of a TSTC cable will be very beneficial in this application.

Optimization of a TSTC conductor design and termination methods should be investigated. We have developed various types of TSTC cables other than one used for this test [2]. Other options, such as (1) Stack first, then twist, then clad. (2) Stack and then embed in helical open grooves on a structured conduit, could be considered for future investigations.

By using helium or hydrogen gas with cryo-cooling system without liquid nitrogen it will be possible to operate at lower temperature such as 50 K to 20 K. In this way the cable capacity can be dramatically increased. Such gas-cooling power-cable system operated below 77 K should be evaluated for specific railway applications. To take subway into account, for example, nitrogen cooling is not preferable since nitrogen gas is heavier than air and could cause suffocation in case of a leak. In the subway application helium gas system should be considered. We will continue to develop and promote superconducting cables, all with the aim of realizing a cable applicable for railway use.

References

[1] Tomita M, Muralidhar M, Suzuki K, Fukumoto Y, Ishihara A, "Development of 10 kA high temperature superconducting power cable for railway systems," *J. Appl. Phys.*, **111**, no. 6, 63910, 2012.

[2] Takayasu M, Minervini V J, Bromberg L, Rudziak K M, and Wong T, "Investigation of twisted stacked-tape cable conductor," *Adv. Cryo. Eng.*, **58**, Plenum, N.Y., 273-280, 2012.

[3] Takayasu M, Mangiarotti J F, Chiesa L, Bromberg L, and Minervini V J, "Conductor Characterization of YBCO Twisted Stacked-Tape Cables," *IEEE Trans. Appl. Superconduct*, **23**, no. 3, 4800104, 2013.

[4] Tomita M, Suzuki K, Fukumoto Y, Ishihara A, Muralidhar M, "Next generation of prototype direct current superconducting cable for railway system," *J. Appl. Phys.*, **109**, no. 6, 63909, 2011.

[5] Takayasu M, Chiesa L, and Minervini V M, "Low field orientation effect on REBCO tape performance and its power cable application," *IEEE Trans. Appl. Superconduct*, **25**, no. 3, 5400305, 2015.

[6] Tomita M, "Development of Superconducting cable for Railway System," *J. Cryo. Super. Soc. Jpn.*, **48**, No. 11, 562-568, 2013.

IOP Conf. Series: Materials Science and Engineering **102** (2015) 012025 doi:10.1088/1757-899X/102/1/012025

Numerical and Experimental Investigation of the Electromechanical Behavior of REBCO Tapes

N C Allen[1], L Chiesa[1] and M Takayasu[2]

[1] Mechanical Engineering Department, Tufts University, Medford, MA 02155, USA
[2] Plasma Science and Fusion Center, MIT, Cambridge, MA 02139, USA

Email: nathaniel.allen@tufts.edu

Abstract. To fully characterize the electromechanical behavior of a Twisted Stacked-Tape Cable (TSTC) it is important to understand the performance of the individual REBCO tapes under various loading conditions. Numerical modeling and experimentation have been used to investigate the electromechanical characteristics of two commercially available REBCO tapes (SuperPower and SuNAM). Tension and combined tension-torsion experiments on single tapes have been continued, from prior preliminary studies, to characterize their critical current behavior and mechanical strength. Additionally, structural finite element analysis was performed on single tapes under tension and combined tension-torsion to investigate the strain dependence of the critical current. The numerical results were compared to the experimental findings for validation. The SuNAM experimental data matched the numerical model very well while the SuperPower tape experienced degradation at lower stress and strain than predicted in the model. The Superpower tape also displayed greater variability in critical current between different samples as compared with the SuNAM tape.

1. Introduction

The Twisted Stacked-Tape Cable (TSTC) is one method for cabling flat REBCO tapes and may be a viable option for implementing high temperature superconductors in various applications including power transmission, fusion and high-energy physics [1]-[2]. The TSTC method consists of a stack of flat REBCO tapes twisted along the axis of the stack; therefore, to fully characterize the electromechanical behavior of a TSTC conductor it is important to first determine the performance of the individual REBCO tapes under their operational loading conditions.

Several studies are available on the characteristics of REBCO coated conductors under various loading conditions including torsion [3]-[7], tension [8]-[13] and combined tension-torsion [7], [12]-[13]. These studies provide good insight into the general characteristics of REBCO tapes. Considering REBCO tapes have varying dimensions and layered compositions between manufacturers, it is important to characterize specific REBCO tapes being used for TSTC conductors as done in this work.

In this work the electromechanical behavior of two commercially available REBCO tapes from SuperPower and SuNAM have been investigated under tension and combined tension-torsion loading. Combined tension-torsion loading is of particular interest for TSTC conductors due to the twisted nature the design. Torsion originates from the twisting of the REBCO tapes for AC loss reduction; while tension occurs from differential thermal contraction and hoop forces within a magnet coil.

Tension and combined tension-torsion experiments in self-field at 77 K have been conducted and the results were evaluated using a numerical finite element model which was developed in tandem.

IOP Conf. Series: Materials Science and Engineering **102** (2015) 012025 doi:10.1088/1757-899X/102/1/012025

Together the numerical and experimental data provides unique insight into the strain dependence of the critical current for these two REBCO coated conductors.

The layered composite architecture of the SuperPower and SuNAM REBCO tapes that were used are shown in figure 1. Overall both tapes are composed of a superconducting film, buffer layers, silver layers, a non-magnetic substrate and a copper stabilizer. The SuperPower tape has a Hastelloy C276 substrate and a Zr-doped REBCO superconductor while the SuNAM tape uses a Stainless Steel substrate and a GdBCO superconductor. The dimensions and specifications of these tapes are provided in table 1. It should be noted that the width, thickness values reported are the mean values calculated from 50 measurements taken over a 350 mm long sample length. The reported critical currents are the average value from 7 SuNAM and 11 SuperPower samples tested and discussed in this paper. These calculated mean values all fall within the range provided from the manufacturers.

Table 1. Tested REBCO tapes and their measured characteristics

Manufacturer	**SuperPower**	**SuNAM**
Type	SCS4050-AP	SCN04150
Processing	IBAD-MOCVD	IBAD
Width	4.027 ± 0.057 mm	4.062 ± 0.008 mm
Thickness	0.092 ± 0.001 mm	0.144 ± 0.001 mm
Substrate *(thickness)*	Hastelloy C-276 *(50 μm)*	Stainless Steel *(100 μm)*
Stabilizer *(total thickness)*	Electroplated Cu *(40 μm)*	Electroplated Cu *(40 μm)*
Critical Current (77 K & self-field)	112 ± 3 A	229 ± 6 A

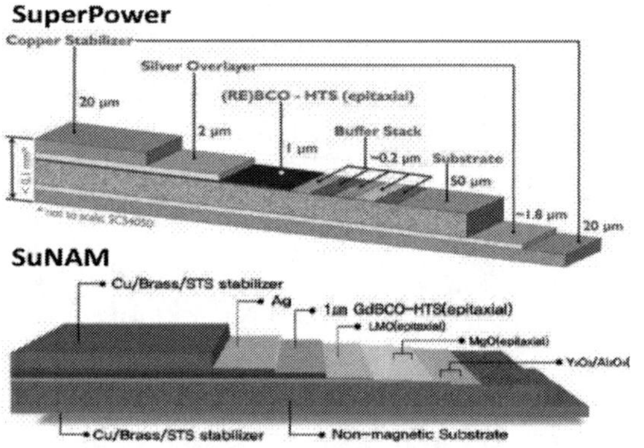

Figure 1. REBCO tape architectures of SuperPower [14] top and SuNAM [15] bottom.

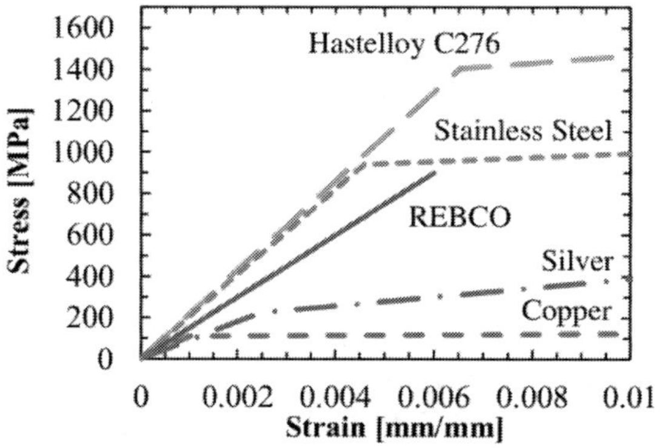

Figure 2. Bilinear material properties inputted into numerical ANSYS® model [14-24].

2. Numerical modeling of REBCO tapes

Structural finite element analysis (FEA) using ANSYS® was done to investigate the electromechanical behavior of REBCO coated tapes under tension and combined tension-torsion loads. The FEA model was used to determine the stress and strain characteristics (elongation and load) as well as the torque magnitude of the tapes under load. The strain profile from the FEA was paired with an analytical model to predict the critical current of the sample [7]. The FEA model was customized to the exact multi-layered architecture of each type REBCO tape (SuperPower and SuNAM) so that detailed results for each constituent layer of the composite tapes can be obtained.

The REBCO tape was modeled using 3D 8-node SOLSH190 structural solid-shell elements with built in layered composite capabilities. The tape was meshed with one element through its thickness and the appropriate layer definitions and properties were defined using shell section commands (i.e. each layer of the tape {substrate, REBCO, etc} is defined within the finite element and all interlayer interactions are calculated in the element formulations). Sufficient mesh density through the width and length of the tape was chosen. The tensile load was applied to the ends of the tape using a surface pressure load while the rotation was applied to the ends of the tape using pilot node constraints.

The material properties for each constituent layer of the composite REBCO tapes were defined in ANSYS® using bilinear stress-strain curves. The bilinear curves are defined by a modulus of elasticity, yield strength and tangent modulus and are shown in figure 2. Care was taken to use the most accurate material properties for the substrate and copper because those materials govern the overall strength of the tapes. The buffer and REBCO superconducting layers have brittle ceramic characteristics and were modeled with linear elastic properties. For consistency, the same REBCO, copper and silver properties were defined for both tapes.

A single REBCO tape under pure tension was initially simulated to determine the mechanical stress-strain trend of the model. The bilinear material properties were modified to produce the best fit to the experimental stress-strain curve (see section 3.2 for more details). Once the material properties were accurately defined producing a model with the same mechanical characteristics as the real REBCO tape, the model was used to determine the strain dependence of the critical current for both REBCO tapes utilizing an analytical model described in [1]. The critical current degradation under pure tension for an untwisted tape was first analyzed followed by the critical current behavior under combined tension-torsion loading in order to investigate electromechanical characteristics of a twisted tape. The results from FEA modeling are compared to the experimental data in section 4.

3. Experimental Details

3.1. Experimental test probe

The tension and combined tension-torsion tests were conducted using the same experimental test probe used in prior preliminary studies [7], [25]. The tests probe is shown in figure 3. The probe is designed to simultaneously measure the axial load, elongation, torque and critical current of REBCO tapes. The probe was used to test both 100 mm and 300 mm long samples. The samples were mounted using stainless steel compression style sample holders. The current was supplied via flexible 12 mm wide high current SuperPower REBCO tapes. This was done so that the current leads did not influence the torque or axial load measurements. Two pairs of voltage taps were used to record the voltage during critical current measurements.

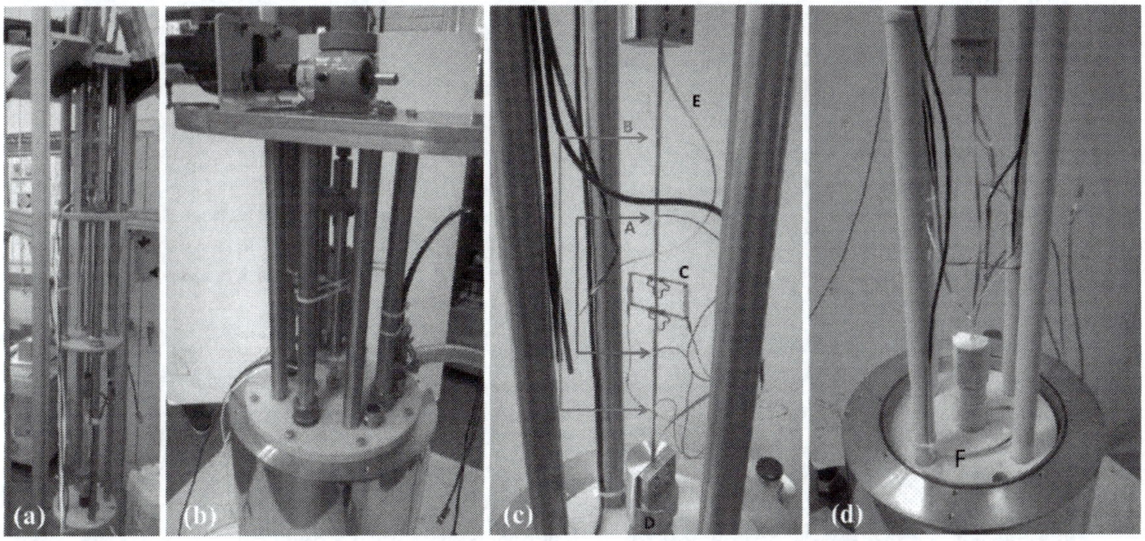

Figure 3. (a) Experimental test probe highlighting stainless steel structure (b) probe inside cryostat (c) Tension sample showing sample holders [D], current leads [E], extensometers [C] and voltage taps [A, B] (d) Torsion-tension sample after being removed from cryostat showing torque meter [F].

Axial displacement was applied outside the cryostat through a pull rod by a 1/40 hp permanent magnet DC electric motor (Bison 011-190-0702) and a 10 kN anti-backlash machine screw actuator (Duff-Norton M5500-124). The axial displacement of the sample was measured outside the cryostat by a 15 mm stroke linear variable differential transformer (LVDT, Omega LD621-15). Double extensometers (Nyilas-type) were used to calibrate the LVDT sensor during pure tension measurements [26]. The extensometers were not used for the tension-torsion tests due to the twist in the sample. The torsion/twist was applied to the top of the sample via the pull rod while the bottom was held by a low capacity reaction torque sensor (RTS-10 Transducer Techniques). The tensile load was measured outside the cryostat by an 1112 N (250 lb) donut style load cell (Futek LTH-350).

3.2. Mechanical stress-strain measurements

In order to define the appropriate material properties for the composite REBCO tapes in the numerical finite element model, the stress-strain behavior of the tapes was measured. The experimental stress-strain results can then be used to validate the stress-strain curve produced by the model allowing the strain dependence of the critical current in the tapes can be investigated. The mechanical stress-strain tests were done in a bath of liquid Nitrogen at 77 K using the experimental test probe defined in the previous section. A high precision load cell mounted on the pull rod at the top of the probe recorded the tensile stress while the strain of the sample was measured by Nyilas-type extensometers mounted directly on the REBCO tape and also by the LVDT mounted in the room temperature zone at the top of the probe. After the probe was cooled down in the bath of liquid Nitrogen the sample was slowly strained at a rate of roughly 3 mm/min while the load and displacement were simultaneously measured. After the onset of yielding occurred the strain was released and the sample was unloaded.

3.3. Critical current measurements

For both tension and combined tension-torsion tests, two pairs of voltage taps were used to determine the samples critical current. The primary pair of voltage taps (denoted by A in figure 3(c)) was located in the center of the sample. The secondary pair of voltage taps (denoted by B in figure 3(c)) spanned twice the distance covering the majority of the sample length. The distance between voltage taps varied with the length of the sample being tested. For 100 mm long samples, the primary and secondary voltage tap pairs spanned 40 and 80 mm long respectively. For the longer 300 mm samples, the primary and secondary voltage tap pairs spanned 100 and 200 mm long. The two sets of voltage taps were used to identify any critical current variation that may occur along the length of the sample.

During the critical current measurements, the current is slowly increased while two nanovoltmeters simultaneously record the voltage between the two voltage tap pairs. The low noise characteristic of the REBCO tapes and the accuracy of the nanovoltmeters allowed for repeatable precise determination of the critical current and n-value. The critical current presented in this work was calculated from the lower electric field criterion of 100 $\mu V\ m^{-1}$. The upper electric field criterion used for SuperPower tapes was 500 $\mu V\ m^{-1}$ while it was 200 $\mu V\ m^{-1}$ for SuNAM tapes, which were more prone to burning due to the larger currents they carry. A characteristic voltage versus current plot for Superpower and SuNAM samples is shown in figure 4 along with the electric field criterion for a tap span of 100 mm.

4. Results

4.1. Stress-Strain Characteristics

The stress-strain curve for SuperPower and SuNAM REBCO tapes at 77 K is shown in figure 5. The solid curves in the figure denote the experimental stress-strain results while the dashed lines indicate the stress-strain result from the structural FEA model of the layered composite tapes. The experimental tests for both tapes produced very repeatable stress-strain behavior and the curves shown in figure 5 are the average stress-strain curves from 6 tests of each tape. The bilinear material properties in the numerical model (shown in figure 2) were adjusted to produce the best fit to the experimental data, which are shown in figure 5.

The SuperPower and SuNAM tapes have different dimensions, substrates and superconducting films and were found to have different mechanical stress-strain characteristics. The SuNAM tape was found to have a 20% greater modulus of elasticity compared to SuperPower but was also found to have a lower yield stress; about 700 MPa compared to around 820 MPa for SuperPower. That being said, because the SuperPower tape has a smaller overall cross-section it actually yields at a lower tensile load (~300 N) compared to the SuNAM tape (~400 N). The SuperPower tape was also found to have a tangent modulus of nearly double that of SuNAM.

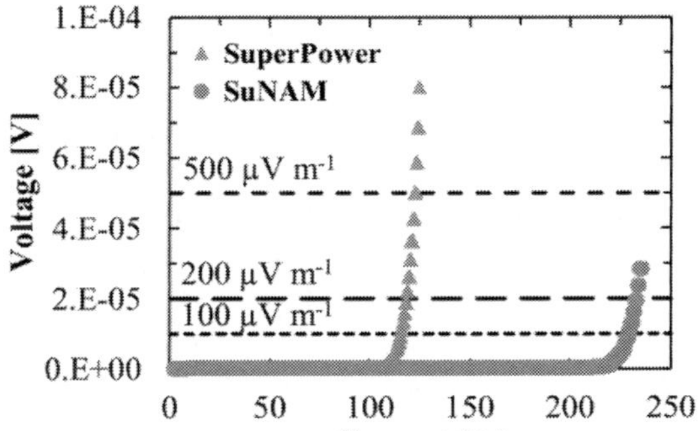

Figure 4. Characteristic voltage vs current plot for Superpower and SuNAM tapes. Criterion defined for a 100 mm long voltage tap.

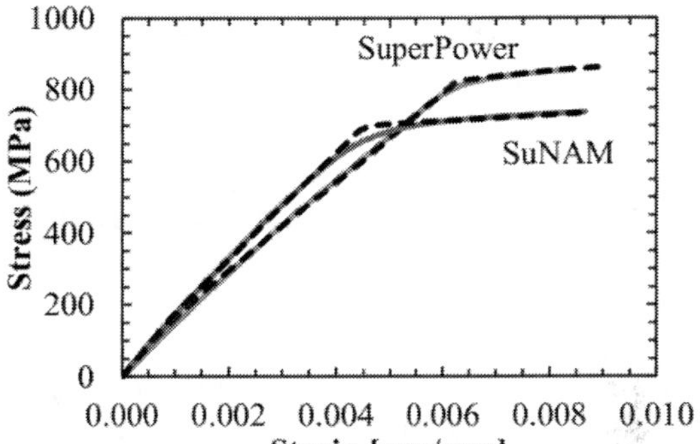

Figure 5. Experimental and numerical (dashed) stress-strain curves for SuperPower and SuNAM tapes at 77 K.

4.2. Normalized critical current behavior under pure tension

The critical current behavior of SuperPower and SuNAM tapes under pure tension is shown in the following figures. Figure 6 plots the normalized critical current versus applied tensile stress while figure 7 plots the normalized critical current versus applied tensile strain. The dashed lines in the plots indicate the predicted critical current determined from the numerical FEA strain results and the analytical model given in [1]. The predicted onset critical current degradation from the finite element model closely matches the yield strength of the tape.

The critical current trend under tension was found to be very different between the two REBCO tapes. The critical current degradation of SuNAM tape closely matched the numerical data and experienced degradation near the yield point of the tape (~700 MPa). Each SuNAM sample experience relatively close and repeatable results whereas the SuperPower tape displayed much greater variability between each sample tested. The critical current for the SuperPower tape was seen to degrade between 500-600 MPa and 0.45-0.55% strain. A few SuperPower samples did have a noticeably higher stress at the onset of critical current degradation (~700 MPa) however that was still much lower than the ANSYS® prediction based on the yield strength of the tape.

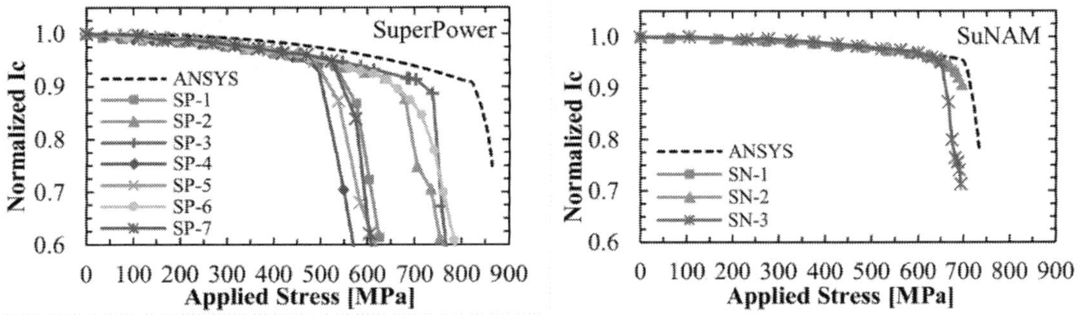

Figure 6. Normalized critical current versus applied tensile stress plots for straight samples.

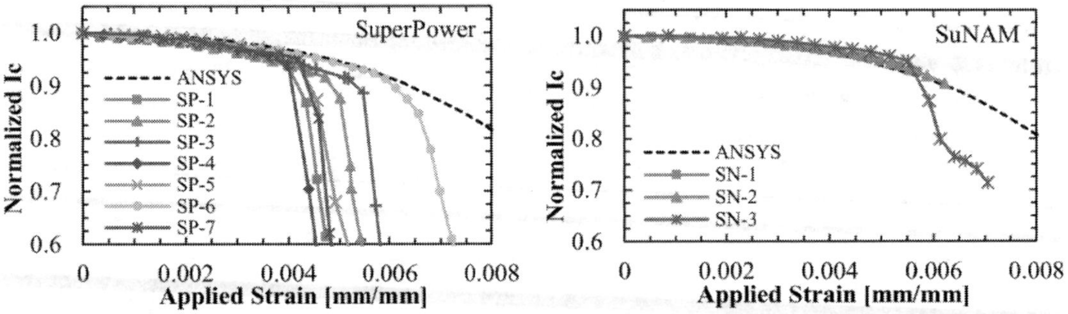

Figure 7. Normalized critical current versus applied tensile strain plots for straight samples.

All samples tested from both manufacturers were taken off of the same spool of tape and were prepared and mounted in the same manner. Care was taken during the preparation and mounting to keep everything consistent between samples. Considering this and the fact that the SuNAM tape had more repeatable data, we conclude that the variability in the SuperPower samples is real and may be explained by a microscopic slipping/shearing phenomena between layers inside the tape that causes critical current degradation before the yield point of the sample. A deeper analysis of the tapes looking at their microstructure is required. SuNAM tape may not have experience such a phenomena because it is 60% thicker and uses a different fabrication process.

4.3. Normalized critical current behavior under combined tension-torsion

The combined tension-torsion results for SuperPower and SuNAM tapes are displayed in the following figures. Figure 8 plots the normalized critical current versus applied tensile stress while figure 9 plots the normalized critical current versus applied tensile strain. The dashed lines again indicate the critical current trend predicted from the numerical FEA strain results. The tension-torsion tests were conducted first for a 200 mm twist pitch which is the twist pitch being used for current TSTC conductors [2]. Tests were also carried out for a 150 mm twist pitch to identify the influence of additional twist. The experimental critical current was normalized to the initial value for the twisted samples (200 and 150 mm twist pitch), although the initial critical currents of the twisted samples saw no difference compared to the straight samples.

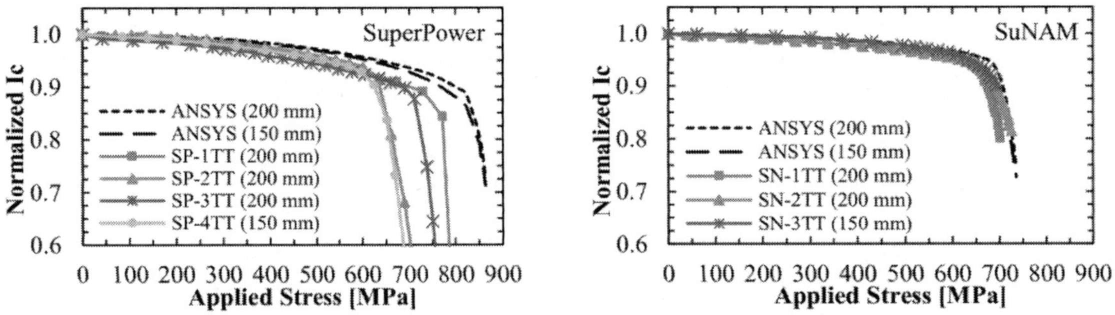

Figure 8. Normalized critical current versus tensile stress plots for combined tension-torsion samples. Numerical results are shown by dashed lines and the sample twist pitch is indicated in parenthesis.

The numerical results under combined tension-torsion shown in figure 9 also indicate a very similar critical current trend compared to the pure tension model. This result was confirmed by the experimental results, which also showed that the combined tension-torsion samples saw critical current degradation around the same location with the tensile samples. The SuNAM samples for 200 and 150 mm twist pitch were almost identical and both showed degradation around 700 MPa which again is at the yield point of the tape same as in the tension samples. The tension-torsion SuperPower samples at 200 mm twist pitch again showed a higher degree of variability as was found in the tension samples. The SuperPower samples saw an onset of degradation between 700 and 800 MPa, which matches the "best" samples from the pure tension results. Those results seem to indicate a slight improvement in critical current performance in the SuperPower tension-torsion samples but more data would be necessary to validate.

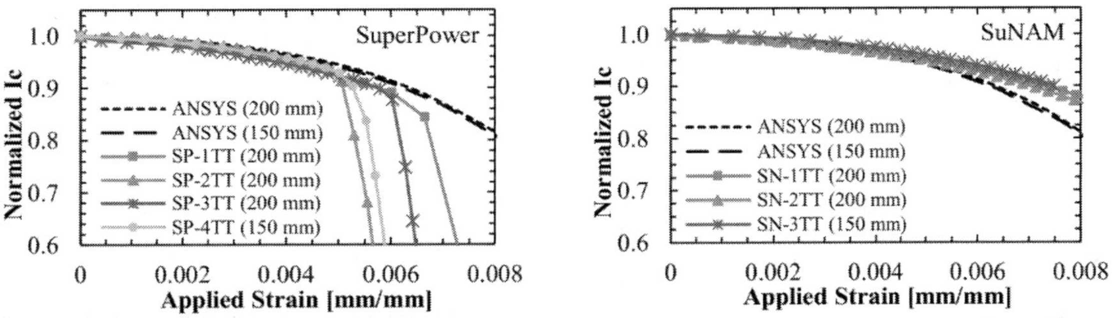

Figure 9. Normalized critical current versus tensile strain plots for combined tension-torsion samples. Numerical results are shown by dashed lines and the sample twist pitch is indicated in parenthesis.

4.4. Torque behavior under combined tension-torsion

As the tensile load was applied to twisted REBCO tapes the torque was concurrently measured. Figure 10 below is a plot of the change in torque versus the axial tensile stress applied to the twisted sample. For the most part the experimental results show good agreement with the torque performance predicted by the FEA model. The results from both REBCO tapes indicate an increasing torque with increasing applied tension. Additionally, both tapes indicate that for a shorter twist pitch the change in torque under tensile load is greater. For the same tensile load the SuperPower tape experienced less change in torque compared to the SuNAM tape. Those results are consistent with previously presented data that showed the SuNAM tape had higher torque under pure torsion [3]. The SuNAM tape experienced a pronounced plateau of the torque at high stress (near its yield point), which was not observed in the SuperPower results.

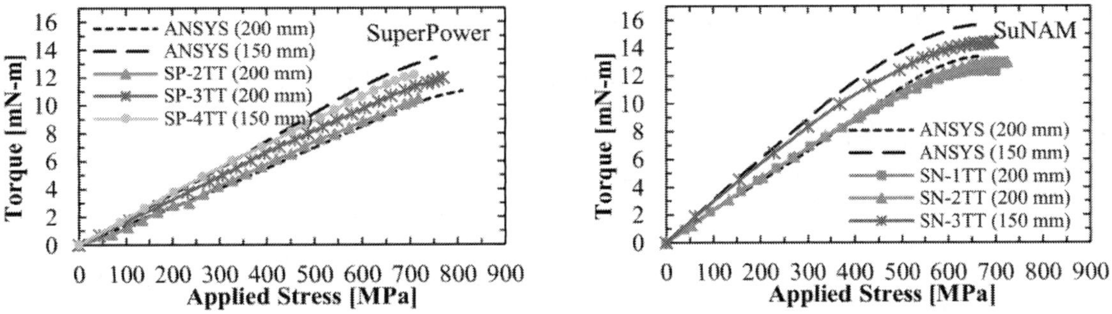

Figure 10. Change in torque versus tensile stress plots for combined tension-torsion samples. Numerical results are shown by dashed lines and the sample twist pitch is indicated in parenthesis.

5. Discussion and conclusion

In this work, we analyzed the electromechanical behavior of two commercially available REBCO tapes (SuperPower and SuNAM) under tension and combined tension-torsion loading. The characteristics of the tapes were investigated through experimentation in self-field at 77 K and by structural finite element analysis. The finite element model was initially validated by adjusting the

IOP Conf. Series: Materials Science and Engineering **102** (2015) 012025 doi:10.1088/1757-899X/102/1/012025

material properties until the numerical results matched the measured experimental stress-strain curves. Following this, the numerical FEA strain results were fit to an analytical model that predicted the critical current behavior of the tapes under tension and combined tension-torsion. These results were then evaluated against the experimental results from both REBCO tapes.

For tension, the SuNAM experimental data matched the numerical model very well and saw a degradation in critical current beginning at the onset of plastic deformation in the tape. The SuperPower tape experienced degradation at lower stress than numerically predicted and displayed greater variability between samples as compared with SuNAM. These behavior is still under investigation. Overall, TSTC conductors should be designed to remain well below the yield strength of the REBCO tapes to avoid irreversible critical current degradation.

The combined tension-torsion critical current results were found to be almost identical to the pure tension results for both tapes indicating that the combined tension-torsion has minimal effect on the critical current performance for the two tested twist pitch (200 and 150 mm). This indicates that the twist of a TSTC conductor, for a twist pitch greater than 150 mm, should not influence the critical current behavior of that TSTC conductor under axial tension. The torque characteristics of the REBCO tapes under combined tension-torsion were also investigated and were found to linearly increase with applied tension. The numerical FEA results were found to be in good agreement with the measured torque data of both manufactures' tapes.

References

[1] Takayasu M *et al* 2012 *Supercond. Sci. Technol.* vol 25 no 014011
[2] Takayasu M *et al* 2015 *IEEE Trans. Appl. Supercond.* (submitted HTS4Fusion Special Issue) "Present Status and Recent Developments of the Twisted Stacked-Tape Cable (TSTC) Conductor"
[3] Chiesa L *et al* 2013 *IEEE Trans. Appl. Supercond.* vol 23 no 3
[4] Shin HS, Dizon J, Kim T, Ha D, and Oh S 2007 *IEEE Trans. Appl. Supercond.* vol 17 no 2
[5] Weiss K, Goldacker W and Nannini M *IEEE Trans. Appl. Supercond.* vol 21 no 3
[6] Takao T *et al* 2007 *IEEE Trans. Appl. Supercond.* vol 17 no 2 pp 3513–3516
[7] Allen NC, Chiesa L and Takayasu M 2014 *IEEE Trans. Appl. Supercond.* vol 25 no 3
[8] Cheggour N *et al* 2003 *Appl. Phys. Lett.* vol 83 no 20 4223–4225
[9] Van der Laan DC *et al* 2010 *Supercond. Sci. Technol.* vol 23 no 7 072001
[10] Barth C, Mondonico G and Senatore C 2015 *Supercond. Sci. Technol.* vol 28 no 045011
[11] Dedicatoria M, Shin H, Ha H, Oh S and Moon S 2010 *Supercond. Cryo.* vol 12 no 4
[12] Ilin K *et al* 2015 *Supercond. Sci. Technol.* vol 28 no 055006
[13] Majkic G *et al* 2009 *IEEE Trans. Appl. Supercond.* vol 19 no 3
[14] Hazelton DW 2013 *CHATS-AS Workshop* Cambridge MA USA
[15] "SuNAM 2G HTS Wire Catalog" http://www.i-sunam.com/home/en_company,4,1,4,4
[16] Reed R and Mikesell R 1967 (*Colorado: National Bureau of Standards*)
[17] Brechna H, Hartwig G and Schauer W 1971 *Institut fur Experim. Kernphysik*
[18] Sugano M *et al* 2011 *Supercond. Sci. Technol.* vol 24 no 075019
[19] Clickner C *et al* 2006 *Cryogenics* vol 46 pp 432-438
[20] Cheggour N *et al* 2003 *IEEE Trans. Appl. Supercond.* vol 13 no 2
[21] Verdier M, Huang H, Spaepen F, Embury J and Kung H 2006 *Phil. Mag.* vol 86 pp 5009–16
[22] Goretta K *et al* 1995 *Pro of 4th World Congress on Supercond.* vol 3290 pp 633–638
[23] Kanchana V *et al* 2006 *J. Phys. Condens. Matter* vol 18 pp 9615-9624
[24] Ledbetter H and Ming L 1991 *J. Mater. Res.* vol 6 pp 2253–5
[25] Bader D 2012 *Mechanical Engineering Masters Thesis* Tufts University
[26] Nyilas A 2005 *Supercond. Sci. Technol.* vol 18 pp S409-15

Optimized and practical electrical joints for CORC type HTS cables

T Mulder[1,2], A Dudarev[1], D C van der Laan[3], M G T Mentink[1], M Dhallé[2] and H H J ten Kate[1]

[1] CERN, Geneve, Switzerland
[2] University of Twente, Enschede, the Netherlands
[3] Advanced Conductor Technologies (ACT), Boulder, USA

Email: tim.mulder@cern.ch

Abstract. Within CERN the development of REBCO-CORC (Conductor On Round Core) type cables is pursued in view of possible application in future detector and accelerator magnets. An important issue is the design and qualification of terminations for connecting CORC cables mutually or to bus-bars. A termination design is envisaged that combines a simple manufacturing process with a lowest possible joint terminal resistance in the few nΩ range at 4.2 K, first for a single CORC cable and subsequently for CORC based Cable-in-Conduit Conductors. The investigation concerns the effect of tapering the CORC cable within the joint to form a staircase like geometry, which allows current to pass more directly from the copper joint casing to the inner REBCO layers of the CORC cable. Simulations have shown a substantial decrease in joint resistance at operating current in the case both CORC cable and joint casing are tapered. The CORC cable and new joint were tested at CERN. In this paper, some details of the new joint design, fabrication process, and model are presented and the results are summarized.

1. Introduction

The CORC (Conductor On Round Core) cable is a new REBCO tape based cable technology. The CORC cable is a flexible round cable that comprises many REBCO tapes wound in multiple layers around a cylindrical wire [1]. CERN is pursuing the development of CORC type REBCO cables in view of possible application in future detector and accelerator magnets. CORC is especially interesting for detector magnets because its design is simple and robust, the cable has a high Tc and cable geometry allows manufacturing of a Cable-in-Conduit type of Conductor of several CORC strands. Furthermore, the cable does not require heat treatment and has good mechanical properties [2]. An important issue in the development is the design and qualification of low-resistive and practical joint terminals. These joint terminals can be used to connect a CORC cable to a busbar or to connect two CORC cables mutually.

In the past, several CORC joint terminals were designed that where either bulky [3] or injected current only via the outer layer into the cable. The electrical contact to the inner layers is poor, due to the Hastelloy substrate within the tape, leading to a relatively high joint resistance and bad current sharing. At CERN a new joint terminal is designed by which current is injected in each layer of the cable as opposed to just the outer layer. This results in a significant decrease in resistance at high currents near to the critical current of the cable. Complimentary, a custom-built Finite Element Model (FEM) is developed to help understanding the behavior of the joints. The model is clearly able to demonstrate the improvements when comparing the old and new designs, and is used to evaluate and optimize different terminal configurations.

2. Cable and new joint terminal

A 12 meter long CORC cable was manufactured for CERN by ACT for characterization, joint development and the production of the first 6-around-1 REBCO-CORC cable Cable-in-Conduit Conductor. This cable actually is (in 2015) the longest CORC cable ever produced. The cable comprises 38 SCS4050 (4 mm wide and 96 μm thick) Superpower tapes wound in 12 layers around an aluminum core. The aluminum core has a diameter of 4 mm and is increased to 5 mm with five layers of copper tape. The total diameter of the cable is 7.6 mm. This cable is used to test the fabrication method and the characterization of joint terminals.

2.1. Joint terminal design

In the past, the joint terminal comprised a copper casing in which the CORC cable was inserted, after which it is filled with solder material. The solder material is able to flow between the layers since the tapes spring out after the sleeve around the cable is removed. The new joint terminal design features a change to the CORC geometry that leads to a low-resistive path to the inner REBCO layers. The path, created by trimming several layers of tapes into shorter sections, exposes the layers underneath. The assembly is then filled with a solder material, thereby creating a low-resistive path from casing to cable. In this configuration current can flow directly into the lower layers without passing the Hastelloy substrate of the upper layers. Automatically, it follows that the current is distributed more evenly over all layers. The resistance can be optimized for the specific operating conditions of the joint by varying the number of trimmed layers, the length of the trimmed sections and the overall length of the joint terminal. In addition, the copper casing can be machined to create a similar staircase-like geometry that matches the shape of the trimmed cable. This geometry causes an increase of the copper to solder ratio in the joint terminal and thus, a lower joint resistance can be achieved. A schematic view of the original and new design are presented in figure 1.

2.2. Joint production

The tapes in the cable are trimmed by precisely cutting the individual tapes with a sharp precision knife. After cutting the tape, it tends to spring out and thus has to be held in place. This can be achieved by wrapping a thin copper wire tightly around the trimmed layer. The wire can be removed while inserting the trimmed cable section into the casing of the joint terminal. However, for casings with a staged bore closely following the trimmed section this is impossible, since the tapes spring out in the casing and get stuck at the edge of the first machined section. Our solution is to keep the tapes together with a bandage of solder as can be seen in figure 2. The solder adds a few μm to the diameter of the trimmed cable. On

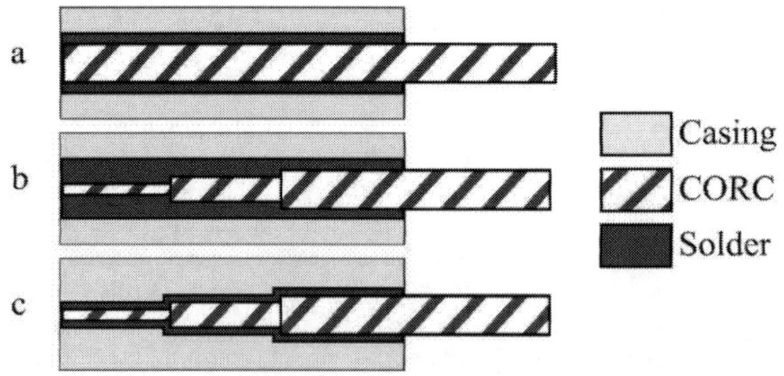

Figure 1. Schematic view of (a) the original joint terminal design, (b) the new joint design with a trimmed CORC cable and (c) the new joint design with a trimmed CORC cable and a staged casing.

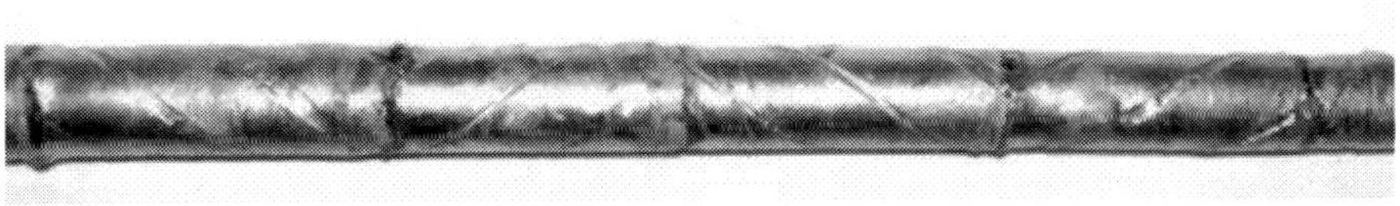

Figure 2. Five tapered sections of the CORC cable, each 17 mm in length, held together by a bandage of solder.

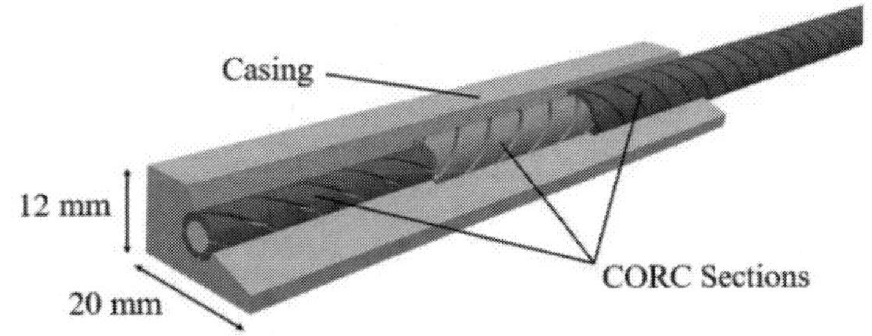

Figure 3. 3D representation of the trimmed CORC cable inside a rectangular copper casing shaped to fit the cable. All voids between the tapes inside the casing are filled with a soldering alloy.

filling the joint casing with solder, the bandage of solder melts and the tapes can spring out as required for achieving minimum resistance.

Two manufacturing methods were tested. In the first method the joint terminal comprises several individual sections each with a different bore diameter. The sections are moved over the cable; each section fitting the different trimmed steps of the cable. The sections are soldered together at the time the casing is filled. The second method uses a single casing covering the entire joint length by which the stepping is achieved with multi-stage drilling. Prototypes of both variations were manufactured. The multiple sections of the first method is more practical to machine, however the second method facilitates easy solder filling and handling after the cable is inserted. Overall the second method is preferred for future joint terminals. A picture of a block shaped prototype joint terminal is presented in figure 3.

Eutectic Sn63Pb37 is used in all joint terminal samples. This solder material has very good wetting properties, maintains good mechanical properties at cryogenic temperature, and with a melting temperature of 183 °C it does not exceed the 200 °C degrees at which REBCO starts to degrade [5, 6]. The casing is attached to an aluminum block, which is fitted with two controllable cartridge heaters and a thermometer to monitor the temperature. Through small holes present in the top surface, non-acid flux and solder can be injected and they allow gas to escape. Both, joint and aluminum block are inclined to facilitate flow and escape of evaporated flux. When solder starts to come out of the first hole, the hole is closed and the filling continues step by step.

2.3. Fabrication issues

Two issues were encountered during joint manufacturing: delamination of the tapes and small gas pockets inside the joint. Delamination is observed in the terminal after cross-sections of the produced terminals were made. About 5% of the trimmed tapes showed slight delamination of REBCO tape. This is an issue because the copper delaminates from the REBCO layer and thus no longer facilitates an optimal electrical connection between solder and REBCO. Delamination is caused by the soldering temperature working on the Hastelloy layer that wants to spring out while the copper still is firmly attached to the layer below. The Hastelloy layer in the next generation of REBCO tapes will have less thickness and thus less delamination is expected. The second issue concerns small gas pockets observed

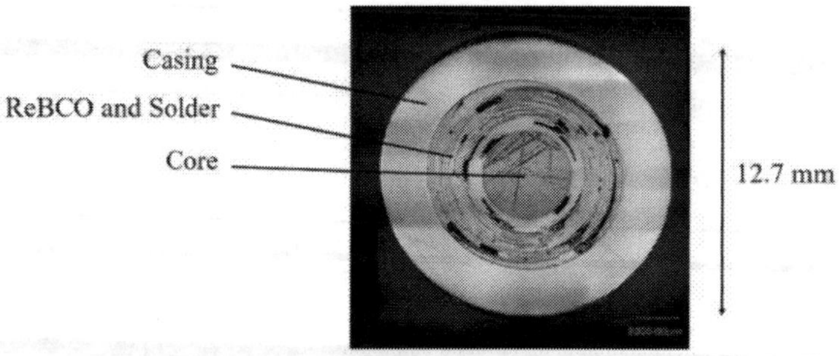

Figure 4. Cross-section of a cylindrical joint terminal with gas pockets (black areas) visible.

between the tapes as shown in figure 4. The number of gas pockets can be reduced by increasing the inclination angle during filling and increasing the number of holes for gas to escape.

2.4. Experimental joint terminals

The two joint terminals are manufactured and mounted on a 1.8 m long CORC cable. This cable and its joints are characterized in the FRESCA cable test facility [6]. The terminals are 200 mm long and follow the new design. The cable ends are trimmed into twelve 17 mm long sections. The joint casing comprises three OFHC copper cylinders with each a different bore diameter that matches the shape of the trimmed CORC cable. Each cylinder has two small holes for solder filling and degassing.

3. Joint model

A custom-built joint model is developed for simulating and evaluating joint terminal configurations. The model is based on a network of resistances that includes both linear resistances (solder and casing) and non-linear resistances (REBCO tapes) and assumes a steady state current. A schematic representation of the resistances network is shown in figure 5.

The current is injected in one terminal and extracted at the other terminal. The current is allowed to redistribute optimally over both terminals. Between the tapes in the present CORC cables there are only pressure contacts. For this reason it is assumed that for short cable sections the current is distributed only in the terminals. The model uses both Kirchhoff's voltage and current laws to solve the system of equations. Following Kirchhoff's voltage law the sum of the electrical potential differences between nodes in a closed network is always zero:

$$\sum_{k=1}^{n} V_k = 0. \tag{1}$$

In addition following Kirchhoff's current law, the sum of the currents that flow into a node is equal to the sum of the currents that flows out of that node:

$$\sum_{k=1}^{n} I_k = 0. \tag{2}$$

The system is solved for both the current between- and voltage over the nodes giving the following equation for all normal conducting sections:

$$(V_2 - V_1) - I_{12}R_{12} = 0, \tag{3}$$

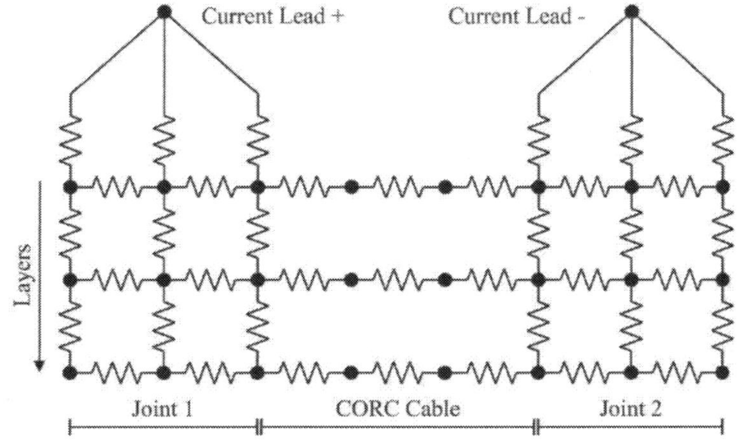

Figure 5. Schematic view of the resistances network used in the novel joint model.

and for all superconducting sections:

$$(V_2 - V_1) - U_0\left(\frac{x(I)I}{I_c}\right)^n - (1 - x(I))IR = 0, \tag{4}$$

where V is the voltage, I is the current, R is the calculated normal-resistance between two nodes and x is a function stating the ratio between current in the REBCO layer and in the stabilizer material, respectively. The resistance between tapes is calculated as the serial resistance of the materials within the tape, with a small strip of solder in parallel, following:

$$\frac{1}{R_{mn}} = \left(x * \frac{\rho_{solder}\ln\left(\frac{r_n}{r_m}\right)}{2\pi L}\right)^{-1} + \left(\sum_{m=1}^{n}(1-x) * \frac{\rho_m \ln\left(\frac{r_{m+1}}{r_m}\right)}{2\pi L}\right)^{-1} \tag{5}$$

Here x is the fraction of solder in parallel with the tapes and r is the radial position of the layer. Since the tape comprises different materials, the serial resistance is calculated across all materials [7, 8]. A uniform current distribution is assumed and therefore the resistance in the angular direction is ignored.

3.1. Simulation results

The model is used to simulate various joint terminal configurations and evaluate their performance. The primary subject of investigation is the difference in joint resistance between a cable that is trimmed and one that is not. The geometry of the CERN CORC cable with 12 layers and 38 tapes is used in the simulation with a joint length of 200 mm and at 4.5 K. The simulated joints are in zero magnetic field and self-field is ignored. In this section the following cases are discussed: the non-trimmed cable, trimmed in 3 sections of each four layers, trimmed in 6 sections of each two layers, trimmed in 12 sections of one layer and trimmed in 12 sections with a matching casing. In figure 6 simulation results for these configurations are presented.

The non-trimmed configuration has the longest length of the outer most layer, this layer has the lowest resistance to the casing and thus this configuration has the lowest resistance at $I \ll I_c$. The simulation clearly shows that different layers in the non-trimmed configuration saturate before current passes to subsequent inner layers. When a layer is nearly saturated, the resistance rises in a step-like

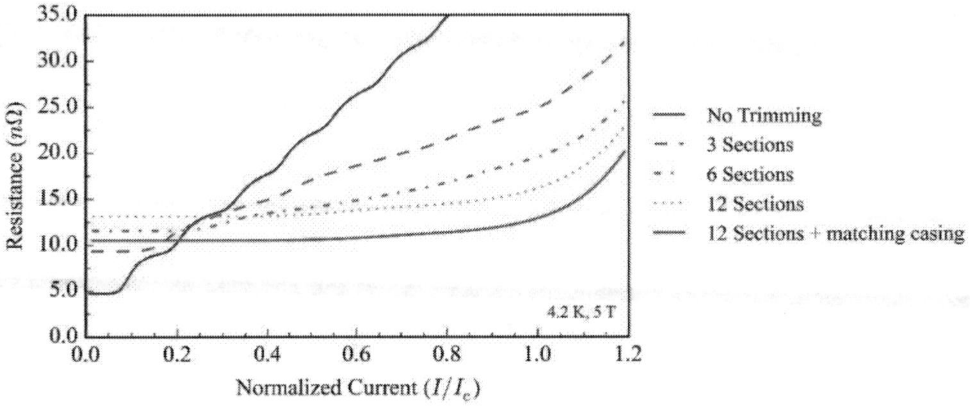

Figure 6. Simulated joint terminal resistance for five trimming configurations.

manner. The simulation shows that trimming the CORC cable increases the initial joint terminal resistance at low currents, but can cause a significant reduction in resistance at currents near I_c compared to the non-trimmed cable. The trimmed cable starts to excel when the current becomes high enough to cause saturation of the outer few layers of the cable. The lowest resistance at nominal current, about 75% of I_c, is obtained by trimming the cable into 12 sections. The joint terminal resistance can be reduced by about 50% by only trimming the cable.

Machining the casing to follow closely the shape of the trimmed cable reduces the joint terminal resistance further by another 15% relative to the non-trimmed cable. In total, the resistance of this type of joint terminal can be decreased by about 65% when following the new joint terminal design. The model also shows that cables with more layers will benefit more strongly from the new design. Furthermore, the terminal resistance scales with length. Increasing the length of the joint terminal to some meters reduces the terminal resistance significantly and long joint lengths are usually no issue in large magnet systems.

4. Experimental

A characterization test of the CORC cable is performed in the FRESCA cable test facility at CERN. The joint performance is examined during the measurement. FRESCA's dipole magnet can provide a background field up to 10 T over homogeneous magnetic field region of 600 mm. The cable is centered in the peak field, both joints are outside the peak-field region and only see self-field. Current is injected in one joint and is carried by the CORC cable over the peak-field region. A stack of three NbTi Rutherford cables is connected to the second joint and return the current to the power supply. The sample is inserted vertically into the bore of the dipole magnet, therefore the joint terminals will be addressed as 'top' and 'bottom' terminal.

The test is performed in an applied magnetic field ranging from 3 to 9 T at 4.5 K. In figure 7 the measured and simulated resistance of the joint at 4.5 K is presented by which the mid-section of the cable is in the dipole field. The critical current of the cable at 6 T is about 11 kA, enabling a large current range in which the resistance can be determined and still with limited effect of the self-field on the joints. A constant, magnetic field independent, joint terminal resistance of 30 nΩ is measured for the bottom joint terminal and a non-linear magnetic field dependent resistance of 20-35 nΩ is measured for the top terminal. Although the top joint terminal does not see the applied field, it does show a large magnetic field dependency and a significant current dependency. The resistance increases up to 85% of I_c and decreases when the current in increased further. This joint terminal behavior is unexplained at this moment.

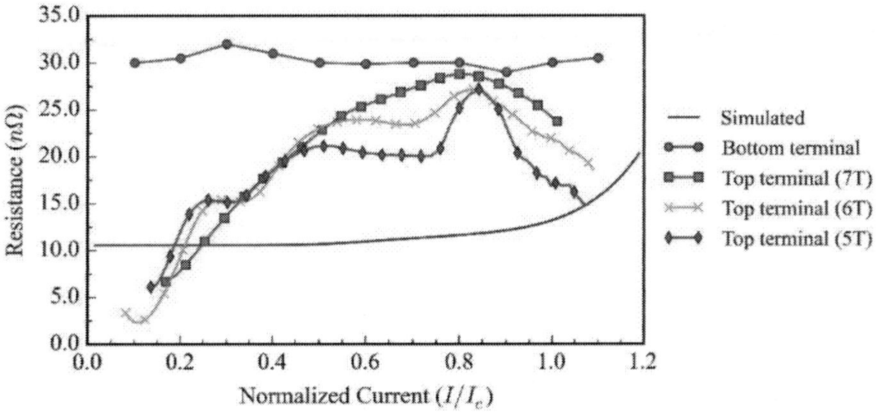

Figure 7. Measured joint terminal resistances of the bottom and top joint terminal at 4.5 K as function of the normalized current where the mid-section of the cable was placed in applied magnetic field. The resistance of the top terminal varies with the applied field, the resistance of the bottom terminal did not. The simulated joint terminal resistance is plotted as function of the normalized current at 4.5 K, 1 T, as the terminals are outside the peak magnetic field region.

During the measurement a large resistance between the bus bar and the CORC joint was observed, implying an imperfect electrical contact. Probably this led to an inhomogeneous current distribution over the joints, possibly causing the unexpected difference in resistance. More measurements are required to fully understand and benchmark the joints. Also an increase in the accuracy of the model is anticipated.

5. Conclusion

A novel joint technique for CORC type cables with many layers has been developed aiming at a drastic reduction of the joint resistance. Simulations have shown a possible reduction in joint resistance up to some 65% for a CORC cable with 12 layers following the new design. A test of the joint was performed at 4.5 K, at full current and in a background magnetic field of up to 9 T. The measured joint resistance is higher, roughly by a factor of 2 or 3, than the simulated values, possibly due to a non-optimal solder splice. More variations of such joints will be tested in the near future and, as the next step the design, will be scaled up for making joints to a 45 kA at 4.5 K and 10 T class CORC based Cable-in-Conduit Conductor.

References

[1] Van der Laan D C 2009 Supercond. Sci. Technol. 22 065013.
[2] Van der Laan D C 2014 presentation at HTS4Fusion Villigen, Switzerland.
[3] Willering G P, van der Laan D C, Weijers H W, Noyes P D, Miller G E and Viouchkov Y 2015 Supercond. Sci. Technol. 28 035001.
[4] Yazaki S, Karasawa A, Kotoyori T, Ishiyama A and Miyahara N 2013 IEEE Trans. Appl. Supercond. 23 4602304.
[5] Lu J, Han K, Sheppard W R, Viouchkov Y L, Pickard K W and Markiewicz W D 2011 IEEE Trans. Appl. Supercond 21 3009.
[6] Verweij A P, Genest J, Knezovic A, Leroy D F, Marzolf J-P and Oberli L R 1999 IEEE Trans. Appl. Supercond. 9 153.
[7] Polak M, Barnes P N and Levin G A 2006 Supercond. Sci. Technol. 19 817.
[8] Ekin J W 2006 Experimental Techniques for Low Temperature Measurements (New York: Oxford University Press) Appendix A6 pp 566-592.

RRR and thermal conductivity of Ag and Ag-0.2 wt.%Mg alloy in Ag/Bi-2212 wires

P Li[1], L Ye[1,2], J Jiang[3], T Shen[1,4]
1. Magnet Systems Department, Fermi National Accelerator Laboratory, Batavia, IL 60510, USA
2. Department of Material Science and Engineering, North Carolina State University, Raleigh, NC 27695, USA
3. Applied Superconductivity Center, National High Magnetic Field Laboratory, Florida State University, Tallahassee, FL 32310, USA

E-mail: tshen@fnal.gov

Abstract. Residual resistivity ratio (RRR) and thermal conductivity of metal matrix in metal/superconductor composite wires are important parameters for designing superconducting magnets. However, the resistivity of silver in reacted Ag/Bi-2212 wires has yet to be determined over temperature range from 4.2 K to 80 K because Bi-2212 filaments have a critical transition temperature T_c of ~ 80 K, and because it is unknown whether the RRR of Ag/Bi-2212 degrades with Cu diffusing from Bi-2212 filaments into silver sheathes at elevated temperatures and to what degree it varies with heat treatment. We measured the resistivity of stand-alone Ag and AgMg (Ag-0.2 wt.% Mg) wires as well as the resistivity of Ag and Ag-0.2 wt.% Mg in Ag/Bi-2212 round wires reacted in 1 bar oxygen at 890 °C for 1, 8, 24 and 48 hours and quickly cooled to room temperature. The heat treatment was designed to reduce the critical current I_c of Bi-2212 wires to nearly zero while allowing Cu loss to fully manifest itself. We determined that pure silver exhibits a RRR of ~ 220 while the oxide-dispersion strengthened Ag-Mg exhibits a RRR of ~ 5 in stand-alone samples. A surprising result is that the RRR of silver in the composite round wires doesn't degrade with extended time at 890 °C for up to 48 hours. This surprising result may be explained by our observation that the Cu that diffuses into silver tends to form Cu_2O precipitates in oxidizing atmosphere, instead of forming Ag-Cu solution alloy. We also measured the thermal conductivity and the magneto-resistivity of pure Ag and Ag-0.2 wt. % Mg from 4.2 K to 300 K in magnetic fields up to 14.8 T and summarized them using a Kohler plot.

1. Introduction

Silver-sheathed $Bi_2Sr_2CaCu_2O$ (Ag/Bi-2212 hereafter) multi-filament round wire has long been considered as a promising candidate for building high field magnets generating fields beyond 20 Tesla. Recently, the critical current density J_c of commercial powder-in-tube (PIT) Ag/Bi-2212 wires was significantly improved, with the best engineering critical current density J_E and the J_c at 4.2 K and 20 T exceeding 700 A/mm^2 and 2800 A/mm^2, respectively [1], [2]. In comparison, the best commercial Cu/Nb_3Sn wires exhibit non-Cu J_c of 3000 A/mm^2 at 4.2 K and 12 T, which decreases rapidly with increasing magnetic fields to 1500 A/mm^2 at 4.2 K and 15 T. The record J_c in Ag/Bi-2212 was achieved by removing porosity in the Bi-2212 filaments in PIT superconducting wires, through densification treatments [1]-[3]. Importantly, such high J_c can be achieved in long-length Bi-2212 wires through overpressure partial melt processing so Ag/Bi-2212 round wire is becoming a magnet-grade conductor [2].

In additional to using pure silver as a stabilizer, commercial Bi-2212 wires use oxide-dispersion strengthening silver alloys to improve their mechanical strength [4], [5]. Popular silver alloys used include Ag-0.2 wt.% Mg, Ag-Al, Ag-Mn, and Ag-Mg-Sb [6] and they typically occupy 25% of the total volume of the conductor. Resistivity ρ and thermal conductivity κ of both silver and silver alloys are important materials properties. For example, they are important for designing the quench protection

[4] To whom any correspondence should be addressed.

circuitry [7], [8] and affect the design of the entire magnet system, because during a magnet quench, the power generation density in a normal zone is given by $(1-\lambda)J_m^2(t)\rho(B,T)$, where λ is the volumetric ratio of superconductor in the metal/superconductor composite, J_m the current density in the metal matrix, and $\rho(B,T)$ the effective resistivity of the metal matrix. Both J_m and $\rho(B,T)$ would depend on the magneto-resistivity of pure silver and Ag-alloys.

One would think that the task is rather simple as one just needs to measure $\rho(B,T)$ of silver and silver alloys used in the industrial production of Ag/Bi-2212 wires. This, however, ignores the influence of Bi-2212 filaments on transport properties of silver and silver alloys. Bi-2212 filaments are embedded in a silver matrix, which again is encased inside an Ag-alloy external sheath, without a reaction barrier. During the partial melt processing, the elements inside Bi-2212 filaments can diffuse into silver, introducing a small amount of impurity atoms and lattice defects to enhance the scattering of electron at low temperatures and therefore raising the residual resistivity at low temperatures. Among all the elements forming Bi-2212, Cu has the highest solubility and a diffusion coefficient of ~2-3x10^{-11} cm^2/s in Ag at ~ 890 °C [9]. However, it is not easy to determine resistivity and thermal conductivity of silver and silver alloys in a reacted Ag/Bi-2212 wire because Bi-2212 filaments in reacted Bi-2212 wires have a T_c of ~82 K and an I_c of several hundred amperes [9]-[12].

In this paper, we report the electrical resistivity and thermal conductivity of Ag and Ag-0.2 wt.% Mg (AgMg hereafter) alloy used in industrial production of Ag/Bi-2212 conductors. These properties were tested in temperature from 4.2 K to 300 K and in magnetic field up to 14.8 T. We also developed a new protocol for determining the residual resistivity ratio (RRR) of Ag/Bi-2212 wires and its dependence on Cu-loss and heat treatment parameters.

2. Experiments

2.1. Magneto-resistivity and thermal conductivity measurement of stand-alone Ag and AgMg alloy

As the first step, this study determined the $\rho(B,T)$ of silver and silver alloys used in industrial production of Ag/Bi-2212 wires. Samples of the silver and sliver-magnesium alloy (Ag with 0.2 wt.% Mg) were supplied by Oxford Superconducting Technology (OST) as round wires with a diameter of 0.55 mm. Samples were heat-treated using a standard partial-melt-processing in 1 bar flowing pure O_2 to remove the influences of cold deformation and to introduce the MgO precipitates. The resistivity of the heat-treated Ag and AgMg alloy samples were measured in a Quantum Design Physical Property Measurement System (PPMS) using a standard four-probe method with activation current up to 100 mA as a function of temperature from 4.2 K to 300 K and of magnetic field up to 14.8 Tesla. The thermal conductivity of these heat-treated Ag and AgMg wires was measured from 300 K to 4 K using the thermal-transport-option (TTO) of PPMS. The measurement was calibrated using a nickel sample from Quantum Design.

2.2. Resistivity measurement of silver in heat-treated Ag/Bi-2212 wires

We also determined the resistivity of Ag and Ag-0.2 wt.% Mg in Ag/AgMg/Bi-2212 wires using a new protocol, which takes into consideration of possible effects of Cu loss on resistivity of silver. Samples were reacted at 890 °C, at which Bi-2212 filaments are in melt state and contain a mixture of an amorphous liquid and two solid crystalline phases (a Cu-free phase (CF) and an alkaline-earth cuprate phase (AEC)), and then rapidly cooled down to room temperature. The majority of Bi-2212 filaments in these samples are expected to be filled with liquid, AEC, and CF, and contain a few Bi-2212 whiskers, and therefore the I_c of these samples is close to zero. The dwelling time at T_{max}=890 °C was varied from 1 hour to 48 hours to amplify the possible effect of Cu-loss on sheath resistivity, as Cu-loss is expected to be most active when Bi-2212 filaments are in the melt state. Two types of wires were studied. The first wire is PMM101108-1 (37x18 architecture, ϕ0.8 mm) fabricated by OST with a Ag:AgMg:Bi-2212 ratio of 0.5:0.25:0.25; this wire represents the widely fabricated commercial Ag/Bi-2212 wires. The second wire, PMM140721 (85x6 architecture, ϕ0.8 mm), has a Ag:AgMg:Bi-2212 ratio of 0.33:0.42:0.25 and represents wires with higher mechanical strength with a larger volume of

AgMg. After a standard partial melt processing in 1 bar flowing O_2, the I_c of PMM101108-1 at 4.2 K, self-field is 425 A whereas the I_c of PMM140721 is 450 A whereas after 100 bar overpressure processing, the I_c at 4.2 K, self-field of PMM101108-1 increases to 870 A, producing a J_E(4.2 K, and 20 T) of approximately 650 A/mm^2. For simplicity, PMM101108-1 and PMM140721 are named type "A" and "B" in the rest of this paper.

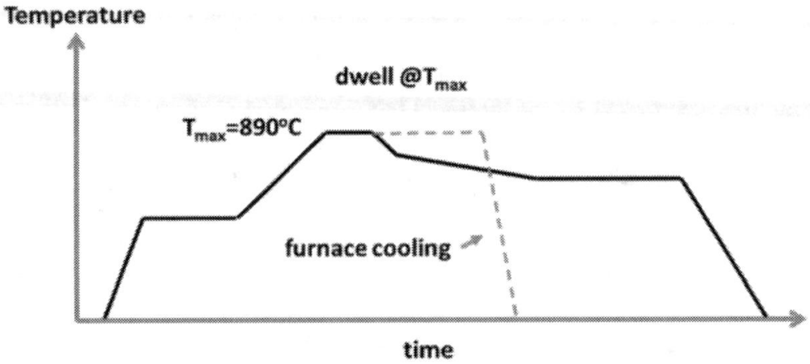

Figure 1. The heat treatment procedure used to generate I_c-depressed samples for sheath resistivity measurements. The beginning part of this procedure is similar to the standard partial-melt-processing (PMP): the temperature ramps from room temperature to 820 °C at 160 °C /hour, dwells there for 2 hours and then ramps to 890 °C at 50 °C/hour. The dashed-line marks where this procedure differs from the standard partial-melt-processing (PMP): the slow cooling (10 °C/hour from 890 °C to 880 °C and 2.5 °C/hour from 880 °C to 830 °C) and the rest of the PMP are replaced with a fast furnace cooling and the dwelling time at T_{max}=890 °C varies from 1 hour to 48 hours.

2.3. Microscopy observation of Cu-loss

To determine at what temperatures Cu loss occurs and where Cu migrate, the surface and cross-sections of various heat-treated Ag/Bi-2212 conductors were examined using a scanning electron microscope JOEL-5900 (SEM) equipped with energy dispersive x-ray spectrometers (EDS).

3. Results

3.1. Resistivity and thermal conductivity of Ag and AgMg

Figure 2 presents $\rho(B, T)$ plots of Ag and Ag-0.2 wt. % Mg wires heat treated using a standard partial melt processing in 1 bar flowing O_2. For pure silver wire, the measured zero-field resistivity at 300 K, ρ(300 K, 0 T), is 16.14 nΩ m and that at 4.2 K, ρ(4.2 K, 0 T), is 0.0754 nΩ m, which gives a *RRR* ratio, defined as ρ(300 K, 0 T)/ρ(4.2 K, 0 T), of 214. For Ag-0.2 wt. % Mg wire, ρ(300 K, 0 T) is 19.7 nΩ m and ρ(4.2 K, 0 T) is 3.84 nΩ m, giving a RRR of 5.1. Magnetic field impacts the resistivity of Ag significantly in temperature range from 4.2 K to 50 K. At 4.2 K, applying a 14.8 T field increases the resistivity of Ag by a factor of five, whereas for temperature higher than 70 K, the increase in resistivity is less than 15%. The resistivity of AgMg is not sensitive to magnetic field and only increases by 10% with field increasing from 0 T to 14.8 T at 4.2 K and the field-induced resistivity increase is negligible at higher temperatures. Using all the $\rho(B, T)$ data from this study, a Kohler plot is generated by plotting the magnetic field–induced resistivity change, $[\rho(B,T)-\rho(B=0,T)]/\rho(B=0,T)$, against $B*[\rho(B=0,273\ K)/\rho(B=0,4.2\ K)]$, as shown in Figure 2 (C). Both the Ag and AgMg results converge to the reference data reported in Ref [13] well. The thermal conductivity κ of Ag and AgMg wires is shown in Figure 2 (D). κ of Ag has a maximum of ~2700 W/Km at ~15 K and is ~500 W/Km at room temperature. κ of AgMg is much lower than Ag and slowly increases with temperature, which is common in silver with low RRR and silver alloys. The thermal conductivity of Ag decreases significantly with increasing magnetic field, showing a maximum of ~1600 W/Km at 9 T whereas the thermal conductivity of AgMg changes little with increasing magnetic field to 9 T.

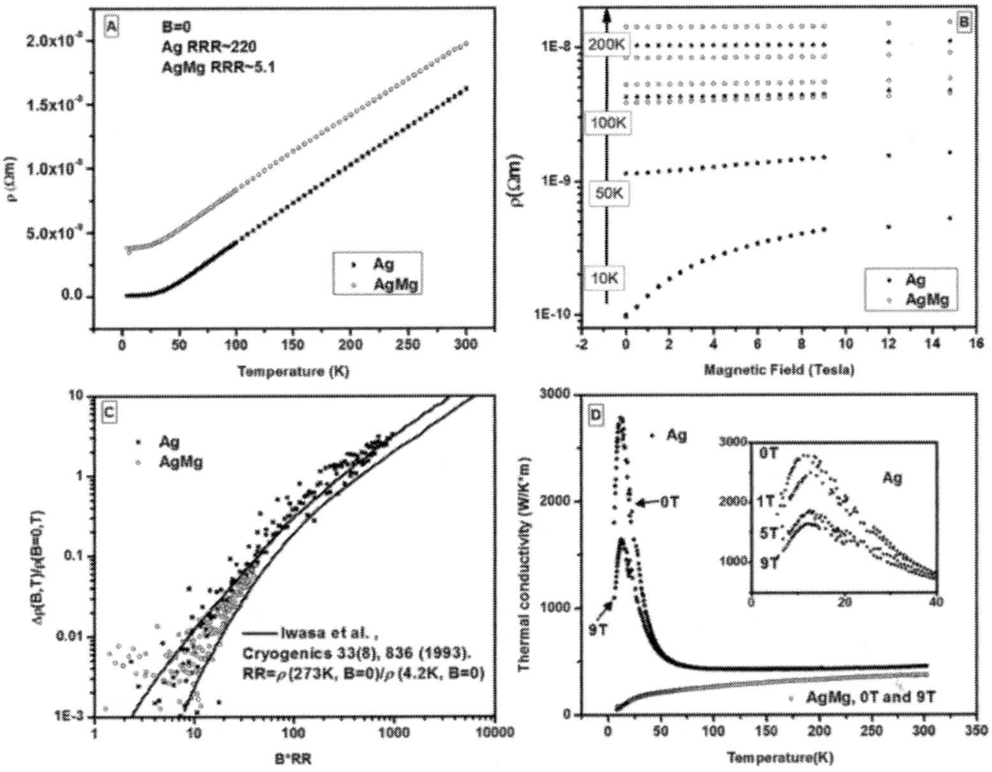

Figure 2. Resistivity and thermal conductivity of heat-treated Ag and AgMg (0.2 wt.% Mg). (A). $\rho(T)$ of pure Ag and Ag-0.2 wt.% Mg in zero magnetic field over 4.2 – 300 K. (B): Dependence of ρ_{Ag} and $\rho_{Ag\text{-}0.2\,wt.\%\,Mg}$ on magnetic fields at 10 K, 50 K, 100 K and 200 K. (C): Kohler plot generated from all the $\rho(B, T)$ data obtained with the data reported in Ref [13] shown as two lines. (D): Thermal conductivity of Ag and AgMg as a function of temperature and magnetic field. The inset of (D) shows the low temperature (< 40 K) thermal conductivity data of Ag at 0 T, 1 T, 5 T and 9 T.

3.2. Observation of Cu-loss in Ag/Bi-2212 conductors

We observed clear evidences of Cu loss in two types of conductors. Figure 3 shows the back-scattering electron images of the surfaces of heat-treated Ag/Bi-2212 dip-coated tapes whereas the figure 4 for heat-treated Ag-0.2 wt.% Mg/Ag/Bi-2212 round wires. The sheath of dip-coated tapes is pure silver and the outmost sheath of round wires is Ag-0.2 wt.% Mg. On the surfaces of all these conductors, Cu_2O particles are found. Meanwhile, EDS was not able to detect Cu signal in the Ag matrix, within the detection limit 0.1% of EDS. It is important to note that these Cu-rich particles can appear even without heating Ag/Bi-2212 above the partial melting point of Bi-2212. As shown in Figure 4, such Cu-rich particles can be found on round wires heated to 800 °C and dwelled for 70 hours.

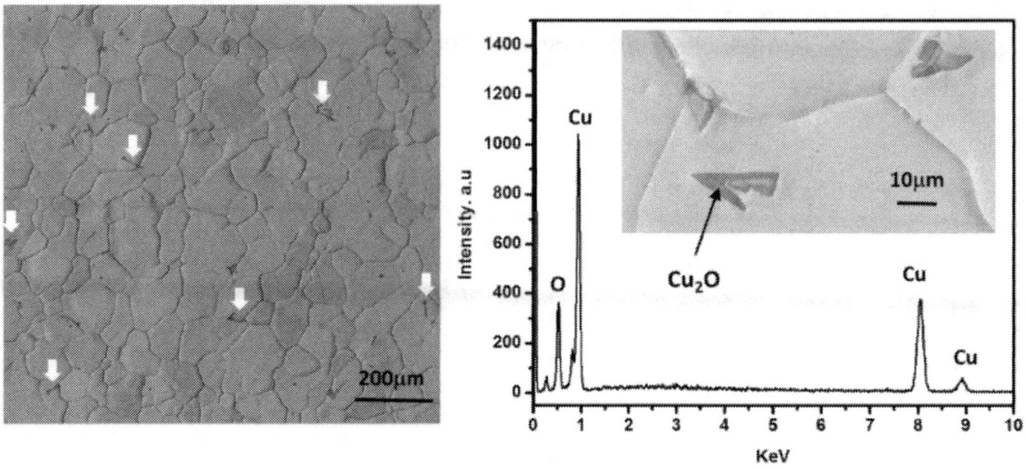

Figure 3. Left: SEM image of the Ag sheath surface of a Ag/Bi-2212 dip-coated tape after the PMP heat treatment, with some Cu-rich particles marked by the arrows. Right: high magnification image of several Cu-rich particles (the dark grey particles) and the EDS spectrum of one of them, indicating that its composition is Cu_2O.

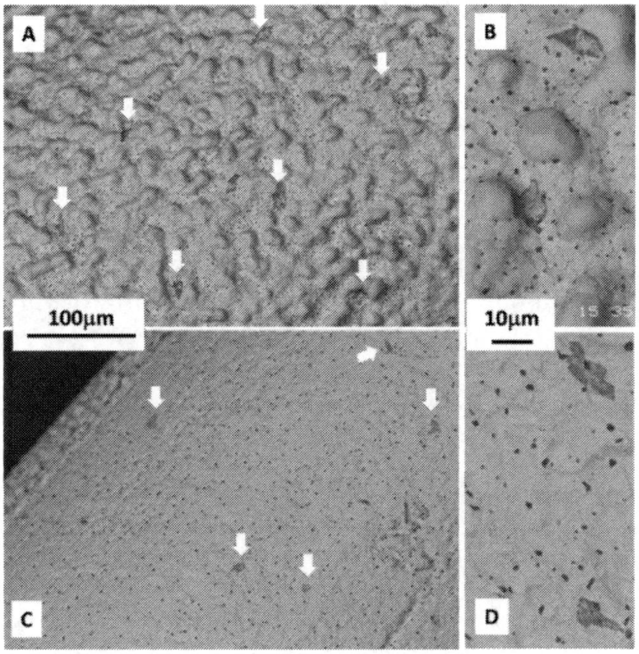

Figure 4. The AgMg Surface of the heat-treated conductor A, whose external outer sheath is AgMg. Samples in (A) and (B) were fully heat-treated using the standard PMP. Some Cu-rich particles are marked by the arrows in (A) and (B) shows two Cu-rich particles (dark grey) with higher magnification. The smaller, black particles are MgO. Samples in (C) and (D) were heated at 800 °C for 70 hours. Cu-rich particles are also evident in (C) and (D).

3.3. Sheath resistivity of I_c-depressed wires

We measured the *V-I* curves of samples reacted at 890 °C for 1, 8, 24, and 48 hours and fast cooled to room temperature using the standard four-probe method at 300 K, 77 K, and 4.2 K. The results are summarized in Table 1. At 4.2 K, all samples exhibited a rather linear *V-I* characteristics with *I* going up to 30 A, indicating that the Bi-2212 filaments carried I_c of less than 1-3 A. Interestingly, the resistivity of the sheathes, including both that of Ag and that of Ag-0.2wt%Mg, did not significantly increase with the time at 890 °C; rather it decreases by 17% and 29% for the conductor A and the conductor B, respectively.

Table 1. Resistivity of the I_c-depressed samples. The bold numbers are from direct measurements and the other are from calculation.

Wire	Time at 890 °C (hour)	ρ_{sheath}(300 K) (10^{-8} Ωm)	ρ_{sheath}(4 K) (10^{-11} Ωm)	ρ_{Ag}(4 K) (10^{-11} Ωm)	RRR of Ag
"A"	1	**~1.72**	**7.23**	4.84	351
	8		**6.98**	4.67	363
	24		**6.46**	4.33	392
	48		**5.97**	4.02	436
"B"	1	**~1.80**	**18.6**	8.16	198
	8		**17.6**	7.75	209
	24		**14.2**	6.24	260
	48		**13.3**	5.85	277

4. Discussion

Figure 3 and Figure 4 showed that for Ag/Bi-2212 wires either in melt state or in powder state, Cu can dissolve into silver and diffuse through Ag and Ag-0.2 wt.% Mg to form Cu_2O particles on their surfaces, providing another strong evidence that Cu-loss is common in reacted Ag/Bi-2212 wires in addition to those presented in [6], [11], [12], [14]. This is consistent with studies of AgCu alloy [15], which shows that Cu tends to segregate and form precipitates under oxidizing atmosphere, instead of forming a uniform solid solution alloy.

Using the resistivity of silver sheathes in two wires at 4.2 K in Table 1, we calculated the resistivity of pure silver by assuming: (i) All electrical current flows in silver and Ag-Mg, with the Bi-2212 filaments' I_c being less than 3 A; (ii) Ag and Ag-Mg form a parallel circuit; (iii) the RRR of the Ag-0.2 wt. % Mg is less than 5. The results were also presented in the Table 1. The calculated ρ_{Ag}(4.2 K) was in the range of 4.02-4.84x10^{-11} Ωm and 5.85-8.16x10^{-11} Ωm, close to or smaller than that of the stand-alone Ag sample (7.3x10^{-11} Ωm), yielding RRR ranging from 198 to 436. This is reasonably consistent with RRR=220 obtained from the stand-alone sample, suggesting that heat treatment and Cu-loss have very limited effect on the resistivity and RRR of Ag. This is perhaps because the formation of Cu_2O particles in O_2 environment minimizes the Cu atoms that stay as foreign atoms in Ag lattice, and perhaps also because the Cu content is quite low. We estimated that the total amount of Cu diffusing from Bi-2212 filaments into Ag is less than 0.1 at.% of Cu in Ag, based on the fact that the number density of Cu_2O particles is one or two orders lower than that of the MgO particles, shown in Figure 4.

5. Conclusion

We measured the $\rho(B,T)$ and $\kappa(B = 0, T)$ of Ag and Ag-0.2 wt.% Mg over the temperature range 4.2 - 300 K for magnetic induction up to 14.8 T. We found strong microstructural evidences that Cu in Bi-2212 filaments dissolves into silver and diffuses through Ag and Ag-0.2 wt.% Mg both when Bi-2212 filaments are in the melt state and when they are in the powder state. We proposed a simple protocol for determining the RRR of Ag in the Ag/Bi-2212 wires, which involves calculating the resistivity of Ag after measuring the *E-J* curves at 4.2 K of Ag/Bi-2212 wires reacted at 890 °C for 48 hours and cooled to room temperature; such heat treatment allows to suppress the I_c of Bi-2212 filaments while maintaining or increasing Cu loss. The measurement and analysis indicated that the RRR of Ag surprisingly does not degrade with an extended heat treatment of 48 hours at 890 °C. This is perhaps explained by the fact that Cu diffuses through Ag to form Cu_2O precipitates on the surface. The RRR of Ag-0.2 wt.% Mg is determined to be ~5 and the RRR of Ag is determined to be 200-440.

6. References

[1] Jiang J, Starch W L, Hannion M, Kametani F, Trociewitz U P, Hellstrom E E and Larbalestier D C 2011 *Supercond. Sci. Technol.* **24**, 082001
[2] Larbalestier D *et al.* 2014 *Nat. Mater.* **13** 375
[3] Shen T, Ghosh A, Cooley L and Jiang J 2013 *J. Appl. Phys.* **113** 213901
[4] Cheggour N, Lu X, Holesinger T, Stauffer T, Jiang J and Goodrich L 2012 *Supercond. Sci. Technol.* **25** 015001
[5] Li P, Wang Y, Godeke A, Ye L, Flanagan G and Shen T 2015 *IEEE Trans. Appl. Supercond.* **25** 1
[6] Navarro R 2000 *Supercond. Sci. Technol.* **13** 147
[7] Weijers H *et al.* 2010 *IEEE Trans. Appl. Supercond.* **20** 576
[8] Shen T, Ye L, Turrioni D and Li P 2015 *Supercond. Sci. Technol.* **28** 075014
[9] Majewski P, Aubele A, Fahr T and Aldinger F 2001 *Physica C* **351** 62
[10] Majewski P, Aubele A and Aldinger F 2000 *Physica C* **341–348** 517
[11] Wesolowski D E, Rikel M O, Jiang J, Arsac S and Hellstrom E E 2005 *Supercond. Sci. Technol.* **18** 934
[12] Shen T, Li P, Jiang J, Cooley L, Tompkins J, McRae D and Walsh R 2015 *Supercond. Sci. Technol.* **28** 065002
[13] Iwasa Y, McNiff E J, Bellis R H and Sato K 1993 *Cryogenics* **33** 836
[14] Shen T, Jiang J, Kametani F, Trociewitz U P, Larbalestier D C and Hellstrom E E 2011 *Supercond. Sci. Technol.* **24** 115009
[15] Guo J, Lewis J, Goretta K and Schwartz J 1995 *J. Appl. Phys.* **78** 4596

Acknowledgements

This work was funded by the Office of High Energy Physics (OHEP) of the U.S. Department of Energy (DOE) through Fermi Research Alliance (DE-AC02-07CH11359) and a U.S. DOE Early Career Award to TS. Work at the National High Magnetic Field Laboratory (NHMFL) was supported by the DOE-OHEP under grant number DE-SC0010421, by the National Science Foundation under NSF/DMR-1157490, and by the State of Florida. We thank Dr. Hanping Miao and Dr. Yibing Huang with the Oxford Superconducting Technology for supplying us the Ag and Ag-0.2 wt.% Mg samples used in this study, the U.S. Conductor Development Program of DOE for supplying us the wire PMM140721, and Dr. Jun Lu with NHMFL for discussion.

IOP Conf. Series: Materials Science and Engineering **102** (2015) 012028 doi:10.1088/1757-899X/102/1/012028

Nanoparticle Ag-enhanced textured-powder Bi-2212/Ag wire technology

J N Kellams[1,3], P McIntyre[1], N Pogue[2] and J Vandergrifft[1]
[1] Department of Physics and Astronomy, Texas A&M University, College Station, 77843-4242, TX, USA
[2] Paul Scherrer Institute, Villigen, Switzerland

E-mail: jnkellams@tamu.edu

Abstract. A new approach to the preparation of cores for Bi-2212/Ag wire is being developed. Nanoparticle Ag is homogeneously dispersed in Bi-2212 fine powder, and the mixture is uniaxially compressed to form highly textured, cold-sintered core rods. The rods can be assembled in a silver matrix, drawn to form multifilament wire, and restacked and drawn to form multifilament wire. Preliminary studies using tablet geometry demonstrate that a non-melt heat treatment produces densification, grain growth, intergrowth among grains, and macroscopic current transport. The status of the development is reported.

1. OPIT Bi-2212 wire

The state-of-art fabrication method for Bi-2212 multi-filament round wire is the oxide-powder-in-tube (OPIT) process. In the OPIT process an approximately phase-pure fine powder of Bi-2212 is loaded into silver tubes, the loaded tubes are drawn, stacked, and re-drawn to make multi-filament wire [1]. The wire is cabled and wound onto its final form, and the windings undergo a partial melt (PM) heat treatment during which the Bi-2212 powder is fully melted (~887 C) and then gradually re-crystallized and annealed. The PM heat treatment leads to several difficulties:

- Macro-voids are formed when the void spaces among the powder particles coalesces in the melt liquid.
- The melt liquid etches along the grain boundaries of the silver matrix,
- The recrystallized Bi-2212 grains have limited connectivity.

Overpressure processing at ~5-10 MPa during the formation heat treatment dramatically improves performance [2,3]. The overpressure process could be a practical challenge for the large windings required for collider dipoles and insert windings of solenoids. We have sought a process by which both high density and connectivity could be achieved without melting the Bi-2212 powder cores. The present work is the latest in a sequence of efforts to this purpose.

2. Textured Bi-2212 powder

A fine powder of Bi-2212 can be highly textured by uniaxially pressing the powder to form (4 x 4 x 152) mm square-cross-section rods [4]. The uniaxial pressure causes the a-b plains of the micaceous powder particles to align. The goal is to develop a process to achieve connectivity amongst the

[3] To whom any correspondence should be addressed.

IOP Conf. Series: Materials Science and Engineering **102** (2015) 012028 doi:10.1088/1757-899X/102/1/012028

textured grains of such a Bi-2212 core, without melting it, so that the above problems associated with liquid formation can be averted.

Fine powder Bi-2212 is loaded into the female aperture of a forming die shown in figure 1. The male and female dies are assembled and loaded into a hydraulic press, and the powder is compressed to 140 MPa. Since Bi-2212 is micaceous, the particles naturally align when compressed and the powder core attains ~80% texture, measured using X-ray diffraction [4]. The density of the green-state powder core was measured to be approximately 80% of single-crystal density.

Figure 1. a) Female die form; b) Entire die assembly loaded into hydraulic press.

The square-cross-section textured-powder (TP) Bi-2212 core is then inserted into a silver tube and then extruded and drawn to fine wire. The TP core drew down to approximately 0.4 mm, reducing by 20% per pass, without any sausaging or breaks. A heat treatment development was undertaken to seek conditions that would produce connectivity among Bi-2212 grains. Remarkable growth of the a-b planes of all grains was observed as shown in figure 2, but there was little evidence of grains growing into adjacent grains (by twisting or twinning). Current transport measurements were made on monocore wire samples, and no significant evidence of macro-scale current transport was observed [4].

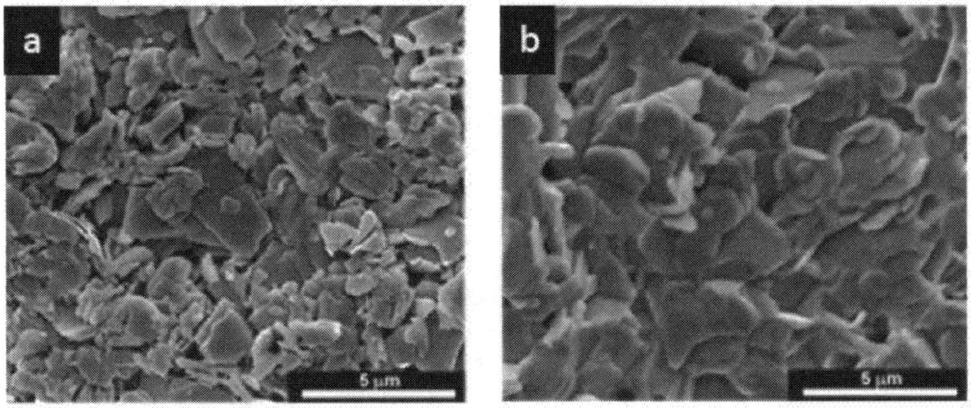

Figure 2. SEM images of TP cores before (a) and after (b) a non-melt heat treatment at 870 C for 2 hours.

3. Nano-Ag-enhanced TP cores

The properties of Bi-2212 grains at temperatures near melt are strongly affected where there is an interface between a Bi-2212 grain and a silver grain [5]. The melt temperature at the interface is lowered, and when a melt is recrystallized grains align along the silver interface. Magnetic imaging studies have shown that in conventionally processed OPIT wire the majority of current flows near that silver interface.

We conjectured that those effects could be enhanced if silver nanoparticles were dispersed homogenously throughout the Bi-2212 fine powder of a core. The benefits of the Ag/Bi-2212 interface would then be realized everywhere inside the core and not just near its boundary with the silver matrix.

A process has been developed to homogenously mix silver nanopowder with Bi-2212 to form enhanced textured-powder (ETP) cores. Initial studies of microstructure and heat treat studies were made using a pellet die to compress 2 cm diameter round pellets of the ETP powder mixture. Figure 3 shows that homogenous dispersion was achieved using high-energy acoustic mixing for both silver powder with 20 nm (Figure 3a) and 100 nm (Figure. 3b) mean particle size.

Figure 3. SEM images of green-state silver enhanced Bi-2212 pellets, with silver sizes of 20 nm (a) and 100nm (b).

Initial heat treatment studies showed excellent grain growth, interconnectivity of Bi-2212 grains, and the absence of alkaline earth cuprates (AEC) and copper free (CF) regions in the maximum temperature domain of 876 C – 878 C for 48 hours. When the pellets were also annealed at 845 C for 48 hours after the maximum temperature no parasitic phases were found. When the mixture was fully melted at a maximum temperature of 881 C AEC and CF regions became very common (Figure 4). A resistance test was performed using a 4-point contact array in a pellet sample. It showed supercurrent transport over ~1 cm scale [6].

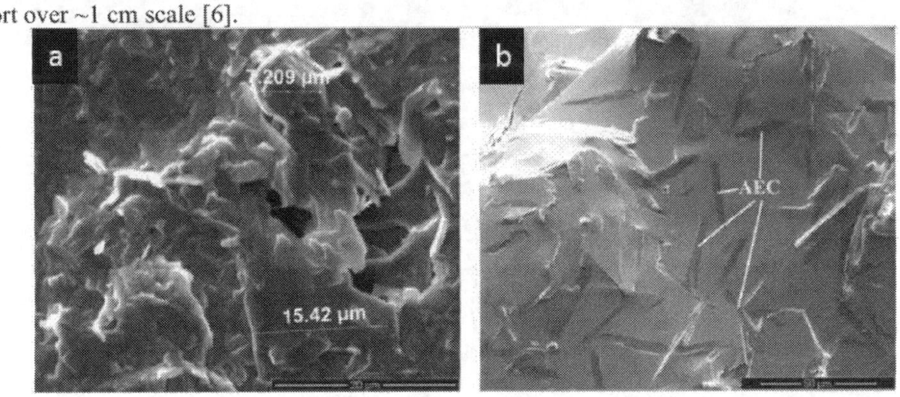

Figure 4. a) ETP core heat treated at 878 C, no AEC, CF regions, or parasitic phases found. b) ETP core heat treated at 881 C, fully melted, AEC and CF regions common.

4. Future Work

Square-cross-section ETP cores are being prepared similarly to the TP cores described above. Four ETP cores are being made using two sizes and two weight concentrations of silver nanopowder (Table 1). The cores are loaded into a silver alloy tube in an inert atmosphere, and the billet will be drawn down to a sub-millimeter monofilament wire. Figure 5 shows an ETP core being loaded into the silver tube using an alignment block to minimize handling of the extremely fragile cores. Wire specimens of each composition will be wound on ceramic barrels, and a systematic heat treat development will be undertaken. Current transport will be measured and microstructure will be characterized for all samples.

Table 1.

Mixture	Silver/Weight %	Silver particle size
Large 3 %	3%	80-100 nm
Large 5 %	5%	80-100 nm
Small 3 %	3%	20 nm
Small 5 %	5%	20 nm

5. Conclusion

A method has been developed for homogeneously dispersing nanopowder Ag into Bi-2212 fine powder. The powder mixture has been formed into highly textured cores using uniaxial pressing. It has been found that a non-melt heat treatment produces excellent grain growth and interconnectivity, no parasitic phases. A pellet configuration was tested and macroscopic superconducting transport was observed.

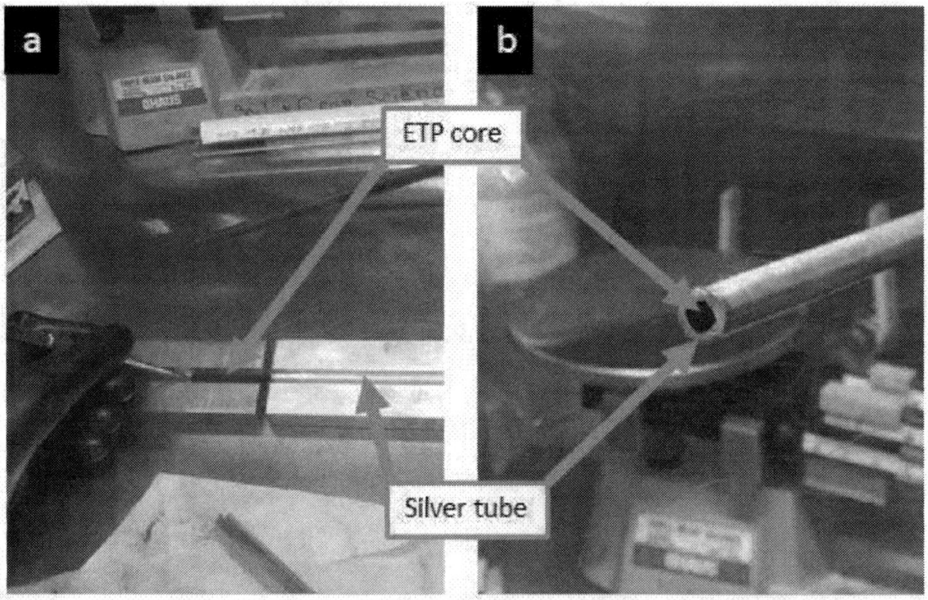

Figure 5. An ETP core being loaded into a silver alloy tube.

6. References

[1] Miao H *et al.*, ‘Development of round multifilament Bi-2212/Ag wires for high field magnet applications’, IEEE Trans. Appl. Superconduct. 15, 2, 2554 (2005).

[2] Jiang J *et al.*, 'Doubled critical current density in Bi-2212 round wires by reduction of the residual bubble density', Superconductor Science and Technology 24, 082001 (2011).
[3] Hellstrom E *et al.*, 'Optimizing 2212 with overpressure processing', WAMHTS-1 – Hamburg, May 21-23, 2014: http://indico.cern.ch/event/308828 /session/9/contribution/32/material/slides/6.pdf
[4] Damborski K *et al.*, 'Transport properties of high green core density Bi-2212 textured-powder conductors', AIP Conf. Proc. 1574, 232 (2014): http://scitation.aip.org/content/aip/proceeding/aipcp/10.1063/1.4860629
[5] Hasegawa T *et al.*, 'The effect of Ag on the formation of Bi2Sr2Ca Cu2Oy thick film', Physica C222, 111-118 (1994).
[6] McIntyre P *et al.*, 2015 'Texture-Powder Bi-2212 Ag Wire Technology Development', *Proc. Int. Particle Accelerator Conf. (Richmond VA, USA)*

IOP Conf. Series: Materials Science and Engineering **102** (2015) 012029 doi:10.1088/1757-899X/102/1/012029

Synthesis of nanoscale magnesium diboride powder

D K Finnemore[1] and **J V Marzik**[2]
[1] Ames Laboratory USDOE and Iowa State University, Ames, Iowa, US
[2] Specialty Materials, Inc., Lowell Massachusetts, USA

E-mail: finnemor@ameslab.gov

Abstract. A procedure has been developed for the preparation of small grained magnesium diboride (MgB_2) powder by reacting nanometer size boron powder in a magnesium vapor. Plasma synthesized boron powder that had particle sizes ranging from 20 to 300nm was mixed with millimeter size chunks of Mg by rolling stoichiometric amounts of the powders in a sealed cylindrical container under nitrogen gas. This mixture then was placed in a niobium reaction vessel, evacuated, and sealed by e-beam welding. The vessel was typically heated to approximately 830°C for several hours. The resulting MgB_2 particles have a grain size in the 200 nm to 800 nm range. Agglomerates of loosely bound particles could be broken up by light grinding in a mortar and pestle. At 830°C, many particles are composed of several grains grown together so that the average particle size is about twice the average grain size. Experiments were conducted primarily with undoped boron powder, but carbon-doped boron powder showed very similar results.

1. Introduction

In this work, we report the development of a batch process for the conversion of plasma synthesized boron powder[1] into MgB_2 powder. Early studies of the performance characteristics of MgB2 conductors showed that the critical current density, Jc, of a MgB_2 wire was controlled primarily by grain-boundary pinning[2,3]. Hence, for an *ex-situ* powder-in-tube (PIT) process to fabricate MgB_2 conductors, it is generally preferred to have small grain size for the starting MgB_2 powder in order to give a dense array of grain boundaries in the final wire.

2. Experimental

In the procedures reported here to convert plasma synthesized boron powder into MgB_2 powder, the boron powder and the magnesium were handled in a glove box under dry nitrogen. Samples were transported in sealed glass jars filled with N_2. The plasma synthesized boron powder was mixed with millimeter size chips of Mg by rolling stoichiometric amounts of the elements in a Mason jar under dry N_2 gas for about 10 minutes. This mixture was then placed in a cylindrical Nb reaction vessel (28 mm OD x 150 mm long x 1 mm wall thickness). An end cap was affixed and the tube was evacuated prior to closing the tube by e-beam welding. The vessel was heated to 830°C for 24 h in a horizontal furnace under an argon atmosphere. A tubing cutter was used to open the reaction vessel near the weld, and the powder was poured into a Mason jar for storage in the glove box. The powder tended to form agglomerates up to a centimeter in size but these could be broken up by light grinding in a mortar and pestle.

As shown in figure 1, most of the grains are in the 100 to 800 nm range for a sample reacted at 830°C for 24 h. A sputter coating of iridium (~2 nm) was applied to highlight the shape and size of the particles for examination in the scanning electron microscope (SEM). At the reaction temperature some grains grew together to make elongated particles (figure 1). Figure 2 illustrates a particularly large particle where the grains linked up into a particle ~2000 nm long.

IOP Conf. Series: Materials Science and Engineering **102** (2015) 012029 doi:10.1088/1757-899X/102/1/012029

Figure 1. Scanning-electron Microscope micrograph illustrating the grain size and connectivity for MgB_2 reacted at 830°C for 24 h.

Figure 2 A micrograph showing a particularly large particle where several grains have grown together for powder reacted at 830°C for 24 h.

A rough estimate of the distribution of both the particle size and the grain size distribution was made by taking many of micrographs at 15,000x and 50,000x and taking the long dimension on the micrograph as a measure of the particle size. Typically the largest 124 of the particles were used in the viewing fields of four micrographs, so these measurements ignored the smallest grains and particles. The results are shown in figures 3a & 3b. Measurements of the average grain size have a lot of uncertainty but the average grain size seems to diminish from about 600 nm for a 48 h reaction at 830°C to about 300 nm for an 8 h reaction at 830°C.

We also prepared three 50g batches of $Mg[B_{0.98}\ C_{0.02}]_2$. Carbon was added by introducing CH_4 to the gas stream in the plasma synthesis of the boron powder. The results were indistinguishable from the work shown above for the undoped boron.

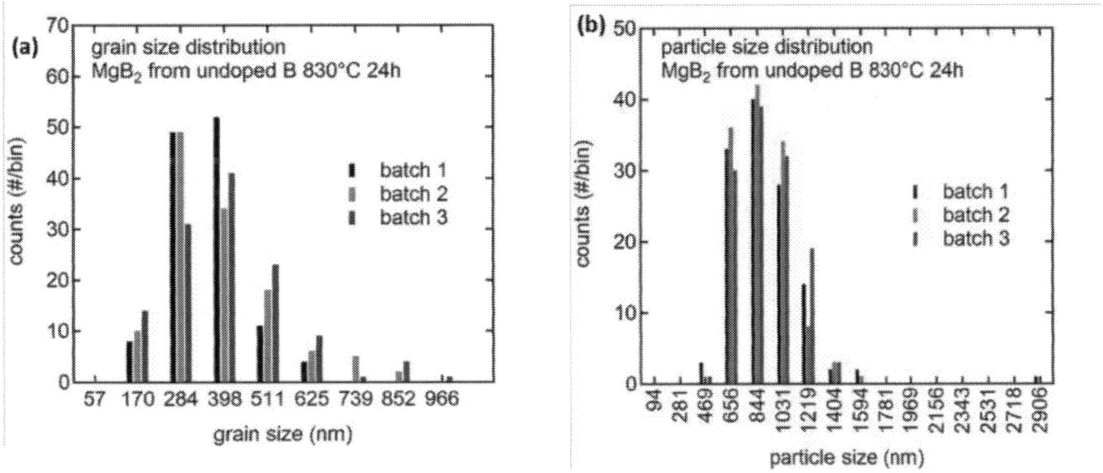

Figure 3. (a) Grain size and (b) particle size and distribution estimated from 3 different 50 g batches of MgB_2 powder as measured for 124 grains or particles from each batch. The grain or particle size average length is labeled for the bin being counted. Note 1 particle is shown at 2906 nm in (b).

If the reaction temperature was raised and the reaction time extended, the particles got larger and more interconnected as illustrated by the micrograph in figure 4 for a sample reacted at 860°C for 48 h.

Figure 4. SEM micrograph of MgB_2 powder reacted at 860°C for 48 h.

Figure 5. Micrograph of MgB_2 particles grown at 830°C for 8 h.

If the reaction time at 830°C was reduced to 8h, the average particle size was distinctly smaller as shown in figure 5.

Figure 6 shows a quantitative measure of the average grain size for different reaction times at 830°C. Four micrographs were taken at 50,000x, the largest linear dimension of each grain was used as a measure of the grain size, and the average of 124 of the largest particles was used for each reaction time.

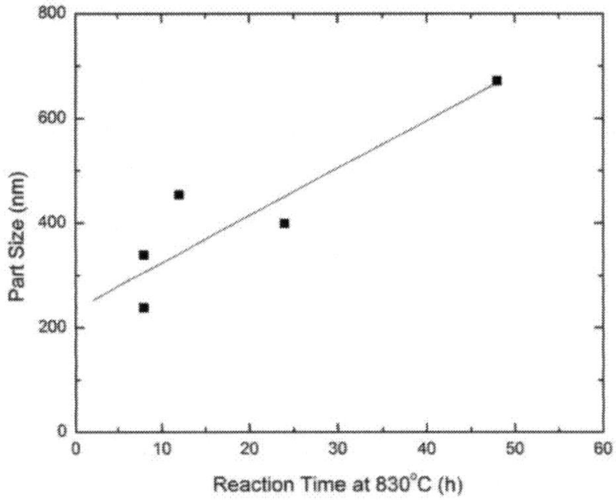

Figure 6. Plot showing the average grain size as a function of reaction time.

As the reaction time was reduced, there was an increasing chance that some higher borides such as MgB_4 or some MgB_7 formed in the resulting powder, indicative of an incomplete reaction between Mg and B. For the shortest time reported here of 8 h at 830°C, we do not see any regions of MgB_4 or MgB_7 in the SEM nor in the X-ray diffraction pattern of the powder in figure 7 below. If there were diffraction lines from MgB_4 or MgB_7, they were below the level of detection indicating that that they must be at least less than a couple of percent.

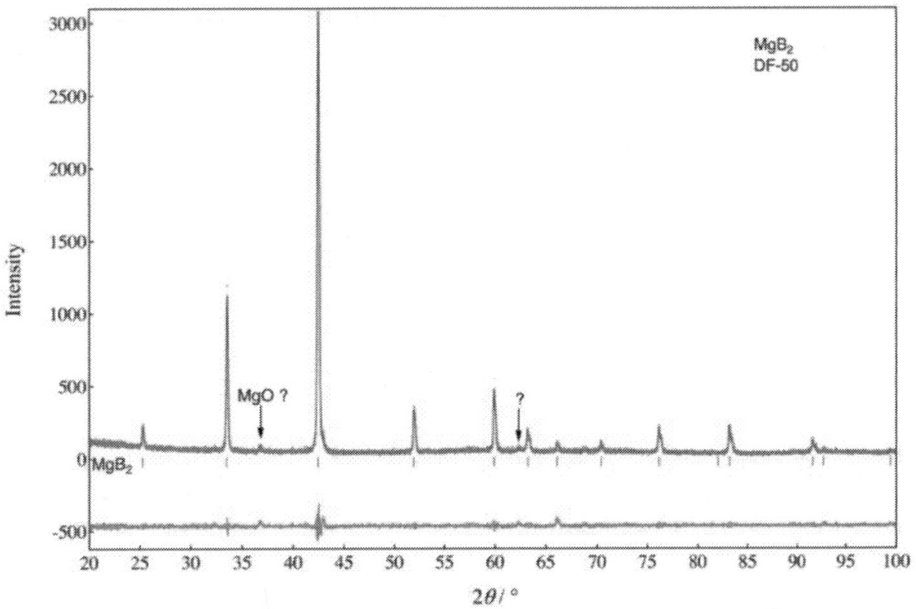

Figure 7. X-ray diffraction pattern for MgB_2 powder for a reacted at 830°C for 8 h using Cu-K_α radiation. All the lines above the noise are MgB_2 except two, one marked as a probable MgO line and one not identified as indicated.

In carrying out this study, we attempted several variants on the above procedure, each of which had some shortcomings. In our early exploration of this problem, we had hoped to use an iron reaction vessel because the cost is relatively low compared to the likely alternatives of niobium or tantalum. We found, however, that the superconducting transition temperature of the MgB_2 powder that was synthesized in an Fe vessel consistently had a transition temperature, T_c, of about 2 K lower than the T_c of similar powder prepared in a Ta or Nb reaction vessel (figure 8).

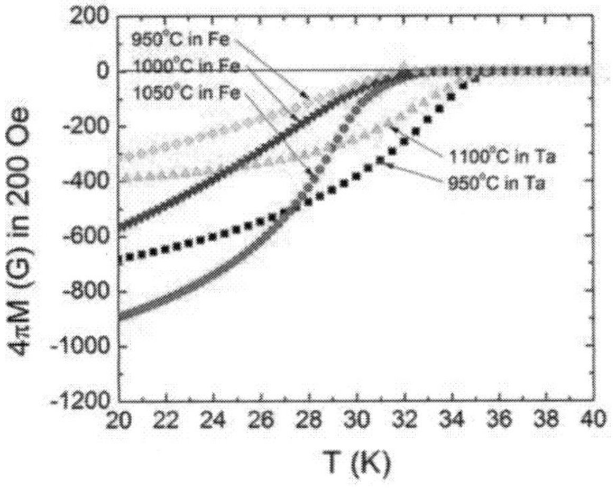

Figure 8. Comparison of T_c for MgB_2 powder grown in Fe or Ta reaction vessels

Figure 9. MgB_2 particles from reaction in an Armco Fe reaction vessel held at 970° C for 18 h.

In addition, the Fe also seemed to promote the growth of large hexagonal grains of MgB_2 as shown in figure 9. Often grains as large as 5 micrometers were found and the particles tended to form as platelets rather than equiaxed particles. The vapor pressure of Fe at 1000° C is 10^{-6} torr but that seemed to suffice to alter the grain growth. With these hints of difficulty, the reaction vessel material was changed to Nb and no further work was done with Fe.

Hence, for most of the work reported here, the experiments were done in cylindrical Nb reaction vessels. The typical batch size was approximately 50 grams. In a second variant to the process, the Nb cans were arc-welded shut in an argon atmosphere of half to three quarters of an atmosphere instead of e-beam welding in a vacuum. We found however, that the presence of argon slowed the diffusion of Mg vapor through the boron packed powders and made the conversion of the plasma synthesized boron powder to MgB_2 rather slow and erratic. The reaction rates seemed to depend on the details of the packing of the mix of Mg and B powder. When the Nb cans were e-beam welded shut in a vacuum, the rate and the completeness of the reaction seemed much less dependent on the quality of the packing of the reactants and the resulting grain size of the MgB_2 varied systematically with time and temperature. When the Ar was removed by using e-beam welding of the Nb vessel in a vacuum, the process of conversion of plasma synthesized boron to MgB_2 became quite regular.

3. Conclusions

Plasma synthesized boron powder with particle size in the 20 nm to 300 nm range, can be transformed into MgB_2 powder with particle size in the 100 nm to 1000 nm range by reacting the boron powder with magnesium vapor in an evacuated niobium reaction vessel at 830°C for about 24 h . Two alternate procedures gave unsatisfactory results. First, our attempts to use iron reaction vessels gave a systematically lower T_c and resulted in a more plate-like growth habit than powders grown in Nb. Second, our attempts to seal the reaction vessel under argon by arc welding resulted in erratic and slow growth rates for the MgB_2. Conversion rates to MgB_2 were faster and more regular if the Nb vessel was sealed under vacuum with e-beam welding. The presence of argon in the reaction vessel slowed the diffusion of magnesium vapor through the boron packed powders and slowed the reaction of magnesium with the boron powder.

4. Acknowledgements

Part of this work was funded by NIH SBIR Phase II grant #5R44EB007139-03. Ames Laboratory is operated for the U.S. Department of Energy by Iowa State University under Contract No. DE-AC 02-07CH11358.

5. References

1) Marzik J V, Suplinskas R J, Wilke R H T, Canfield P C, Finnemore D K, Rindfleisch M, Margolies J, and Hannahs S T, 2005 *Physica C* **423** 83
2) Larbalestier D C *et al.*, 2001 *Nature* **410** 186
3) Wilke R H T, Bud'ko S L, Canfield P C, Finnemore D K, Suplinskas R J, and Hannahs S T, 2005 *Physica* C **424** 1

Structure and properties of oxygen-containing thin films and bulk MgB_2

T Prikhna[1,6], A Shapovalov[1], W Goldacker[2], M Eisterer[3], A Kozyrev[1], V Shaternik[4], V Boutko[5], A Gusev[5], H W Weber[3], M Karpets[1], T Basyuk[1], V Sverdun[1], V Moshchil[1], M Belogolovskiy[4] and N Sergienko[1]

[1] V. Bakul Institute for Superhard Materials of the National Academy of Sciences of Ukraine, 2, Avtozavodskaya Str., Kiev 07074, Ukraine

[2] Karlsruhe Institute of Technology (KIT), 76344 Eggenstein, Germany

[3] Atominstitut, Vienna University of Technology, Stadionallee 2, 1020 Vienna, Austria

[4] G.V. Kurdyumov Institute for Metal Physics of the National Academy of Sciences of Ukraine, 03680 Kyiv, Ukraine

[5] Donetsk Institute for Physics and Engineering named after O.O. Galkin of the National Academy of Sciences of Ukraine, R. Luxemburg str.72, Donetsk-114, 83114, Ukraine

E-mail: [1] prikhna@mail.ru, [2] wilfried.goldacker@kit.edu, [3] eisterer@mail2.ati.ac.at, [4] shaternik@mail.ru, [5] boutko@teor.fti.ac.donetsk.ua

Abstract. A structural Auger spectroscopy study of MgB_2 thin (~140 nm) oxygen-containing polycrystalline films produced by magnetron sputtering and 99% dense MgB_2 bulks synthesized at 2 GPa allows us to conclude that j_c of MgB_2 depends to a high extent on the amount and distribution of oxygen in the material matrix. j_c reached 7.8-2.7 MA/cm^2 below 1T at 20 K in the films and 0.3-0.9 MA/cm^2 (depending on the boron used) in the bulks. The higher j_c in MgB_2 thin films can be associated with finer oxygen-enriched Mg-B-O inclusions and their higher density in the film structure compared to the bulk. Calculations of the total electron density of states (DOS) in MgB_2, $MgB_{1.75}O_{0.25}$, $MgB_{1.5}O_{0.5}$ and MgBO showed that all the compounds are conductors with metal-like behaviour. The DOS is even higher in $MgB_{1.5}O_{0.5}$ than in MgB_2 and the binding energy is similar. So, the experimentally found presence of some dissolved oxygen in MgB_2 does not contradict its high SC performance. The introduction of a high amount of oxygen into the MgB_2 structure does not dramatically reduce the material's T_c and allows obtaining high j_c as observed in our MgB_2 films and bulks.

1. Introduction

Although MgB_2 is an oxygen-free superconductor, the high affinity of Mg toward oxygen makes it practically impossible to synthesize MgB_2-based materials free of oxygen. Eom et al. [1] established that in thin MgB_2 films (with a c-axis parameter of 0.3547 nm, which was larger than that for bulk material: 0.3521 nm) the substitution of oxygen for boron in the boron layers leads to a lower T_c but to a steeper slope of dH_{c2}/dT both in the parallel and perpendicular magnetic field than that for films with

[6] To whom any correspondence should be addressed.

normal parameters. Also, the authors supposed that additional co-pinning by the non-superconducting MgO particles can contribute to the total pinning force. Birajdar et al. [2] found in MgB_2 grains of tapes oxygen-enriched regions and MgO precipitates of size 15–70 nm. The high O/Mg mole fraction ratio (0.23 for precipitates and 1.64 for an O-rich region) gave the authors of [2] reason to consider the precipitate as being MgO and the O-rich region being a mixture of MgO and $Mg(OH)_2$. Liao et al. [3, 4] using high-resolution transmission electron microscopy (HREM) have shown that the oxygen substitution occurs in the bulk of MgB_2 grains to form coherently ordered $MgB_{2-x}O_x$ precipitates from about 5 up to 100 nm in size and that such precipitates can act as pinning centers, thus increasing the critical current density. These precipitates as assumed in [3, 4] were formed due to the ordered replacement of boron atoms by atoms of oxygen and are of the same basic structure as the MgB_2 matrix, but with composition modulations, and no difference in the lattice parameters between the precipitates and the matrix can be detected in conventional electron diffraction patterns. However, extra satellite diffraction spots were seen in some directions implying a structural modulation nature of the precipitates. The precipitates have the same orientation as the MgB_2 and the replacement of boron by oxygen makes the precipitates stronger in electron scattering. The periodicity of oxygen atom ordering depends on the concentration of oxygen atoms in the precipitate and first of all occurred in the (010) plane. A long-term exposure to oxygen at high temperatures results in the transformation of $Mg(B,O)_2$ precipitates to MgO with a little change in precipitate sizes. The density of the precipitates of sizes 10–50 nm has been $10^{15}/cm^3$, and of size 50–100 nm has been $10^{14}/cm^3$ [4]. The unit cell structure of the inclusions essentially did not differ from that of the MgB_2, which, in parallel with the low X-ray atomic scattering factor of B, can be the explanation of the difference in results between X-ray and SEM-EDX examinations of MgB_2-based materials.

Our previous SEM–Auger study [5] showed that the MgB_2 matrix of the MgB_2 material synthesized from Mg:2B at 2 GPa with high j_c contained some dissolved admixture of oxygen, but much more admixture of oxygen was present in the separate oxygen-enriched inclusions (if synthesis temperature was about 900-1100 °C) or nanolayers (if the synthesis temperature was about 600-800 °C). So, it was concluded that the MgB_2 material consists of the MgB_2 phase (mainly) with small amount of dissolved oxygen and a MgO phase with dissolved boron. This can be the reason why in x-ray diffraction pattern just MgB_2 and MgO have been observed. The presence of higher magnesium borides MgB_x ($x>4$), seen by SEM and shown by Auger analysis to have a very low oxygen content, was not possible to determine by x-ray due to their fine dispersion in the structure and the large number of low symmetry atoms in the unit cell what essentially reduce the intensities of the x-ray reflections as compared to those of the MgB_2.

The present study compared structure and properties of superconducting oxygen-contained MgB_2 bulk and thin-film materials in connection with calculated densities of states for pure MgB_2 and for compounds with different oxygen substitution for boron.

2. Experimental

The bulk MgB_2 materials were synthesized from Mg and B taken in the Mg:2B stoichiometry at high quasihydrostatic pressure (2 GPa) and 800 or 1050 °C for 1 h (in contact with BN). We used Mg turnings Technical Specification of Ukraine 48-10-93-88 and commercially available precursor B powders from HC Starck. We used two boron powders; boron-1 (designated as B-1) with a grain size <5 μm, 0.66 wt% O, 0.31 wt% C, 0.48 wt% N, and 0.32 wt% H; and boron-2 (designated as B-2) with a grain size of 4 μm, 1.5 wt% O, 0.47 wt% C, 0.40 wt% N, and 0.37 wt% H. Boron and magnesium were mixed and milled in a high-speed planetary activator with steel balls for 1-3 minutes.

The thin MgB_2 films were deposited by DC magnetron sputtering under 1 Pa Ar pressure on 8×8×0.2 mm^3 sized (001) sapphire substrates at room temperature, which subsequently were annealed at 600-650 °C for 5 min in Ar under 10 Pa. For deposition we used a MgB_2 target synthesized by hot pressing at 30 MPa, 800 °C, 1 h using boron powder B-2.

The microstructure of the materials was characterized by x-ray structure analysis (with Rietveld refinement) and by SEM with microprobe x-ray and Auger (JAMP–9500F) analysis. The

Augerspectroscope JAMP–9500F is combined in the same device with a high resolution SEM. It is possible to change the SEM microprobe analyzer for the Auger probe and perform etching of the 1×1 mm^2 analyzed material surface area by Ar ions in the studies. JAMP–9500F offers the possibility to analyze quantitatively small volumes of about 10 nm in diameter and 1 nm in the depth. In situ etching can remove surface oxides that form between the time the sample is polished and analyzed, allowing pristine MgB_2 surfaces to be analyzed, as well as the composition of nano-inclusions and nano-layers.

The critical current density, j_c, was estimated from magnetization measurements in an Oxford Instruments 3001 vibrating sample magnetometer (VSM) using the Bean model. The connectivity was estimated from the difference in resistivity at 40 K and 300 K measured by the four probe technique. The amount of the superconducting (SC) shielding fraction was calculated from the ac susceptibility at 5 K with a numerical correction for the demagnetization of the actual sample geometry.

3. Results and discussions

The difference in SC behaviour of MgB_2 bulk materials depending on synthesis temperatures can be characterized as an increase of critical current densities j_c in low and medium magnetic fields with increasing synthesis temperature and an increase of j_c in high magnetic fields with decreasing synthesis temperature (see, for example, figure 1 a). Such behaviour has been observed for materials produced from different types of initial borons without and with additions such as Ti, Zr, Ta, SiC. The typical X-ray diffraction pattern of MgB_2 synthesized at 1050 °C using boron B-1 (sample I in table 1) is presented in figure 1b. Sample I showed the highest j_c at 20 K of 0.9 and 0.7 MA/cm^2, at 0 and 1 T respectively, see figure 1a). Sample I contains a well crystallized MgB_2 phase and admixing MgO and BN phases (Table 1). Sample II made with boron B-2, which contained more oxygen, was also made at 1050 °C, and contains the same phases (table 1), but the MgO concentration is higher. Sample II had j_c at 20 K of 0.4 and 0.3 MA/cm^2 at 0 and 1 T, respectively. It also had 99% density, 80% connectivity, and contained 94% of shielding fraction.

It should be mentioned that materials synthesized at lower temperature (800 °C) contain approximately the same amount of MgO as that prepared at 1050 °C using the same type of boron, but their structure contains some MgH_2 and unreacted Mg. The sizes of the inclusions of unreacted magnesium were rather big and the inclusions were randomly distributed so they cannot be effective pinning centers in MgB_2. Thus synthesis temperature does not influence general oxygen concentration in MgB_2, but oxygen impurities in the initial boron does.

The detailed study of MgB_2 structures by SEM and Auger showed that no MgO was found in the MgB_2 but Mg-B-O nanolayers "L" 10-20 nm thick ("L" is brightest regions in figure 1d) were found if material was synthesized under low temperature (800 °C). Such nanolayers had a stoichiometry near $MgB_{1.2\text{-}2.7}O_{1.8\text{-}2.5}$, as Auger quantitative analysis showed. In the materials synthesized at high temperature (1050 °C) segregation of oxygen occurred and in the MgB_2 matrix separate Mg-B-O inclusions "I" formed (which appear as the brightest regions in figure 1f and had

Table 1. Phases present in bulk MgB_2 (synthesized at 2 GPa, 1050 °C, 1 h using boron B-1 and B-2) and their lattice parameters.

Sample I Mg:2B (boron B-1): 91 mass% (90 vol.%) MgB_2, 5 mass% (5 vol.%) MgO, and 4 mass% (5 vol.%) BN				
MgB_2		MgO	BN	
a, nm	*c*, nm	*a*, nm	*a*, nm	*c*, nm
0.30844	0.35215	0.42227	0.25060	0.66830

Sample II Mg:2B (boron B-2): 77 mass% (82 vol.%) MgB_2, 13 mass% (10 vol.%) MgO and 10 mass% (8 vol.%) BN				
MgB_2		MgO	BN	
a, nm	*c*, nm	*a*, nm	*a*, nm	*c*, nm
0.30844	0.35236	0.42202	0.25040	0.66862

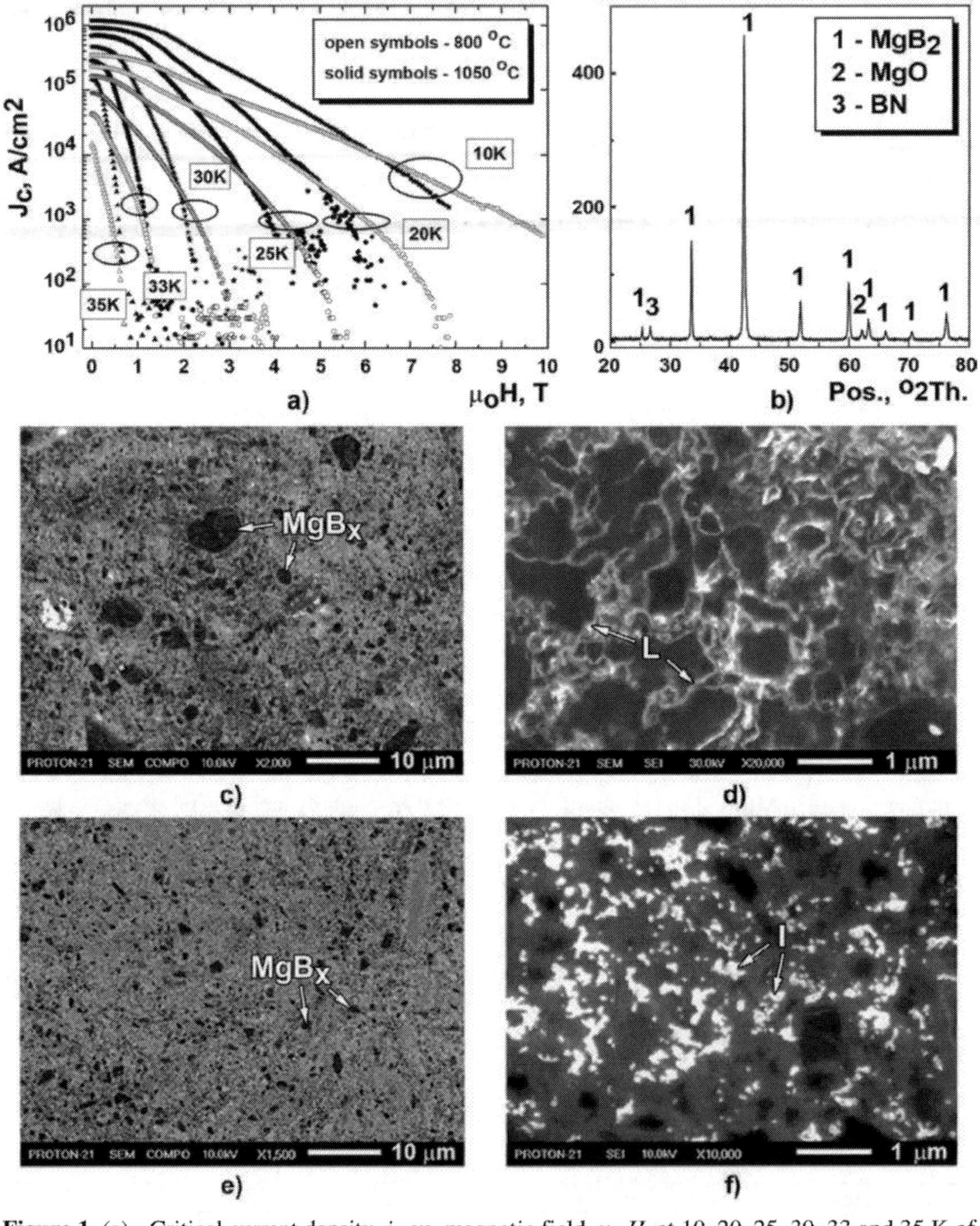

Figure 1. (a) - Critical current density, j_c, vs. magnetic field, $\mu_o H$, at 10, 20, 25, 30, 33 and 35 K of bulk MgB_2 materials synthesized at 2 GPa for 1 h from Mg:2B(1) at 800 °C (open symbols) and 1050 °C (solid symbols); (b) - x-ray diffraction pattern of bulk MgB_2 synthesized at 2 GPa, 1050 °C for 1 h from B-1; (c-f) – structures (after surfaces etching by Ar) in BEI (c, e) and SEI (d, f) SEM modes of bulk MgB_2 samples prepared at 2 GPa, 1 h at 800 °C from B-1 (c, d) and 1050 °C from B-2 (e, f).

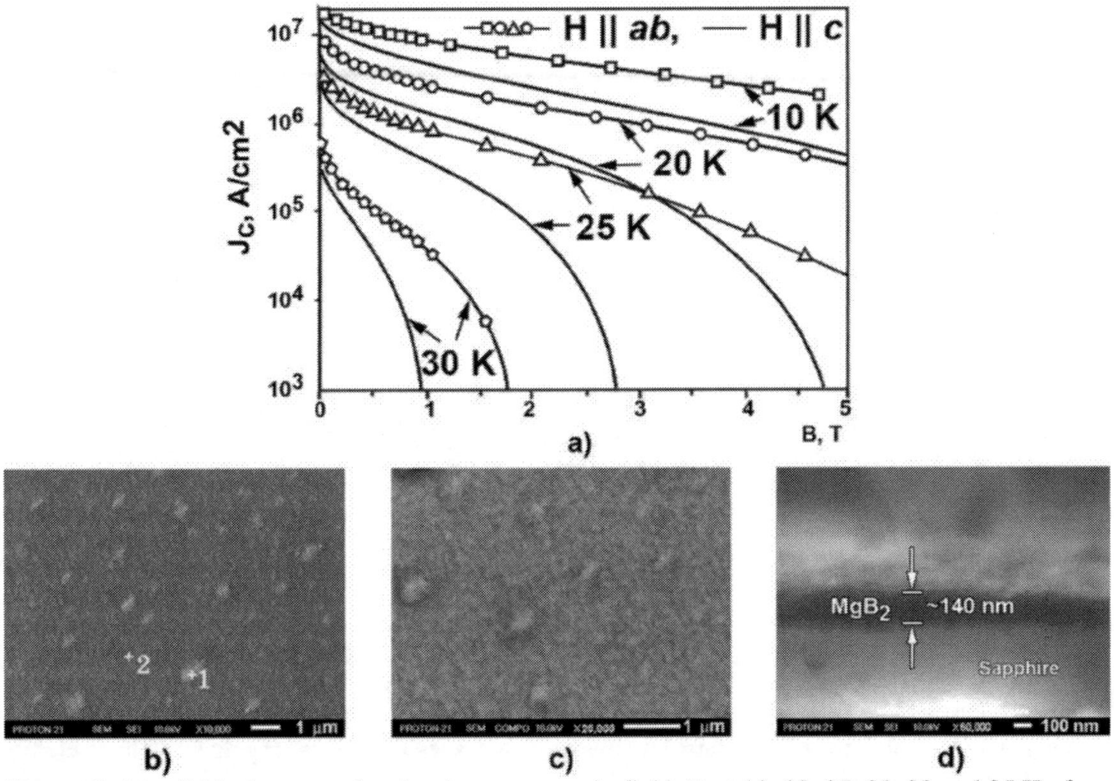

Figure 2. (a) - Critical current density, j_c, vs. magnetic field, B, at 10, 20, 25, 30, 33 and 35 K of thin MgB_2 film (for the cases when the field is parallel ($H \| ab$- lines with symbols) and perpendicular ($H \| c$- lines without symbols) to the substrate surface); (b-d) – structures (after surface etching by Ar) of MgB_2 film surface (b, c) and edge (d) in SEI (b, d) and BEI (c) SEM modes (oxygen enriched areas look brighter).

stoichiometry near $MgB_{0.6\text{-}0.8}O_{0.8\text{-}0.9}$). So, synthesis temperature can affect oxygen distribution in the MgB_2 structure. Besides this, a decrease of the sizes of inclusions of higher magnesium borides MgB_x ($x \approx 11\text{-}13$, which Auger analysis showed having stoichiometry near $MgB_{11\text{-}13}O_{0.2\text{-}0.3}$) takes place if the synthesis temperature increases (which are the darkest regions in figures 1c-f).

Auger spectra obtained from the MgB_2 matrix (appears grey) of the bulk sample presented in figure 1f showed stoichiometry near $MgB_{2.2\text{-}1.7}O_{0.4\text{-}0.6}$ from multiple analyses after multiple etching steps, which provided a depth profile. The quantitative Auger analysis allows an excellent localization of the electron beam and Ar etching in the chamber during analysis, so the oxidized layer that formed between the time the samples was polished and when it was analyzed was removed from the surface of the material before the analysis so the presence of oxygen in high amount in the "bright" nanolayers and inclusions as well as some small amount of oxygen in the MgB_2 matrix is from the sample synthesis. We cannot specify the exact stoichiometry of the oxygen-enriched layers or inclusions, but can determine that the relative amount of oxygen in their composition is high.

Figure 2 present j_c dependences (figure 2a) with the magnetic field in two perpendicular directions for the thin (140±10 nm) MgB_2 film and the correlated SEM SEI (figure 2b, d) and BEI (figure 2c) images. The SEM images were taken in two perpendicular directions. j_c in the film at 20 K ranged from 7.8 to 2.7 MA/cm^2 for 0 and 1 T, respectively. Auger studies gave approximate compositions at points 1 and 2 in figure 2b of $MgB_{3.6}O_{2.4}C_{0.5}$ and $MgB_4O_{1.2}C_{0.8}$, respectively. Thus the brighter colour

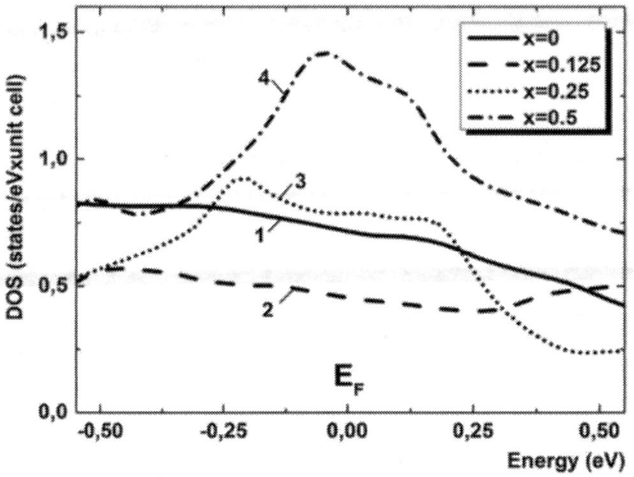

Figure 3. Calculated density of states (DOS) for MgB_2 (curve 1) and $Mg(B_{1-x}O_x)_2$: $MgB_{1.75}O_{0.25}$ (curve 2), $MgB_{1.5}O_{0.5}$ (curve 3) and MgBO (curve 4).

appears mainly due to a higher oxygen concentration (approximately two times) and figure 2c demonstrates the presence in the thin film of a high density of very fine oxygen-enriched inhomogeneities. The Auger quantitative analysis is very sensitive to the presence of light elements and for a correct quantitative estimation a reference spectrum should be used for the respective compound, which is not available, even for MgB_2. So, the present estimations were done using reference spectra of other boron-, magnesium-, oxygen- and carbon-containing substances. Because of this we do not know the exact stoichiometry, but using the same reference spectra we can estimate to what extent the relative amount of each element in the material. It should be mentioned that oxygen-enriched inhomogeneities marked by "1" in figure 2b appeared due to non-uniform evaporation of magnesium (because of eruption of big Mg clusters) and these cannot be the relevant pinning centers due to their rather big size and sparse distribution. The pinning can be due to the smaller, "bright" oxygen-enriched nano-inclusions, the density of which in the material is extremely high (figure 2c).

Figure 3 presents calculated total density of electronic states in $Mg(B_{1-x}O_x)_2$ compounds: pure MgB_2, $MgB_{1.75}O_{0.25}$, $MgB_{1.5}O_{0.5}$ and MgBO. To analyze the Mg-B-O compounds, we have applied the density functional theory [6] based on a full-potential linearized augmented plane wave method with the generalized gradient correction to exchange-correlation potential [7] using WIEN2k [8]. In the nearest vicinity of the Fermi energy E_F corresponding dependencies are not constant (figure 3) and their changes with x are non-monotonous. But, in any case, all studied compounds are conductors with a non-zero density of states at the Fermi level. Large density of delocalized states proves the metallic type of charge transport in them. Table 2 compares the total densities of states and related binding energies E_b in the studied Mg-B-O compounds. The numerical study of possible lattice distortion effects in the oxygen-enriched compounds showed that, for $x = 0.125$ and 0.25, deviations from the unperturbed lattice are indeed small, while it is not so for $x = 0.5$. In our opinion, this indicates the presence of an upper metastable boundary for the Mg-B-O system when the oxygen content increases. It should be noted that a rise of DOS near Fermi level with oxygen doping $x \geq 0.25$ (figure 3) increases the total energy (table 2) and can be therefore energetically unfavourable. A comparison of E_b shows that the appearance of $Mg(B_{1-x}O_x)_2$ with hexagonal symmetry in the case of comparatively low oxygen content ($x \approx 0.25$) thus seems possible from this viewpoint. According to Auger estimation the superconductor had near $MgB_{2.2-1.7}O_{0.4-0.6}$ stoichiometry.

Table 2. Total DOSs and binding energies, E_b, in the $Mg(B_{1-x}O_x)_2$ compounds.

x	0	0.125	0.25	0.5
total DOS	0.72	0.43	0.96	1.36
E_b, Ry	1.12	1.10	1.06	0.94

The concentrations of oxygen in the Mg-B-O inclusions ("I") and nanolayers ("L") in bulk MgB_2 (figure 1d, f) are rather high and most probably they have a structure which is close to MgO (solid solution of boron in MgO) and they can thus be effective pinning centres. The formation of nanostructural inhomogeneities with different oxygen content in magnesium diboride seems to be one of the main factors for improving the transport characteristics in MgB_2 and the presence of some oxygen in the MgB_2 matrix is not an obstacle for attaining high SC currents.

4. Conclusions

The comparison of nanostructures of MgB_2 oxygen-containing thin film (with thickness of ~140 nm) and highly dense oxygen-containing bulk materials (synthesized under 2 GPa pressure) demonstrating a high superconductive performance gave us reasons to conclude that the critical current density, j_c, of MgB_2–based materials to a high extent depends on the distribution of admixture oxygen what can be regulated by manufacturing conditions and by the synthesis temperature, in particular. The higher j_c of MgB_2 thin films as compared to bulk was explained by the higher density of finer oxygen-enriched Mg-B-O inhomogeneities in the film structure. The results of calculations of the electron density of states in MgB_2, $MgB_{1.75}O_{0.25}$, $MgB_{1.5}O_{0.5}$ and MgBO demonstrated that all the compounds are conductors with metal-like behavior. The stoichiometry of the matrix of bulk MgB_2 materials having high SC performance was about $MgB_{2.2-1.7}O_{0.4-0.6}$ (according to Auger quantitative analysis), i.e. very near to the composition $MgB_{1.5}O_{0.5}$ with one of the calculated low binding energies. The introduction of a high amount of oxygen into the MgB_2 material's structure does not dramatically reduce the material's T_c and allows obtaining a high j_c.

Acknowledgement

This work was in parts performed within the German-Ukrainian project HICONSU (FKZ: 01DK13031) supported by German Ministry of Education and Research (BMBF) and within Austrian-Ukrainian project UA 01/2015. The content and the results are the opinion of the authors.

References

[1] Eom C B, Lee M K, Choi J H, Belenky L J, Song X, Cooley L D, Naus M T, Patnaik S, Jiang J, Rikel M, Polyanskii A, Gurevich A, Cai X Y, Bu S D, Babcock S E, Hellstrom E E, Larbalestier D C, Rogado N, Regan K A, Hayward M A, He T, Slusky J S, Inumaru K, Haas M K and Cava R J 2001 *Nature* **411** 558.

[2] Birajdar B, Braccini V, Tumino A, WenzeT, Eibl O and Grasso G 2006 *Supercond. Sci. Technol.* **19** 916

[3] Liao X Z, Serquis A C, Zhu Y T, Huang J Y, Civale L, Peterson D E, Mueller F M and Xu H 2003 *Journal of Applied Physics* **93** 6208

[4] Liao X Z, Serquis A, Zhu Y T, Huang J Y, Civale L, Peterson D E, Mueller F M, Xu H *F cond-mat/* 0212571

[5] Prikhna T A, Eisterer M, Weber H W, Gawalek W, Kovylaev V V, Karpets M V, Basyuk T.V. and Moshchil V E 2014 *Supercond. Sci. Technol.* **27**, No 4 044013

[6] Parr R G and Yang W 1989 Density-Functional Theory of Atoms and Molecules, *Oxford Univ. Press*.

[7] Perdew J P, Burke S and Ernzerhof M Phys. 1996 *Rev. Lett.* **77** 3865

[8] Blaha P, Schwarz K, Madsen G K H, Kvasnicka D and Luitz J 2001 WIEN2K, An augmented plane wave + local orbitals program for calculating crystal properties. *Techn. Univ. Wien.*

IOP Conf. Series: Materials Science and Engineering **102** (2015) 012031 doi:10.1088/1757-899X/102/1/012031

Assessment of critical factors affecting the performance of trapped field magnets using thin film superconductor tapes

Kavita Selva[1], Xiao-Fen Li[2], Goran Majkic[2] and Philippe Masson[2]

Clear Lake High School, 2929 Bay Area Blvd, Houston, TX 77058, USA

Department of Mechanical Engineering and Texas Center for Superconductivity University of Houston, 4800 Calhoun Rd, Houston, TX 77204-4006, USA

E-mail: 0001084090@students.ccisd.net

Abstract. Trapped field magnets have been investigated using Zr-added (Gd,Y)Ba-Cu-O thin film superconductor tapes. Trapped field profiles were examined both experimentally and by simulation over a temperature range of 30 to 77 K. A good match is found between experimental and simulation trapped field data at 77 K and 65 K, but higher trapped field values were obtained experimentally at lower temperatures. Trapped field values up to 1.55 T were measured in an 11 mm thick stack of 55 μm thick superconductor tapes arranged in a crisscross fashion with three tapes per layer. A substantial increase in the trapped field values was found by simulation, in tape stacks made with 20 μm thick tapes. Using such thin tapes, trapped field values of nearly 2 T is predicted in ~ 13 mm thick tape stacks at 77 K.

1. Introduction

High power wind turbines with ratings of 5 MW or more are preferred to maximize the capture of wind energy. For such high power wind turbines, the use of permanent magnet-based direct drive generators provides significant advantages in terms of cost and reliability. The most powerful, conventional permanent magnets contain a significant amount of neodymium, a rare-earth that is limited in supply which affects the development of high power wind turbines [1, 2]. Additionally, the permanent magnet-based generators are heavy which restricts the feasibility of wind turbines in power range of 10 MW. For this reason, there is a strong incentive to develop new types of magnets that have a lower content of rare-earth materials and that can enable lighter-weight, powerful wind turbines [2]. Thin film superconductor tapes or coated conductors contain only a small amount of superconducting material and hence only a miniscule amount of rare-earth materials while exhibiting very high critical current densities [3]. Trapped magnetic fields up to 7 T have been demonstrated at 4.2 K in single stacks of square 12 × 12 mm superconductor tapes [4 - 6].

It is important to achieve high and uniform trapped magnetic fields in superconductor tapes so that they could replace permanent magnets. Previously, trapped magnetic field profiles in arrays of crisscross- and straight-arranged stacks of (Gd,Y)-Ba-Cu-O (GdYBCO) superconducting tapes were investigated [7] since large arrays would be needed if these tapes are to be used as trapped-field magnets in practical applications. It was found that a crisscross arrangement of arrays of stacked superconductor tapes yielded more uniform trapped magnetic field profiles and a lower decay rate of the trapped magnetic field as a function of distance from the tape stack surface, compared to a straight arrangement [7].

The influence of density of nanoscale defects incorporated in the superconductor film on the magnitude and decay of magnetic fields trapped by the tapes has also been investigated [8]. It was found that tapes with 7.5% zirconium addition which resulted in $BaZrO_3$ nanocolumns in the superconductor film exhibited the highest trapped field and the least time dependent decay of the field at 77 K [8]. Hence in this work, crisscross-arranged superconductor tapes made with 7.5% Zr addition were used. The purpose of this work was to determine how the magnitude of the magnetic field trapped by a stacked array of superconductor tapes is influenced by number of tapes in the stacked array, thickness of individual tapes and the operating temperature.

2. Experimental and Simulation

0.055 mm thick, 12 mm wide thin film GdYBCO tapes with 7.5% zirconium addition were used in this study. The tapes consisted of 0.05 mm thick substrate, 1.5 μm thick superconductor film and no copper stabilizer. The tapes exhibited a nominal critical current density of 2.5 MA/cm^2 at 77 K, 0 T. For the trapped field measurements, first, the superconductor tapes were cut into 36 mm long segments and stacked in a crisscross arrangement with three tape segments per layer [7]. A maximum tape stack thickness of 12.98 mm was tested. The superconductor tape stack was cooled in a magnetic field of 1.5 T to 77 K. After turning off the external magnetic field, the trapped field profile was measured over the surface of the tape stack. Details of the trapped field measurements are provided in [8]. The trapped field measurement system consisted of a three- axis linear motion table on which a cryogenic hall probe (Arepoc, model Axis-3, three-axis hall sensor) was mounted. The trapped magnetic field was measured at the center of the superconductor tape stack, 3 mm from the surface of the stack, as a function of time for up to 3 hours, initially at one minute intervals and later in longer intervals. After 16 minutes, the trapped field was measured at increasing distance from the surface of the center of tape stack. Thirty minutes into the measurement, the Hall probe was scanned over a 50 mm × 40 mm area over the sample holder in 1 mm steps to measure and map the trapped magnetic field at 3 mm from the tape stack surface.

Trapped field measurements were also conducted in a second system, over a temperature range of 30 K to 77 K. In these measurements, the superconductor tape stack was cooled in a field of 5 T to 20 K. After the external field was turned off, the trapped field was measured at the center of the tape stack, 1 mm from its surface. Measurements were conducted at 30, 40, 50, 65 and 77 K.

Simulation of the trapped field profiles was conducted by finite element analysis using COMSOL software. When an external magnetic field is applied to a tape, the magnetic flux penetrates from the sides towards the center of the tape. When the external field is removed, the flux in the tape decreases from the sides of tape. The slope of the magnetic field profile is determined by the critical current density of the tape [9]. Figure 1a is a schematic of the full penetration current flow that is established in the tapes to maintain the trapped field. The figure shows two layers of tape each with three tapes are arranged next to each other – in a horizontal orientation in the first layer and in a vertical orientation in the second layer. The net current density direction is distribution is determined through vector addition resulting in a profile shown in Figure 1b. Since the trapped magnetic field depends on the critical current density (J_c) and because J_c of a tape depends on the trapped field in the adjacent tape it is exposed to, the simulation model uses the engineering current density, J_e, as a function of magnetic field as its primary input; In the current 3D model, the peak field is assumed the same for all the tapes and no flux creep is considered. Since the temperature is an important parameter evaluated in this study, J_e characteristics as a functional of magnetic field at different temperatures from 30 to 77 K were used. The critical current data of 7.5%Zr-added GdYBCO tapes of SuperPower were used [10].

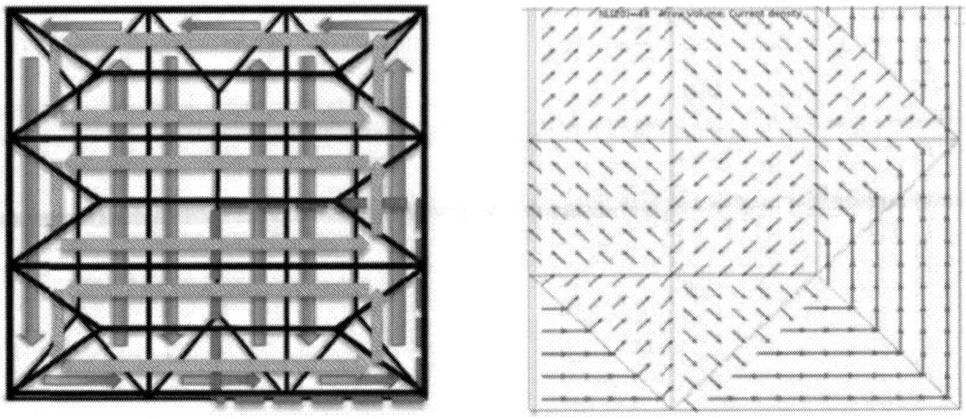

Figure 1. (a) Current density distribution in two layers of crisscross-arranged tapes. Solid arrows and hatched arrows indicate current flow direction in the three vertically-arranged tapes and the three horizontally-arranged tapes (b) Net current density direction from COMSOL model of two layers of crisscross-arranged tapes (because of symmetry, only ¼ of the stack, shown within the square with dashed lines in (a) is modeled)

3. Results and Discussion

A trapped field profile measured 3 mm above the surface of a stack of 236 layers of crisscross-arranged tapes, at 77 K is shown in Figure 2. It can be seen that a crisscross arrangement of superconductor tapes results in nine peaks in the trapped field profile corresponding to the nine overlapping regions formed by crisscrossing of three tapes in each layer. The intensities of the nine peaks are similar, suggesting that a relatively uniform trapped field profile can be obtained over a large area using the crisscross configuration. The valleys in the trapped-field profile correspond to the regions with overlapping edges of each tape. The maximum trapped magnetic field is 0.26 T, 3 mm above the center of the tape stack.

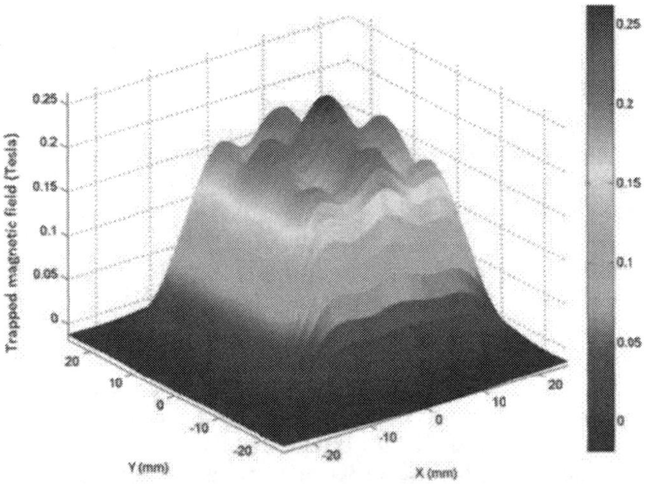

Figure 2. Trapped magnetic field profile at a distance of 3 mm above a stack of 236 layers of crisscross-arranged superconductor tapes made with 7.5% zirconium addition.

Trapped field profile results from the simulation of a stack of 128 layers of crisscross-arranged tapes are shown in Figure 3a. It can be seen that the simulated profile qualitatively matches very well with experimentally-obtained trapped field profile shown in Figure 3b.

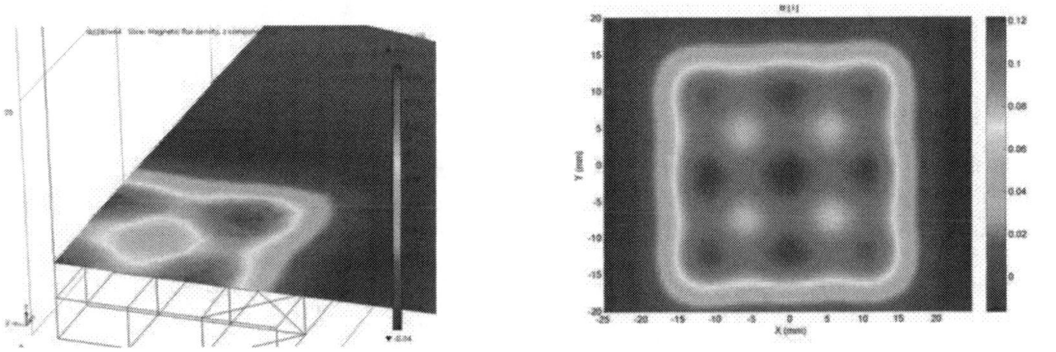

Figure 3. (a) Trapped field profile from simulation of a stack of 128 layers of crisscross-arranged tapes (because of symmetry, only ¼ of the stack is modeled). (b) Trapped field profile from experimental measurements on a stack of 128 layers of crisscross-arranged tapes.

A quantitative comparison of the trapped field values of a 12.98 mm thick stack of 0.055 mm thick tapes (236 layers) obtained from simulation and experiments is shown in Figure 4. A good match is found between the simulation and experimental data. Discrepancy between the trapped field values between those obtained from simulation and experiment may be due to the following reasons: i) experimental data is affected by flux creep. We had reported previously that that flux creep is higher (and trapped field values decay more rapidly with time) in tape stacks with fewer layers [8]; ii) variability in critical current density of tapes used in experiment. A constant critical current density value was used for all tapes in simulation in relation to the peak c-axis field in the stack; iii) difference in the magnetic field dependence of critical current density used in simulation with that of actual tapes used in the experiments.

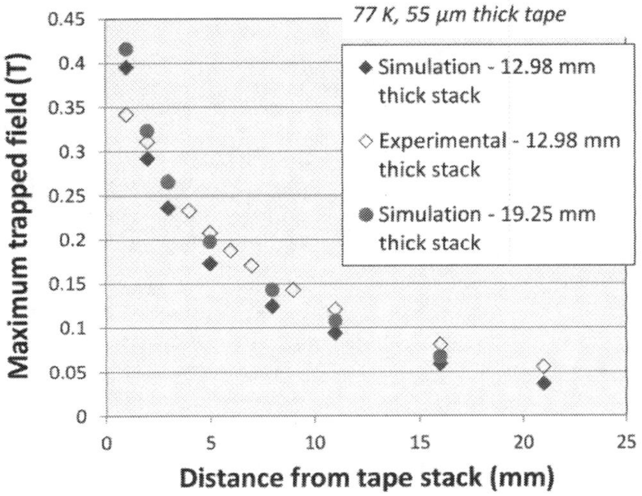

Figure 4. Maximum trapped magnetic field values in stacks of crisscross-arranged 55 µm thick superconductor tapes, obtained from simulation. Experimental data is included for comparison.

Figure 4 also includes simulation data from a thicker (19.25 mm) stack of tapes. No significant increase is seen in the trapped field which indicates that there is no benefit in adding more layers of tapes beyond about 13 mm, using 0.055 mm thick tapes.

Next, the effect of temperature on trapped field profiles with increasing number of layers of 0.055 mm thick tapes was examined by simulation and the results are shown in Figure 5. It is seen in Figure 5 that the maximum trapped field values increase with decreasing temperature from 77 K to 30 K. It is also observed that all temperatures, the trapped field values increase with increasing number of layers of 0.055 mm thick tapes in the stack. Unlike the experimental data that had revealed a linear increase in trapped field values with increasing number of layers [8], the simulation data shows a non-linear (logarithmic) increase in trapped field values with increasing number of layers of tapes in the stack. A trapped field value of nearly 1.4 T was obtained at 30 K which is about six times higher than the value at 77 K.

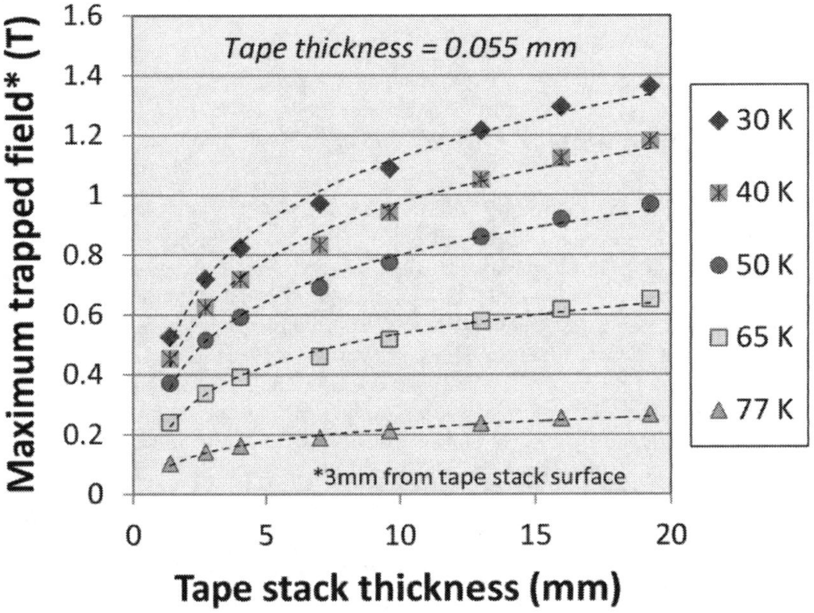

Figure 5. Maximum trapped magnetic field values in stacks of increasing thickness of crisscross-arranged 0.055 mm thick superconductor tapes at different temperatures, obtained from simulation. Dashed lines are logarithmic fit to the data points.

Results from trapped field measurements over a temperature range of 30 to 77 K are shown in Figure 6. As shown in the figure, the experimental trapped field values at 65 K and 77 K match well with simulation data. However, with decreasing temperature, the experimental trapped field values tend to increase linearly whereas a reduced rate of increase is seen in trapped field values from simulation. This finding is similar to the non-linear increase in trapped field values with increasing tape stack thickness in simulation data as seen in Figure 5 compared to the linear increase observed in experimental data [8]. It is seen from Figure 6 that the experimentally-obtained trapped field value reaches 1.55 T at 30 K.

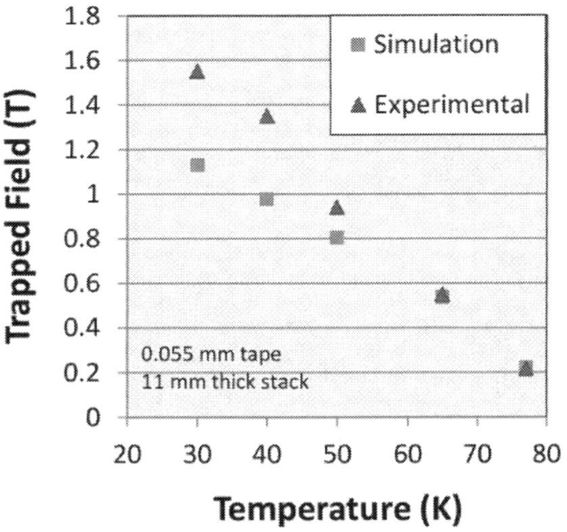

Figure 6. Comparison of maximum trapped fields at different temperatures of 11 mm thick stacks of 0.055 mm thick tapes obtained by simulation and experiments.

Maintaining the temperature at 30 K, the effect of individual thickness of tapes in the stack was investigated. The individual tape thickness was varied from 0.055 mm to 0.035 mm to 0.020 mm. Results are shown in Figure 7. It is found from the figure that at 30 K, the maximum trapped field values increase with decreasing tape thickness from 0.055 mm to 0.020 mm i.e. with increasing number density of tapes in the stack (number of layers of tapes in a given stack thickness). Also, in tapes of all thickness, the maximum trapped field values increase with increasing number of layers of tapes. A trapped field value of nearly 3 Tesla is reached at 30 K.

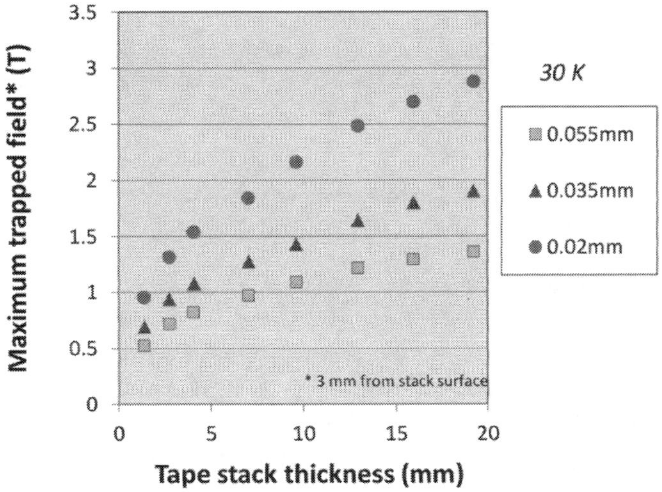

Figure 7. Maximum trapped magnetic field values in increasing thickness of stacks of crisscross-arranged 55 μm, 35 μm and 20 μm thick superconductor tapes at 30 K, obtained from simulation.

A comparison of the maximum trapped magnetic field at 77 K in a 12.98 mm thick tape stack made with 55 and 20 μm thick tapes, obtained from simulation, is shown in Figure 8. It is seen from the figure that the trapped magnetic field increases sharply with decreasing distance from the tape stack surface using the thinner tapes, reaching nearly 2 Tesla, 1 mm from the stack surface. This result clearly shows the benefit of using thinner tapes for maximizing trapped magnetic field.

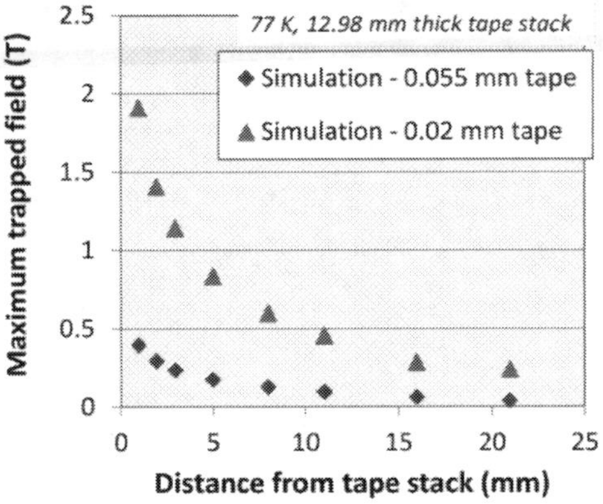

Figure 8. Maximum trapped magnetic field values in stacks of crisscross-arranged 55 μm and 20 μm thick superconductor tapes, obtained from simulation.

Conclusions

The influence of operating temperature, thickness of individual tapes and thickness of tape stacks on the magnitude of magnetic field trapped by crisscross-arranged superconductor tapes have been investigated by simulation. Experimental verification was also conducted with 11 mm thick tape stacks over a temperature range of 30 to 77 K. The results show increased trapped fields with increasing tape stack thickness at all temperatures with 55 μm thick tapes. The trapped fields were found to increase linearly with decreasing temperature in experiments and nonlinearly in simulation. The difference could be due non uniformity in the magnetic field dependence of J_c of long tapes (even though the measured J_c at 77 K, 0 T is uniform) which is not considered in simulation. Additionally, a higher fidelity model that takes flux creep into account needs to be developed. Simulation results showed a drastic improvement in trapped fields when the individual tape thickness is decreased from 55 to 20 μm.

Acknowledgment

Some of the superconductor tapes used in this work were provided by SuperPower Inc.

References

[1] Folger T 2011 *National Geographic* **219,** 136-145.
[2] http://energy.gov/sites/prod/files/DOE_CMS2011_FINAL_Full.pdf
[3] http://www.superpower-inc.com/system/files/SP_2G+Wire+Spec+Sheet_for+web_2013FEC_v2.pdf
[4] Patel A, Hopkins S C and Glowacki B A, 2013 *Supercond. Sci. Technol.* **26,** 032001
[5] Patel A, Filar K, Nizhankovskii V I, Hopkins S C and Glowacki B A, 2013 *Appl. Phys. Lett.* **102,** 102601
[6] Baghdadi M, Ruiz H S, and Coombs T A, 2014 *Appl. Phys. Lett.* **104,** 232602
[7] Selva K and Majkic G, 2013 *Supercond. Sci. Technol.* **26**, 115006
[8] Selva K, Li X-F and Majkic G, 2015 *IEEE Trans. Appl. Supercond.* **25,** 6605005
[9] C. P. Bean, 1962 *Phys. Rev. Lett.* **8,** 250
[10] http://superpower-inc.com/system/files/2014_0812+ASC2014+ICBT+Poster+FINAL.pdf

IOP Conf. Series: Materials Science and Engineering **102** (2015) 012032 doi:10.1088/1757-899X/102/1/012032

Conductance Degradation in HTS Coated Conductor Solder Joints

Edgar R. Canavan, Henning Leidecker, and Lyudmyla Panashchenko

NASA – Goddard Space Flight Center
Greenbelt, MD 20771 USA

Ed.Canavan@nasa.gov

Abstract. Solder joints between YBCO coated conductors and normal metal traces have been analysed as part of an effort to develop a robust HTS lead assembly for a spaceflight mission. Measurements included critical current and current transfer profiles. X-ray micrographs were used to verify proper solder flow and to determine the extent of voiding. SEM of cross-sections with EDS analysis was crucial in understanding the diffusion of the protective silver layer over the YBCO into the solder for different solder processes. The assembly must be stored for an extended period of time prior to final cool-down and operation. Measurements of the joint resistance over the course of months show a significant increase with time. Understanding the interface condition suggests an explanation for the change.

1. Introduction

Spaceflight missions place stringent requirements on the resistance of joints to HTS conductors. The Soft X-ray Spectrometer instrument on JAXA's Astro-H mission will use a micro-calorimeter operating at 50 mK to measure the energy of each incoming x-ray photon with high resolution. The temperature environment of the microcalorimeter is maintained with a cooling chain consisting of two Stirling coolers, a J-T cooler, a superfluid helium tank, and a three-stage Adiabatic Demagnetization Refrigerator (ADR). To reach its 3-year design lifetime, the total design heat load onto the helium tank is 910 μW [1]. Of this, 10 μW is allocated to the superconducting current leads for the ADR solenoids, each of which draw a peak current of 2 Amps. In an off-nominal (failed Stirling cooler) case, the warm end of the superconducting leads must function at temperatures up to 62 K.

To meet these requirements, a set of High Temperature Superconductor (HTS) lead assemblies were designed, built, tested for flight, and delivered to the project [2]. The assemblies used commercial 2G REBCO coated conductor. To minimize thermal conductance, instead of the standard 2–3 μm sputter-deposited silver outer layer, tapes with a silver-gold alloy were used. The tapes were also cut down to 1 mm width. At the cold (warm) end, transition to NbTi/Cu composite (standard copper) wire were made through solder joints to silver-plated, copper traces on a printed circuit board (PCB). After passing electrical, thermal, and vibration tests, the assemblies were put in storage for nearly a year. Just prior to shipment to Japan for integration into the flight dewar, the assemblies were tested again, and it was discovered that the joint conductance had seriously degraded. The leading hypothesis for the failure was corrosion due to insufficient rinsing of solder flux. A large effort was

IOP Conf. Series: Materials Science and Engineering **102** (2015) 012032 doi:10.1088/1757-899X/102/1/012032

mounted to build a new set of assemblies. A significant part of this effort was directed toward understanding the factors influencing solder joint conductance and their degradation over time.

2. Sample Preparation

Samples were prepared using either test boards or spare flight boards. Test boards are large (~ 100 x 150 mm) two-layer PCBs, specifically designed for solder joint testing. They accommodate up to eight ~110 mm sections of HTS tape, with up to 16 joints. The traces are of the same ~ 100 μm thick copper, plated with ~ 0.5 μm immersion silver, as are used in the flight boards. The test boards have wide traces for the high current paths and a nearly continuous bottom layer for heat sinking. A pair of voltage tap traces, connected at either end of each solder joint, allowed the voltage across each joint to be measured. The flight PCBs are 3 layer boards with a nearly continuous central layer thermally linked to plated through holes to provide a thermal path for heat generated in the board. As in the test boards, the flight boards have built-in voltage taps on either end of each solder joint.

The HTS material is the same as used in the flight HTS lead assemblies: 2G REBCO tape with a 50 μm thick Hastelloy substrate, sliced down to 1 mm width. The slicing process does leave the edges of the cuprate layer exposed. In 2G tapes, local spots of low J_c limit critical current. Slitting the tape exacerbates this problem, so the I_c of all lengths of tape was measured and only those with $I_c > 28$ Amp used.

Two indium-based solders were used: In3%Ag, and In48%Sn. Both are eutectics. The melting point of In3%Ag is 143 C, and that of In48%Sn in 118 C. While copper-plated 2G conductors are typically soldered with eutectic SnPb, early testing in the project indicated that with these unplated tapes, the silver layer completely dissolves into the tin-rich phase, even with relatively brief exposure to the molten solder. The low soldering temperature of the two indium alloys should slow this reaction. While In48%Sn has a very low melting point, which should also limit damage to the YBCO layer, In3%Ag should have lower solubility of silver, particularly if soldering occurs close to the (eutectic) melting point. In all cases, 50 μm thick solder ribbon was used, cut into preforms 0.5 mm wide and the length of the joint, typically 25mm. Kester RF741 RMA flux was used. In all cases, the HTS tape was oriented so that the side with the superconducting layer was in contact with the solder.

Flight circuit boards are typically soldered using a multi-zone oven. This allows tight control of temperature and temperature rates. The inherently three-dimensional shape of the HTS lead assemblies (see reference [2]) precluded the use of a standard oven, but an oven-based process was developed in attempt to maintain strict control over process parameters. For the test boards, foil heaters mounted directly to bottom applied heat. The flight boards, which were too small to accommodate heaters, were mounted on somewhat larger aluminum plates with heaters. The heavy copper leads that carried current into the boards tended to create cool spots, so a copper clamp with foil heaters was mounted directly on the leads also. To minimize radiative and convective losses, the PCBs were enclosed in foil-lined fiberboard "oven". A constant pressure was applied to the solder joints with tungsten weights. A stack of thin glass slides, separated by narrow strips of Kapton, minimized the effect of tungsten's heat capacity. The average pressure on the joints ranged from 8 to 15 kPa. Diode thermometers mounted directly on the boards measured temperature, and PID temperature controllers provided power to the heaters. In this way, temperature could be ramped at a fixed rate, or held at a given value, and the temperature could be logged at a rapid cadence. A heating rate of ~ 35 K/min, within the range recommended by the solder manufacturer, was easily achieved. The enclosure was opened as soon as the joint had reached peak temperature, and forced air used to cool the joints, again at a rate (120 K/min) within the range recommended by the manufacturer.

As will be shown below, following this standard process for the use of indium alloy solders resulted in consumption of the silver layer over the REBCO. A series of samples using a hand-soldering process were produced with In3%Ag (m.p. 143 C). For these, boards were preheated to approximately 20 C below the melting point using a hot air soldering station, and the solder was melted using a temperature-controlled soldering iron 20 C above the melting point. Soldering was done in a single swift stroke to minimize the time above the melting point.

3. Measurements

3.1. Joint Resistance

Joint resistances, and the resistance of the HTS tape between joints, were measured in a closed cycled cryostat down to ~4 K, and also in liquid nitrogen (LN_2). For tests in LN_2, heavy copper leads were used, and current was supplied by a Cryomagnetics CS-4 current source. Voltages between pairs of taps were read out with Keithley 2000 multimeters. Current was limited to 25 Amps to keep the heat flux from the traces on the boards from exceeding the burnout limit for pool boiling liquid nitrogen. The resistance of each region was determined from the slope of the voltage-current curve. The voltage-current curves of most joints remained linear up to the maximum current (25 Amps).

For measurements in vacuum, resistances were measured using a Linear Research LR-700 AC resistance bridge. The source of the bridge was connected in series through all of the circuits on the board, and the sense lines were connected to all sense lines through a multiplexer. Measurements were made during cool down, and thus the resistance as a function of temperature from 290 K down to 4 K was obtained. For a typical test board, the room temperature values of its 8 joints range from 0.176 to 0.216 Ω. These values drop by approximately a factor of two in cooling to the T_c of the tape, ~ 92 K. At T_c resistance drops roughly 4 orders of magnitude. Figure 1 shows the low temperature region of the resistance – temperature plot for four of the joints. The resistance behaves like a normal metal, dropping linearly with temperature down to ~30 K, and leveling out at low temperature. Two of the joints undergo a second sharp drop in resistance around 5.3 K, significantly below the reported range for the T_c of In48%Sn (7.1 – 7.5 K); two others show no second transition.

3.2. Joint resistance profile

In an attempt to better understand the variation observed in joint resistances, the resistance profile of joints on two of the test boards was measured. For these measurements, the source of an LR-700 resistance bridge was connected across the circuit, and one sense line was connected to the outer voltage tap, the tap at the end of the tape. The other sense line was connected to a sharp probe. The resistance reading was thus zero when the probe was at the end of the tape, and the polarity of the

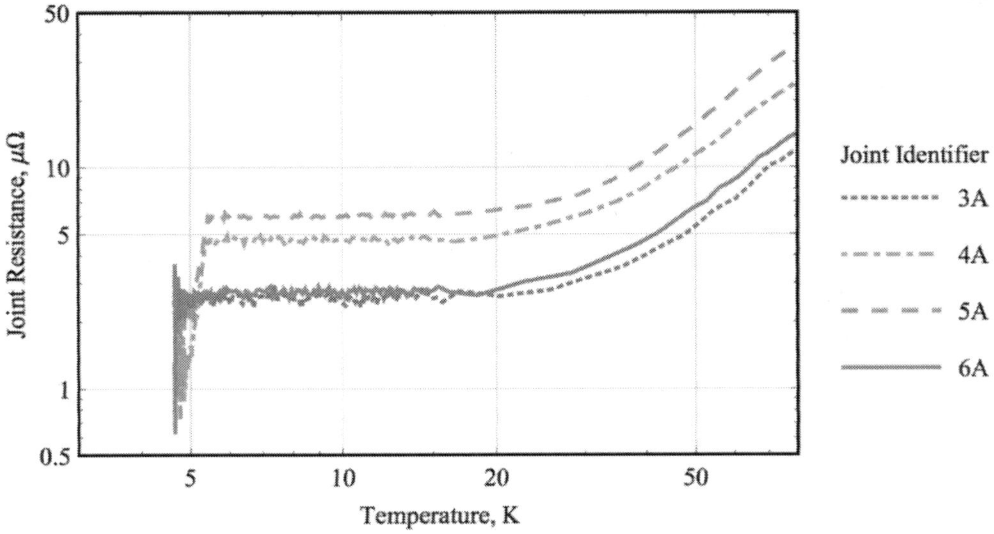

Figure 1. Typical resistance – temperature plot for In48%Sn solder joints on a test board. Below the T_c of the HTS tape, the solder resistivity behaves like a standard metal, dropping linearly with temperature, then levelling out. Samples 4A and 5A are near the centerline of the board; samples 3A and 6A were closer to the edges.

sense lines was chosen so that resistance increased as the probe moved toward the opposite end of the joint. The board was positioned in a shallow container filled so that the surface of the liquid nitrogen was approximately 1 cm above the surface of the board. Prior to immersing it in LN_2, one millimeter scale lines were marked on the board along, and extending somewhat beyond, the solder region. In this way, the position of the probe could be determined to approximately ± 0.25 mm. As mentioned above, the HTS tape is oriented so that the superconducting layer is on the lower side, facing the solder. In these slit tapes, with no silver on the side walls, current flow from the REBCO-coated face to the upper face must pass through the ceramic transition layers between the superconductor and the substrate, which are inherently insulating. For this reason, the probe was only touched to points on the trace of the board next to the solder joint.

The resistance as a function of the position from the end of the HTS tape, $R[x]$, rises rapidly initially, then levels out and asymptotically approaches the full value of the resistance across the joint, R_{full}, which is the value measured across the voltage taps. Equivalently, the difference $R_{\text{full}} - R[x]$ follows an exponential-like drop to zero with increasing x. Figure 2a) shows this difference, scaled by R_{full}, plotted on a semi-log graph versus position. The data roughly follow straight lines. Thus,

$$\frac{R_{\text{full}} - R[x]}{R_{\text{full}}} = 1 - \frac{R[x]}{R_{\text{full}}} \approx e^{-x/\lambda}, \tag{1}$$

where λ is the current transfer length. The solid lines show the least squares best fit to this equation for all 8 joints. Figure 2b) shows the values of λ determined in this way as a function of joint resistance, R_{full}. Note that even the largest value of λ is only ~20% of the physical length of the joint. Note also that λ and R_{full} are reasonably well correlated; the solid line shows the least-squares best linear fit. The inverse of the slope gives $dR_{\text{full}}/d\lambda = 6.8\ \mu\Omega/\text{mm}$. This is of the same order as the trace resistance per length, $dR_{\text{trace}}/dx \sim 8\ \mu\Omega/\text{mm}$, as determined by probe measurements along the trace outside the joint region.

3.3. X-ray imaging

X-ray imaging of solder joints was carried out on most test boards and simulators, as well as all flight solder joints. The flat solder bonds between the trace and the HTS tape are particularly amenable to x-ray imaging. X-ray images reveal variation in solder thickness, areas where it did not flow, and voids.

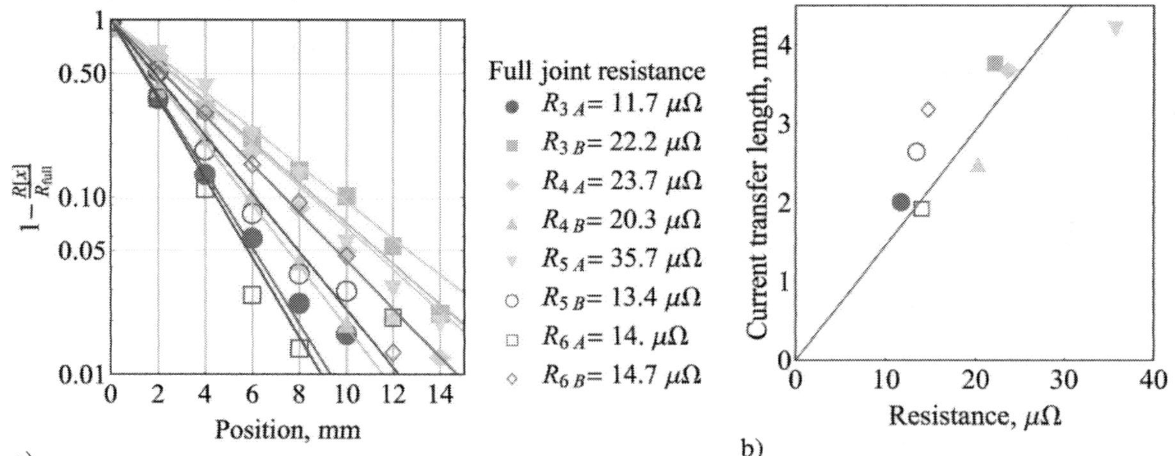

a) b)

Figure 2. a) Resistance profiles of 8 nominally identical 20 mm joint samples. $R[x]$ is the resistance at position x; R_{full} is the resistance across the entire length of the joint. For each sample, the straight line is the least squares best fit to the function $e^{-x/\lambda}$, where λ is the current transfer length. b) Current transfer length as a function of R_{full} for the various joints. Larger joint resistances are correlated with longer transfer lengths. Note that the transfer lengths are a small fraction of the physical length of the joints.

Figure 3 shows typical images for both oven-soldered and hand soldered joints. The level and type of porosity varied significantly from trace to trace, even on traces soldered simultaneously in the oven, and that thus experienced nearly identical temperature excursions. In general however, traces soldered with an iron had smaller voids, and the voids were often more finely dispersed. We believe this difference arises from higher local pressure when soldering by iron, and from the fact that the application point of this pressure moves down the joint, tending to squeeze out gas formed by the interaction of the flux with the oxides on the surface.

3.4. Imaging and elemental analysis of cross-sections

While x-ray imaging is excellent in revealing voids, it is insensitive to cracks parallel to the surface of the interface. For this reason, cross sections were imaged using an optical microscope and a Scanning Electron Microscope (SEM). Cross-sections were cut from the boards with a diamond saw, mounted in epoxy, and polished. Great care must be used to avoid damaging the sample during polishing.

The SEM included the capability for Energy-Dispersive x-ray Spectroscopy (EDS), which analyzes the spectrum of the x-ray emission stimulated by the electron beam as it scans across the sample. EDS thus produces an image that contains the elemental analysis at every pixel. Figure 4 shows SEM/EDS micrographs for two solder joint cross-sections, one produced by oven soldering, the other by hand soldering. The right-hand images in each row are the electron micrographs. The center and left images show the concentration maps of silver and indium, respectively. The indium and silver images have the same scale, so it is possible to identify areas where the silver layers on both the copper trace and the REBCO have dissolved into the solder and formed an indium-silver intermetallic. In both samples, the indium extends all the way to the copper layer, indicating that the thin, ~ 0.5 μm, immersion silver layer has been completely consumed, as has the thicker, 2 – 3 μm silver (actually Ag5%Au) layer on the HTS tape in the oven-soldered sample. This is corroborated by the electron micrographs: areas of indium without high silver concentration are very soft, and trap the grinding media, which show up as dark particles in the image; areas of overlap between indium and silver are hard intermetallics, which do not trap the grit and appear bright. In the hand-soldered sample, the indium does not extend all the way to the superconductor surface, indicating the presence of a thin (0.5 – 1.0 μm) layer of unreacted silver. That most of the silver layer on the HTS is consumed indicates that the indium-silver reaction proceeds very quickly when the solder is molten. As discussed above, in the hand-solder process the board is preheated to about 20 C below the melting point, so the solder melts quickly as the iron passes over it. Swiping the 2 – 3 mm wide tip across the 20 mm long joint region requires about 6 seconds, so any point is molten for only approximately a

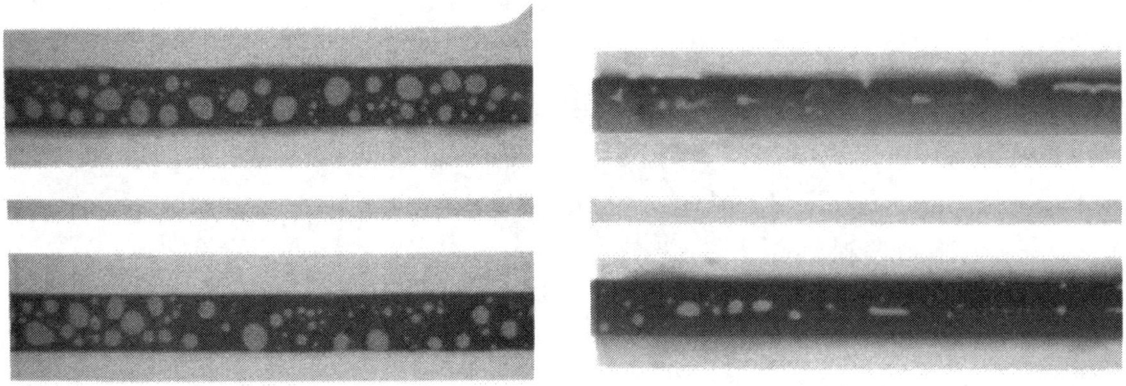

a) Oven soldered sample b) Hand soldered sample

Figure 3. X-ray images of In3%Ag solder joints between copper traces (gray region) and 1 mm wide HTS tape (dark region). In general, oven soldered joints showed higher porosity than hand soldered joints.

IOP Conf. Series: Materials Science and Engineering **102** (2015) 012032 doi:10.1088/1757-899X/102/1/012032

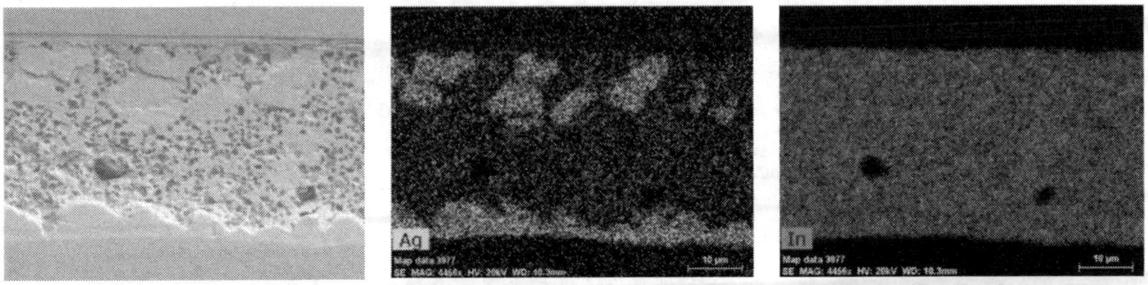

Oven soldered joint: a) Image b) Silver distribution c) Indium distribution

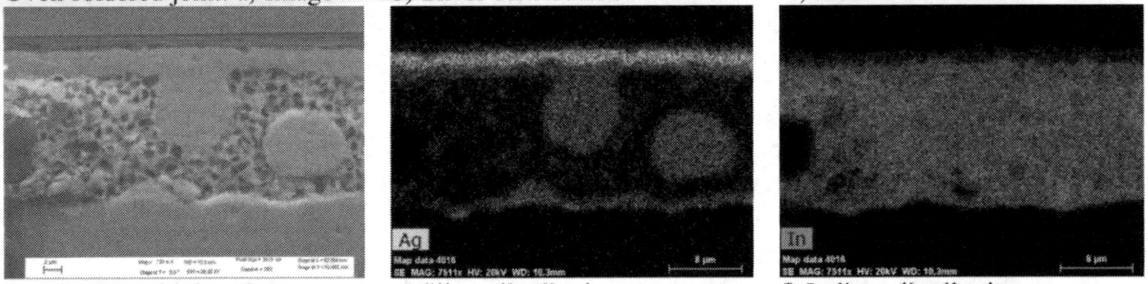

Hand soldered joint: d) Image e) Silver distribution f) Indium distribution

Figure 4. SEM/EDS analysis of In3%Ag solder joint cross sections. Indium and silver maps are scaled the same. Regions bright in both indium and silver are indium silver intermetallic. Note that indium extends all the way to the lower interface (copper) in both samples, indicating the immersion silver layer is completely consumed. In the hand soldered sample, the indium does not extend all the way to the upper (REBCO) interface, showing that some of the silver layer remains intact.

second. By contrast, the board remains above the melting point for about a minute in the oven process.

The fact that the silver layer is almost completely consumed in such a rapid soldering process indicates that avoiding complete consumption is difficult. Figure 5 shows a cross section of the final solder test sample. Even though this was done using nominally the same process as the hand-soldered sample shown in figure 4, in this case the indium has diffused to the upper interface, indicating complete reaction. Figure 5 also shows a region with a small (~146 μm long) crack. The separation occurs primarily between the indium-silver intermetallic and the REBCO layer, although in one ~30 μm region, the REBCO has detached from the Hastelloy substrate, which is the upper gray region. The upper and lower (Hastelloy and copper) interfaces are parallel, indicating the sample has not been distorted in the polishing process. Furthermore, other cross sections from the same sample show a

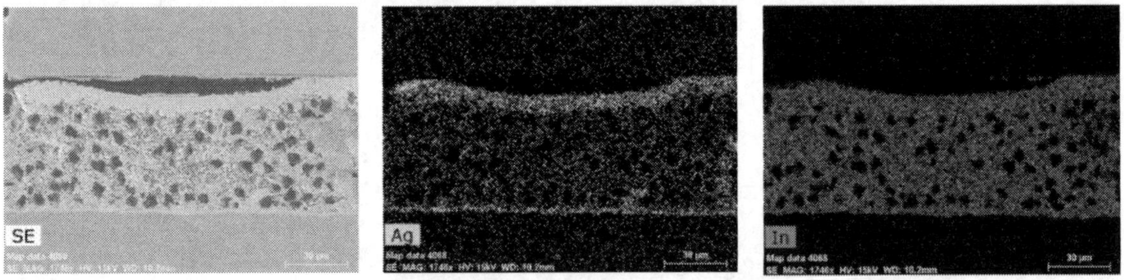

a) Image b) Silver distribution c) Indium dstribution

Figure 5. SEM/EDS analysis of the final In3%Ag solder joint test sample, showing a region around a crack. Note that the indium has diffused to the top boundary, indicating complete conversion of the silver layer to In-Ag intermetallic. Note also that the separation occurs primarily between the REBCO and In-Ag layers, although in one region REBCO has detached from the Hastelloy substrate.

region where the solder has flown into a gap. Such cracks were observed in earlier (hand-soldered) samples, and are likely due to the low peel strength between layers in 2G coated superconductors [3].

3.5. Resistance degradation over time

During the first attempt to fabricate the new flight HTS assemblies, inadequate fixturing prevented sufficient pressure from being applied to the cold end board on one of the assemblies during soldering. X-ray images showed uneven solder flow in some joints. Although the assembly met all requirements in cryogenic testing, the unit was rejected and became available for life testing. It was tested five additional times. In each test, the unit was held in vacuum in the "worst case" temperature configuration, with the warm end board controlled at 62 K, and the cold end at 5.7 K. Current was ramped to 5 Amps, and voltage across every joint, and across the HTS tape was recorded. The V-I curves were all linear, and no voltage appeared across the HTS tape outside the joint, within the error of the measurement. The slope of the V-I curves determines the resistance. As in section 3.1, a multiplexed ac resistance bridge also monitored the resistances on cool-down. The resistances determined by the two techniques agreed. Between the first two measurements, the unit was kept in a nitrogen purge box. In subsequent intervals, the unit was stored in vacuum, except between tests 3 and 4, as discussed below.

Figure 6 show the resistance as a function of time above 273 K since fabrication. For all but one of the joints, the resistance increases with a log-time dependence. This suggests a presumably temperature-dependent chemical reaction at the superconductor-solder interface. Storing the unit in a sealed, nitrogen-filled bag in a freezer at -20 C for a month tested this hypothesis. The small interval between points 3 and 4 is the time required to warm up from 273 K, remove the unit from the cryostat, purge and seal it, and transport it to the freezer after test 3, and to reverse the process before test 4. The resistance change between these two tests is zero within the error of the measurement in some cases, and, where it is not, is accounted for the by the small interval above 273 K. The resistance of one of the joints in the cold end board did not follow the log time behavior, but instead followed a linear increase with time quite precisely for the first five measurements, as figure 6c shows. In the most recent measurement, the resistance fell below this linear trend.

4. Conclusions

Although we have used the term "joint resistance" throughout this paper, it must be kept in mind that this is a practical term for the total resistance across a lap joint that has a normal conductor on one side. A pointed out in section 3.2, $dR_{full}/d\lambda \approx dR_{trace}/dx$, so the joint resistance is primarily the voltage

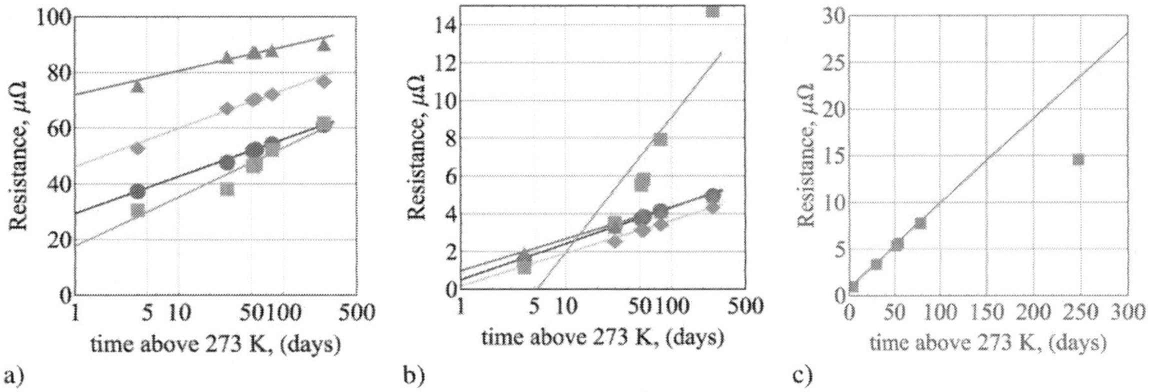

Figure 6. Resistances of the different solder joints on an HTS assembly as a function of time stored above 273 K. a) Resistances on the warm end board, measured at 62 K. b) Resistances on the cold end board, measured at 5.7 K. Note that all joints follow a log-time behaviour except one of the cold end joints. c) This joint followed a linear time behaviour for the first ~80 days.

drop in the copper trace, and is controlled by current transfer length, λ. This length, in turn must be largely controlled by the interface resistance. The interface resistance is determined geometric factors, such as porosity and cracking, and by the intrinsic interface resistivity. We suspect geometric factors, particularly porosity, explain the variance in the data from the simple exponential fit seen in the current transfer plot, figure 2a).

The conductance degradation with time discussed in section 3.5 must almost certainly be due to a slow change in the interface resistivity; it is unlikely a geometric factor, crack growth for instance, would be arrested by storage at -20 C. Ekin, et al. hypothesized [4] and showed [5] that the high oxygen affinity of indium caused the development of an indium oxide layer at the interface of In2%Ag solder applied directly to samples of bulk YBCO. They showed that silver interfaces do not develop oxide layers, and suggested that it be deposited directly on the YBCO, a practice that has become standard. Unfortunately, as shown in section 3.4, indium solder reacts rapidly with the 2–3 μm layer of silver on standard coated 2G superconductors, and can fully consume the silver layer, even if an intentionally rapid solder process is used, leaving a layer of indium-silver intermetallic in contact with the superconductor. For the low soldering temperature used, the intermetallic regions most likely contain both $AgIn_2$ and Ag_2In. For In2%Ag soldered directly to bulk YBCO, Ekin measured interface resistivities of up to 5.5 Ω-mm^2 at 77 K. For the 20 mm^2 joints used in this study, such resistivities would lead to much higher joint resistances than observed. Thus, one might surmise that the tendency to form an oxide layer at the interface with the REBCO is reduced in the intermetallic. However, the low temperature resistance of the oven soldered samples are approximately equal to those of the hand soldered samples. While in the hand soldered cross-sections, some of the silver barrier layer remains intact (figures 4e) and f)), or there is a continuous layer of intermetallic in contact with the superconductor (figures 5b) and c)), in the oven-soldered cross-section, the silver-indium reaction not only goes to completion, but the intermetallic lumps migrate to the middle of the joint, leaving indium solder directly in contact with the superconductor (figures 4b) and c)). This is ostensibly the same configuration as Ekin's In2%Ag/YBCO samples, although in his bulk YBCO samples the grains at the interface are randomly aligned, whereas in the coated superconductor the interface is presumably with a-b planes.

Given the complexities in the use of solder, and the possibilities of continuing chemical reactions, when the use of non-copper clad 2G tape is required, a prudent path may be to develop mechanical joints. Ito and Hashizume [6] have demonstrated mechanical lap joints with resistances comparable to those of soldered lap joints.

Acknowledgements
The authors would like to acknowledge the hard work, dedication and outstanding skill of Tom Hait, Tomomi Watanabe, Diane Kolos, and Michael Sampson, who produced the various samples for this work, usually on a very tight schedule.

References
[1] Mitsuishis I, et al. 2014 *Cryogenics* **64** 189–93
[2] Canavan E R, James B L, Hait T P, Oliver A and Sullivan D F 2014 *Cryogenics* **64** 194-200
[3] Takematsu T, Hu R, Takao T, Yanagisawa Y, Nakagome H, Uglietti D, Kiyoshi T, Takahashi M and Maeda H 2010 *Physica C*
Yanagisawa Y, Nakagome H, Takematsu T, Takao T, Sato N, Takahashi M and Maeda H 2011 *Physica C* **471** 480–5
[4] Ekin J W, Panson A J and Blankenship B A 1988 *Applied Physics Letters* **52**(4) 331–3
[5] Ekin J W, Nelson A J, Swartzlander A B, Kazmerski L L, Panson A J and Blankenship B A 1988 *Applied Physics Letters* **52**(21) 1819–21
[6] Ito S and Hashizume H 2012 *IEEE Transactions on Applied Superconductivity* **22**(3) 6400104

IOP Conf. Series: Materials Science and Engineering **102** (2015) 012033 doi:10.1088/1757-899X/102/1/012033

Temperature dependence of critical current and transport current losses of 4 mm YBCO coated conductors manufactured using nonmagnetic substrate

J Kvitkovic[1], R Hatwar[1,2], S V Pamidi[1,3], S Fleshler[4], C Thieme[4]
[1] Center for Advanced Power Systems, Florida State University, Tallahassee, Florida 32310, USA
[2] Heat Transfer Lab, Mechanical Engineering Department, Johns Hopkins University, Baltimore, Maryland 21218 USA
[3]Department of Electrical & Computer Engineering, FAMU-FSU College of Engineering, Tallahassee, Florida 32310, USA
[4] American Superconductor Corporation, Devens, Massachusetts, USA

E-mail: kvitkovic@caps.fsu.edu

Abstract. The temperature dependence of the critical current and AC losses were measured on American Superconductor Corporation's (AMSC) second generation high temperature superconducting (2G HTS) wire produced by Rolling Assisted Biaxially Textured Substrate (RABiTS) and Metal Organic Deposition (MOD) process. Wires manufactured with two types of substrates were characterized. The magnetic substrate with composition Ni5a%W exhibits a magnetic signature and has non-negligible AC losses in AC power applications. A new nonmagnetic substrate with an alloy composition Ni9a%W has been developed by AMSC to address the AC losses in 2G HTS. The data presented show that the performance of the new conductor is identical to the conductor with magnetic substrate in terms of critical current density. The data on AC losses demonstrate the absence of ferromagnetic loss component in the new conductor and significantly reduced AC losses at low to moderate values of I/I_c. The reduced losses will translate into reduced capital costs and lower operating costs of superconducting electrical devices for AC applications.

1. Introduction

Critical current density and AC losses are two of the most important electrical characteristics of superconducting wires critical for their power applications. Recent advances in the production of long lengths of 2G HTS tapes with high critical current values have made it feasible for deploying HTS cables, fault current limiters, transformers, and energy storage systems. AC losses are one of the major heat loads on the cryogenic systems supporting the power devices, and hence, contribute to the capital investments and operating costs of the overall power system. Many studies were carried out to measure the critical current and transport AC losses on variety HTS tapes to understand the relation between them and to develop theoretical models [1–3]. American Superconductor Corporation (AMSC), one of the major manufacturers of 2G HTS wire uses a technology that is based on Rolling Assisted Biaxially Textured Substrate (RABiTS) for the template on which HTS layer is grown via Metal Organic Deposition (MOD) as described elsewhere [4, 5]. The standard RABiTS technology employs a Ni-W alloy with nominally five atomic percent W (Ni5a%W) as it can be readily produced with a high degree

IOP Conf. Series: Materials Science and Engineering **102** (2015) 012033 doi:10.1088/1757-899X/102/1/012033

of cube texture and can support HTS layers with high critical current densities. Although wire manufactured with Ni5a%W is suitable for many DC applications, the substrate exhibits a ferromagnetic signature which gives rise to non-negligible AC loss under some AC conditions rendering it less desirable for several AC applications. To reduce the AC loss of the wire and address the magnetic behavior of the substrate, a RABiTS process has been developed for an alloy with composition Ni9a%W [6, 7]. This alloy exhibits a relative permeability slightly above one and reversible magnetic behavior at temperatures accessible with liquid nitrogen.

Superconducting cables are typically operated in sub-cooled liquid nitrogen to take advantage of the enhanced critical current densities at lower temperatures and to prevent nitrogen gas bubble formation to ensure dielectric integrity of the cable system. For application of wire with this nonmagnetic substrate to cable systems, it is necessary to understand the critical current and the AC loss characteristics, the relations between the two parameters of the new non-magnetic conductor in the temperature range accessible by liquid nitrogen. In this study we have measured temperature dependence of critical current and transport AC losses on two 2G HTS wires at frequencies 10 -400 Hz in the temperature range of 63 – 78 K. The main emphasis of this study is to compare critical current (I_c) and AC losses of 2G wires produced by AMSC with the original magnetic and newer non-magnetic substrates.

2. Experimental

2.1. Sample description

Two samples of 2G wire used in this study were provided by AMSC. Sample 1 was produced with the Ni5a%W substrate while Sample 2 was manufactured with Ni9a%W substrate. For both samples, the substrate thickness was in the range 62 – 75 μm. HTS layers deposited on the substrates are nominally 1.2 μm thick with both tapes possessing self-field I_c's around 150 – 160 A at 77 K. The HTS insert strip produced on the manufacturing line was slit to 4 mm in width and laminated to 4.4 mm wide brass stabilizers on each face. This wire configuration with 4.4 mm x 150 μm of brass stabilizers on each side has been deployed in numerous power cables.

2.2. Critical current and AC loss measurement apparatus

The sample ends were soldered on to the current leads of the sample holder with a low temperature Indium solder (66%In and 34%Bi). The voltage taps were soldered on the tape axis and the loop width was maintained at three times the half tape width of the HTS tape before they were twisted. Three Lake Shore Cernox temperature sensors were placed below the sample for precise sample temperature measurement. The sample holder was immersed in liquid nitrogen bath in a vacuum tight cryostat that is suitable for pumping on the liquid nitrogen. The temperature of the liquid nitrogen bath in temperature range 63 – 78 K was controlled by regulating the pressure above the bath with a vacuum pump and valve combination. The temperature of the sample was monitored using a Lake Shore temperature controller model 340. The temperature stability at the set point was better than +/- 0.1 K. The DC current–voltage characteristics at different temperatures were measured by the standard four-probe method. A schematic used for the transport current AC loss measurements is shown in Fig. 1. The power amplifier cabinet consists of four blocks of Techron 7782 amplifiers connected in parallel and controlled by a Stanford Research Systems DS360 signal generator. The voltage step-down and isolating transformer was used to achieve the sample current up to 600 A. The transformer also facilitates in minimizing the common mode errors in the measurements.

The actual current through the sample was measured by a low inductance Hall probe current sensor and a digital multimeter Keithley model 2001. The loss voltage was measured using a Lock-in amplifier Stanford Research Systems SR830. The auto-gain and auto-phase procedure was performed at sample current of 5 A using a Rogowski coil output connected to the Lock-in input A. After measuring the phase shift between sample current and Rogowski signal, the phase was changed by 90 degrees to measure the resistive component of the sample loss voltage. The directly measured sample voltage was connected to input A of the Lock-in amplifier. The inductive component of the sample voltage was compensated by a variable voltage, generated in compensation coil, connected input B of the Lock-in amplifier. Lock-in amplifier was set to differential input A-B in float configuration with the time constant of 0.3 s. The AC loss measurements were performed in frequency range 10 - 400 Hz.

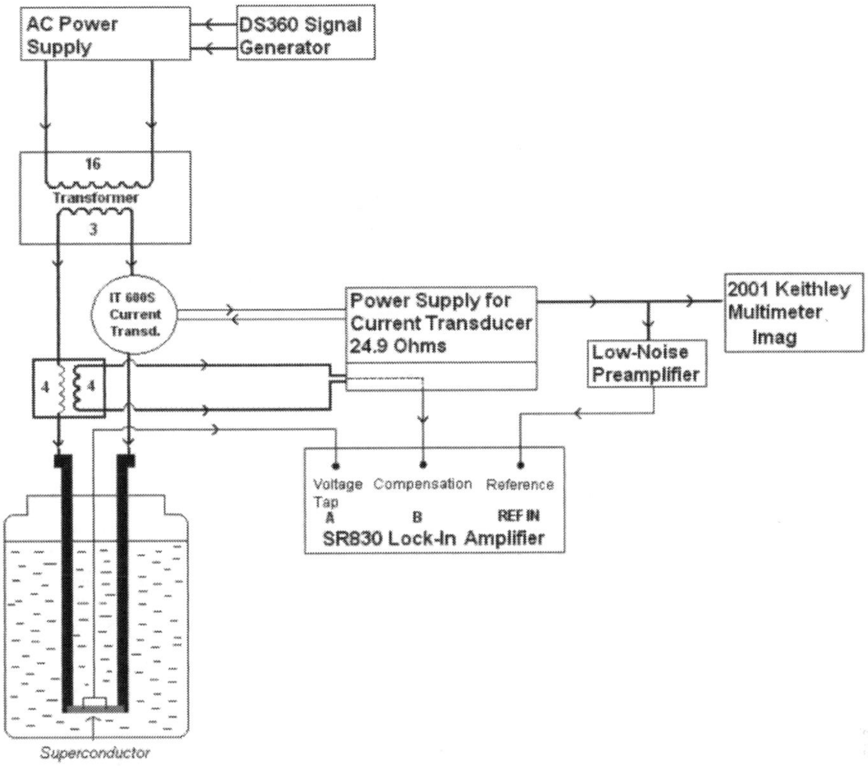

Figure 1. Schematics of the apparatus for transport current loss measurements.

3. Results and discussion

Critical current was measured on samples of 16 cm in length with soldered current contacts of 5 cm long on both sides. The potential taps were placed in the center of the sample with 2 cm between them. The temperature dependence of the critical current in self-field was measured with the electric field criterion of 1 μV/cm. Critical currents of both samples were measured in the temperature range of 63 – 78 K, the results of the critical current measurements are shown in Fig. 2. It can be seen that both samples have approximately the same critical current of 160 A at 77 K and the temperature dependence of I_c of both samples is identical with a slope of ~ -15 A/K. As is typical with 2G HTS, the critical current doubles as temperature is lowered from 77 K to 67 K. We fit the measured data with Onnes formula: $I_c(T) = I_c(0)(1 - T/T_c)$ with $T_c = 88$ K and $I_c(0) = 1310$ A.

The transport current losses per cycle and unit length of Sample 1 and Sample 2 at 78 K as function of the transport current amplitude are shown in Fig. 3. At currents well-below the temperature dependent I_c, the superconductor losses are negligible for Sample 1 and the ferromagnetic loss dominates the loss signal. As current is increased, the superconductor losses increase non-linearly while the ferromagnetic loss tends to saturate. The level of current at which the superconductor losses become significant relative to the ferromagnetic loss depends on the temperature dependent I_c. Loss is a mixture of superconducting loss and a saturated ferromagnetic loss. It can be noticed for Sample 1 that the frequency dependence of AC loss is weak. Sample 2 had ~ 10 times lower losses at 10 A peak.

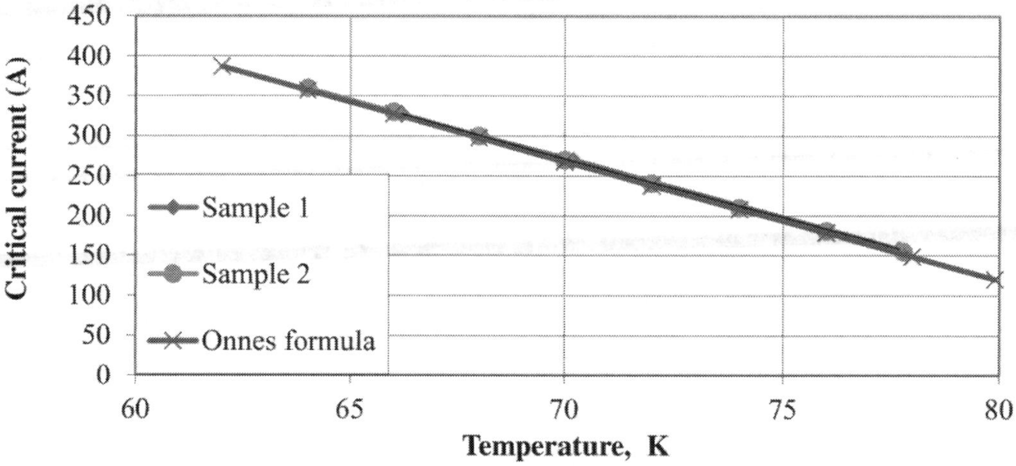

Figure 2. Temperature dependence of critical current of Sample 1 and Sample 2.

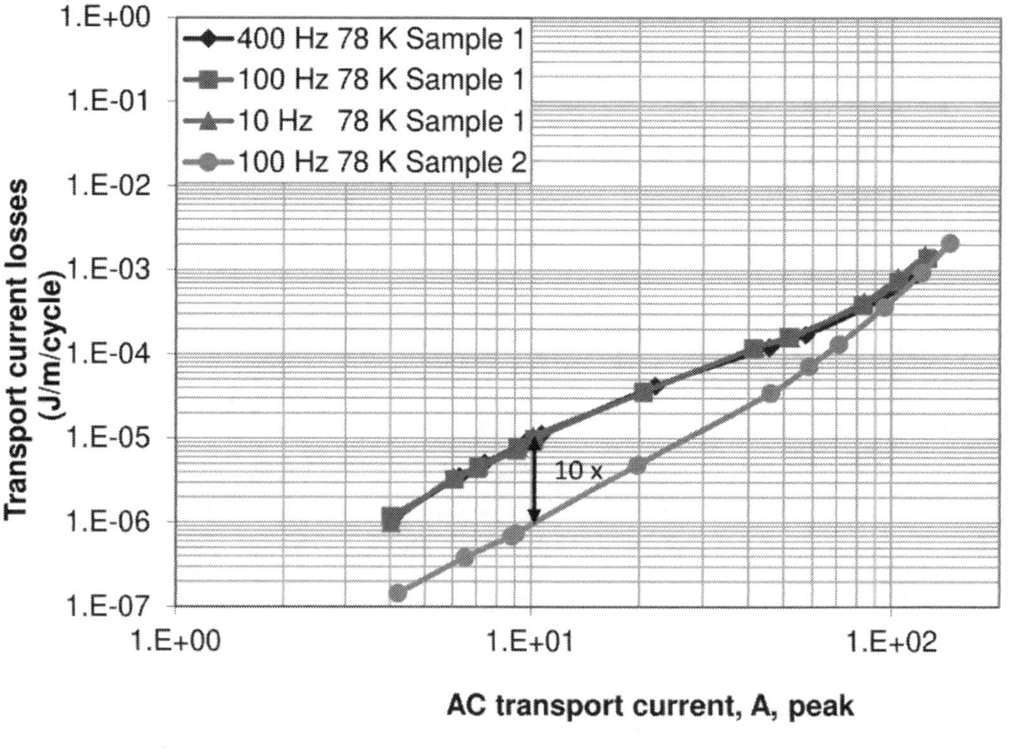

Figure 3. Transport current losses versus transport current of Sample 1 and Sample 2 at 78 K.

To further analyze the impact of the magnetic component to the AC loss in Sample 1, the loss data is normalized by I_c^2 and plotted against normalized transport current (by I_c) in Fig. 4 in the temperature interval of 63 – 68 K at 100 Hz. As seen in the figure, the losses at different temperatures do not scale with normalized current due to the ferromagnetic losses from the magnetic substrate. Ferromagnetic losses do not change with the temperature while the critical current increases with decreasing temperature.

The results of AC loss measurements on Sample 2 at 78 K and 64 K are shown in Fig. 5. The temperature dependence behavior of AC losses in Sample 2 is different than in Sample 1. The measured AC losses are significantly smaller in Sample 2 compared to Sample 1 over the entire transport current interval. The frequency dependence of the losses in Sample 2 at 78 K is weak.

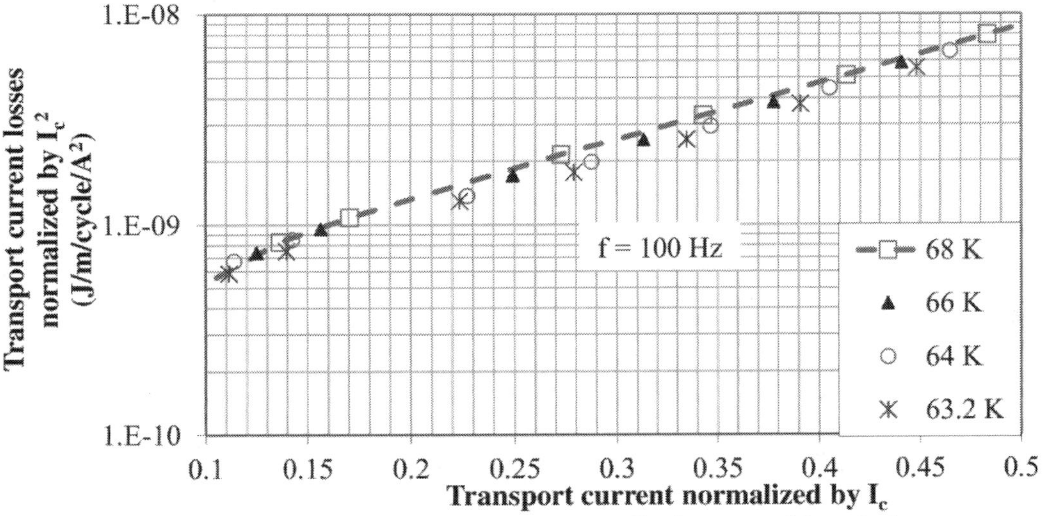

Figure 4. Normalized transport current losses versus normalized current of Sample 1.

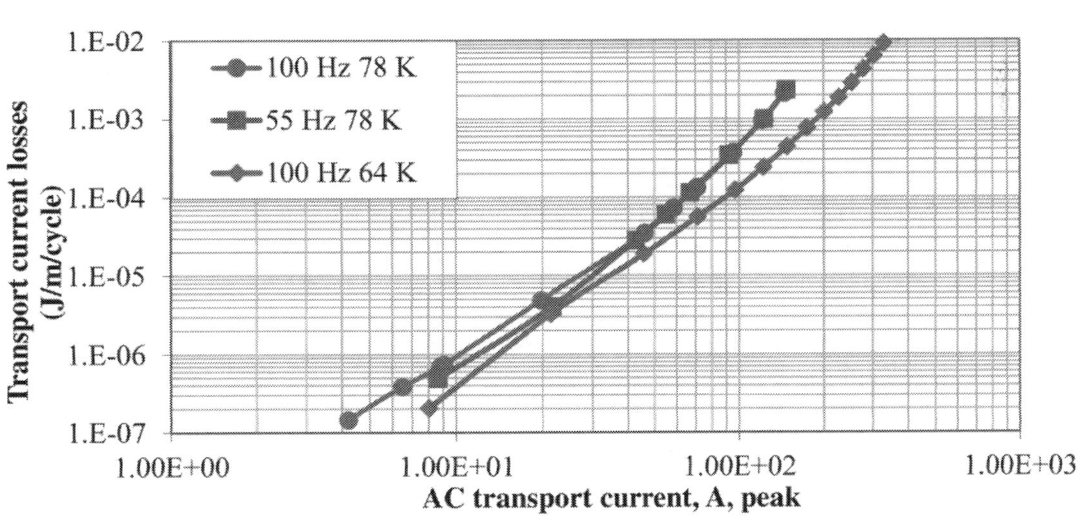

Figure 5. Transport current losses versus transport current of Sample 2 at 64 and 78 K.

Fig. 6 shows the normalized losses versus normalized current of Sample 2 in temperature interval 64 – 78 K at 100 Hz. It can be seen that that the AC loss data at various temperatures scales perfectly. The perfect scaling of the normalized losses clearly indicates that the losses depend just on the current and there is no ferromagnetic loss component present. The loss data of Sample 2 demonstrates that the substrate in Sample 2 is nonmagnetic.

IOP Conf. Series: Materials Science and Engineering **102** (2015) 012033 doi:10.1088/1757-899X/102/1/012033

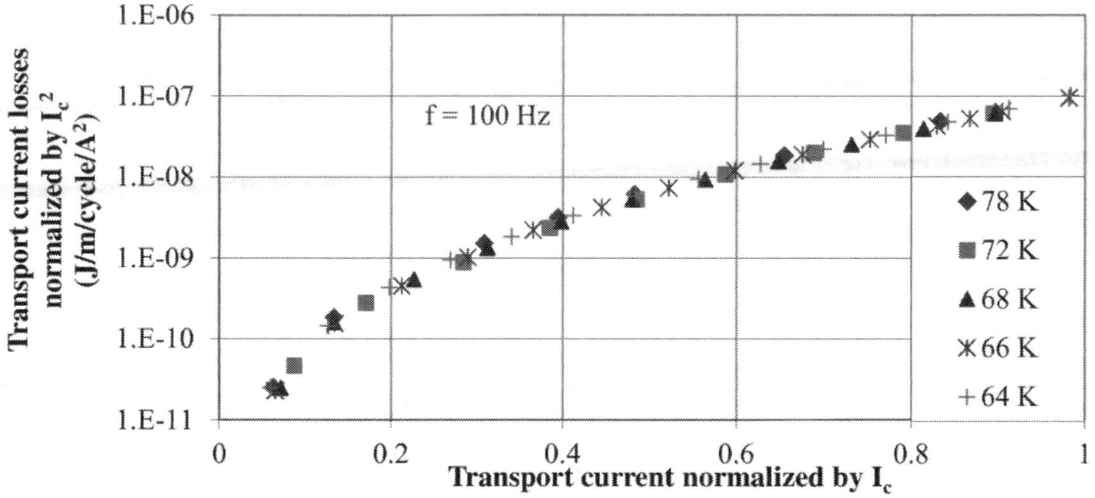

Figure 6. Normalized transport current losses versus normalized current of Sample 2.

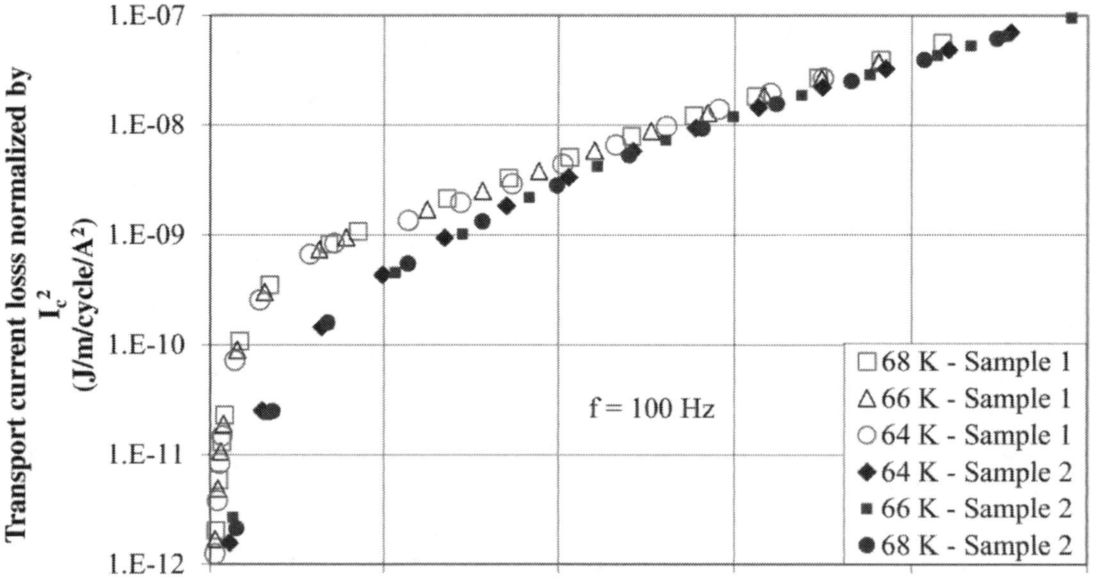

Figure 7. Comparison of normalized transport current losses versus normalized transport current of Sample 1 and Sample 2 at 100 Hz.

Comparison of normalized transport current losses versus normalized transport current of both samples at 100 Hz is presented in Fig.7. One could observe that normalized AC losses of Sample 2 are minimum three times lower for normalized current, I/I_c from 0 to 0.3 while the normalized losses on both samples are almost identical at higher currents.

The data presented demonstrates that the 2G HTS conductor fabricated using the nonmagnetic substrate performs similar to that fabricated from the magnetic substrate in terms of critical current density values and temperature dependence of critical current density. The benefit of the nonmagnetic substrate is clearly demonstrated with lower losses, particularly at the normalized critical current, I/I_c values of < 0.6. Typical superconducting cables and other devices are designed to operate below I/I_c of 0.6 to have operating and safety margin. Hence use of the nonmagnetic substrate significantly reduces the heat load on the cryogenic systems of superconducting cables and other devices for AC applications.

4. Conclusions

Temperature dependence of the critical current and transport current losses were measured on samples from two different batches of 2G HTS wire manufactured by AMSC. One of the samples is fabricated using the magnetic substrate with ferromagnetic properties and the second sample is fabricated with a nonmagnetic substrate. It was demonstrated that the conductor produced with the nonmagnetic substrate performs similar to the conductor with the magnetic substrate in terms of critical current and its dependence on operating temperature. Loss measurements in the temperature range 63 – 78 K have shown that the conductor with the nonmagnetic substrate does not have any magnetic loss component and the AC losses are significantly lower than in the conductor with the magnetic substrate for current values lower than 60 % of the critical current at the given operating temperature. The reductions in the AC losses represent a significantly reduced cryogenic capacity needed and lower operating costs of the cryogenic equipment for power applications.

Acknowledgements
This research was in part supported by the Office of Naval Research under grant number N00014-14-1-0211.

References

[1] Ciszek M, Glowacki B A, Ashworth S P, Campbell A M, Evetts J E 1995 AC Losses of Ag-(Bi,Pb)SrCaCuO-2223 Tapes in Combination of Transverse External Magnetic Field and Transport Current *IEEE Trans. Appl. Supercond.* **5** 709
[2] Majoros M, Ye L, Velichko A V, Coombs T A, Sumption M D and Collings E W 2007 Transport AC losses in YBCO coated conductors *Supercond. Sci. Technol.* **20** S299–S304
[3] Miyagi D and Tsukamoto O 2002 Characteristics of AC transport current losses in YBCO coated conductors and their dependence on distribution of critical current density in the conductors *IEEE Trans. Appl. Supercond.* **12** 1628–31
[4] Fleshler S et al 2009 *Physica C* **469** 1316
[5] Rupich M et al 2010 *Supercond. Sci. Technol.* **23** 014015
[6] Rupich M, Li X, Sathyamurthy S, Thieme C, DeMoranville K, Gannon J and Fleshler S 2013 *IEEE Trans. Appl. Supercond.* **23** 6601205
[7] Fleshler S, DeMoranville K, Gannon Jr J, Li X, Podtburg E, Rupich M, Sathyamurthy S, Thieme C L H, Tucker D and Whitman L 2014 Development Status of AMSC Amperium Wire *J. Phys.: Conf. Ser.* **507** 022005

Institute of Physics
Dirac House, Temple Back
Bristol BS1 6BE UK

ISSN: 1757-8981
ISBN 978-1-5108-1811-8

9 781510 818118